Practical Reliability of Electronic Equipment and Products

ELECTRICAL AND COMPUTER ENGINEERING

A Series of Reference Books and Textbooks

FOUNDING EDITOR

Marlin O. Thurston
Department of Electrical Engineering
The Ohio State University
Columbus, Ohio

1. Rational Fault Analysis, *edited by Richard Saeks and S. R. Liberty*
2. Nonparametric Methods in Communications, *edited by P. Papantoni-Kazakos and Dimitri Kazakos*
3. Interactive Pattern Recognition, *Yi-tzuu Chien*
4. Solid-State Electronics, *Lawrence E. Murr*
5. Electronic, Magnetic, and Thermal Properties of Solid Materials, *Klaus Schröder*
6. Magnetic-Bubble Memory Technology, *Hsu Chang*
7. Transformer and Inductor Design Handbook, *Colonel Wm. T. McLyman*
8. Electromagnetics: Classical and Modern Theory and Applications, *Samuel Seely and Alexander D. Poularikas*
9. One-Dimensional Digital Signal Processing, *Chi-Tsong Chen*
10. Interconnected Dynamical Systems, *Raymond A. DeCarlo and Richard Saeks*
11. Modern Digital Control Systems, *Raymond G. Jacquot*
12. Hybrid Circuit Design and Manufacture, *Roydn D. Jones*
13. Magnetic Core Selection for Transformers and Inductors: A User's Guide to Practice and Specification, *Colonel Wm. T. McLyman*
14. Static and Rotating Electromagnetic Devices, *Richard H. Engelmann*
15. Energy-Efficient Electric Motors: Selection and Application, *John C. Andreas*
16. Electromagnetic Compossibility, *Heinz M. Schlicke*
17. Electronics: Models, Analysis, and Systems, *James G. Gottling*
18. Digital Filter Design Handbook, *Fred J. Taylor*
19. Multivariable Control: An Introduction, *P. K. Sinha*
20. Flexible Circuits: Design and Applications, *Steve Gurley, with contributions by Carl A. Edstrom, Jr., Ray D. Greenway, and William P. Kelly*
21. Circuit Interruption: Theory and Techniques, *Thomas E. Browne, Jr.*
22. Switch Mode Power Conversion: Basic Theory and Design, *K. Kit Sum*
23. Pattern Recognition: Applications to Large Data-Set Problems, *Sing-Tze Bow*

24. Custom-Specific Integrated Circuits: Design and Fabrication, *Stanley L. Hurst*
25. Digital Circuits: Logic and Design, *Ronald C. Emery*
26. Large-Scale Control Systems: Theories and Techniques, *Magdi S. Mahmoud, Mohamed F. Hassan, and Mohamed G. Darwish*
27. Microprocessor Software Project Management, *Eli T. Fathi and Cedric V. W. Armstrong (Sponsored by Ontario Centre for Microelectronics)*
28. Low Frequency Electromagnetic Design, *Michael P. Perry*
29. Multidimensional Systems: Techniques and Applications, *edited by Spyros G. Tzafestas*
30. AC Motors for High-Performance Applications: Analysis and Control, *Sakae Yamamura*
31. Ceramic Motors for Electronics: Processing, Properties, and Applications, *edited by Relva C. Buchanan*
32. Microcomputer Bus Structures and Bus Interface Design, *Arthur L. Dexter*
33. End User's Guide to Innovative Flexible Circuit Packaging, *Jay J. Miniet*
34. Reliability Engineering for Electronic Design, *Norman B. Fuqua*
35. Design Fundamentals for Low-Voltage Distribution and Control, *Frank W. Kussy and Jack L. Warren*
36. Encapsulation of Electronic Devices and Components, *Edward R. Salmon*
37. Protective Relaying: Principles and Applications, *J. Lewis Blackburn*
38. Testing Active and Passive Electronic Components, *Richard F. Powell*
39. Adaptive Control Systems: Techniques and Applications, *V. V. Chalam*
40. Computer-Aided Analysis of Power Electronic Systems, *Venkatachari Rajagopalan*
41. Integrated Circuit Quality and Reliability, *Eugene R. Hnatek*
42. Systolic Signal Processing Systems, *edited by Earl E. Swartzlander, Jr.*
43. Adaptive Digital Filters and Signal Analysis, *Maurice G. Bellanger*
44. Electronic Ceramics: Properties, Configuration, and Applications, *edited by Lionel M. Levinson*
45. Computer Systems Engineering Management, *Robert S. Alford*
46. Systems Modeling and Computer Simulation, *edited by Naim A. Kheir*
47. Rigid-Flex Printed Wiring Design for Production Readiness, *Walter S. Rigling*
48. Analog Methods for Computer-Aided Circuit Analysis and Diagnosis, *edited by Takao Ozawa*
49. Transformer and Inductor Design Handbook: Second Edition, Revised and Expanded, *Colonel Wm. T. McLyman*
50. Power System Grounding and Transients: An Introduction, *A. P. Sakis Meliopoulos*
51. Signal Processing Handbook, *edited by C. H. Chen*
52. Electronic Product Design for Automated Manufacturing, *H. Richard Stillwell*
53. Dynamic Models and Discrete Event Simulation, *William Delaney and Erminia Vaccari*
54. FET Technology and Application: An Introduction, *Edwin S. Oxner*

55. Digital Speech Processing, Synthesis, and Recognition, *Sadaoki Furui*
56. VLSI RISC Architecture and Organization, *Stephen B. Furber*
57. Surface Mount and Related Technologies, *Gerald Ginsberg*
58. Uninterruptible Power Supplies: Power Conditioners for Critical Equipment, *David C. Griffith*
59. Polyphase Induction Motors: Analysis, Design, and Application, *Paul L. Cochran*
60. Battery Technology Handbook, *edited by H. A. Kiehne*
61. Network Modeling, Simulation, and Analysis, *edited by Ricardo F. Garzia and Mario R. Garzia*
62. Linear Circuits, Systems, and Signal Processing: Advanced Theory and Applications, *edited by Nobuo Nagai*
63. High-Voltage Engineering: Theory and Practice, *edited by M. Khalifa*
64. Large-Scale Systems Control and Decision Making, *edited by Hiroyuki Tamura and Tsuneo Yoshikawa*
65. Industrial Power Distribution and Illuminating Systems, *Kao Chen*
66. Distributed Computer Control for Industrial Automation, *Dobrivoje Popovic and Vijay P. Bhatkar*
67. Computer-Aided Analysis of Active Circuits, *Adrian Ioinovici*
68. Designing with Analog Switches, *Steve Moore*
69. Contamination Effects on Electronic Products, *Carl J. Tautscher*
70. Computer-Operated Systems Control, *Magdi S. Mahmoud*
71. Integrated Microwave Circuits, *edited by Yoshihiro Konishi*
72. Ceramic Materials for Electronics: Processing, Properties, and Applications, Second Edition, Revised and Expanded, *edited by Relva C. Buchanan*
73. Electromagnetic Compatibility: Principles and Applications, *David A. Weston*
74. Intelligent Robotic Systems, *edited by Spyros G. Tzafestas*
75. Switching Phenomena in High-Voltage Circuit Breakers, *edited by Kunio Nakanishi*
76. Advances in Speech Signal Processing, *edited by Sadaoki Furui and M. Mohan Sondhi*
77. Pattern Recognition and Image Preprocessing, *Sing-Tze Bow*
78. Energy-Efficient Electric Motors: Selection and Application, Second Edition, *John C. Andreas*
79. Stochastic Large-Scale Engineering Systems, *edited by Spyros G. Tzafestas and Keigo Watanabe*
80. Two-Dimensional Digital Filters, *Wu-Sheng Lu and Andreas Antoniou*
81. Computer-Aided Analysis and Design of Switch-Mode Power Supplies, *Yim-Shu Lee*
82. Placement and Routing of Electronic Modules, *edited by Michael Pecht*
83. Applied Control: Current Trends and Modern Methodologies, *edited by Spyros G. Tzafestas*
84. Algorithms for Computer-Aided Design of Multivariable Control Systems, *Stanoje Bingulac and Hugh F. VanLandingham*
85. Symmetrical Components for Power Systems Engineering, *J. Lewis Blackburn*
86. Advanced Digital Signal Processing: Theory and Applications, *Glenn Zelniker and Fred J. Taylor*

87. Neural Networks and Simulation Methods, *Jian-Kang Wu*
88. Power Distribution Engineering: Fundamentals and Applications, *James J. Burke*
89. Modern Digital Control Systems: Second Edition, *Raymond G. Jacquot*
90. Adaptive IIR Filtering in Signal Processing and Control, *Phillip A. Regalia*
91. Integrated Circuit Quality and Reliability: Second Edition, Revised and Expanded, *Eugene R. Hnatek*
92. Handbook of Electric Motors, *edited by Richard H. Engelmann and William H. Middendorf*
93. Power-Switching Converters, *Simon S. Ang*
94. Systems Modeling and Computer Simulation: Second Edition, *Naim A. Kheir*
95. EMI Filter Design, *Richard Lee Ozenbaugh*
96. Power Hybrid Circuit Design and Manufacture, *Haim Taraseiskey*
97. Robust Control System Design: Advanced State Space Techniques, *Chia-Chi Tsui*
98. Spatial Electric Load Forecasting, *H. Lee Willis*
99. Permanent Magnet Motor Technology: Design and Applications, *Jacek F. Gieras and Mitchell Wing*
100. High Voltage Circuit Breakers: Design and Applications, *Ruben D. Garzon*
101. Integrating Electrical Heating Elements in Appliance Design, *Thor Hegbom*
102. Magnetic Core Selection for Transformers and Inductors: A User's Guide to Practice and Specification, Second Edition, *Colonel Wm. T. McLyman*
103. Statistical Methods in Control and Signal Processing, *edited by Tohru Katayama and Sueo Sugimoto*
104. Radio Receiver Design, *Robert C. Dixon*
105. Electrical Contacts: Principles and Applications, *edited by Paul G. Slade*
106. Handbook of Electrical Engineering Calculations, *edited by Arun G. Phadke*
107. Reliability Control for Electronic Systems, *Donald J. LaCombe*
108. Embedded Systems Design with 8051 Microcontrollers: Hardware and Software, *Zdravko Karakehayov, Knud Smed Christensen, and Ole Winther*
109. Pilot Protective Relaying, *edited by Walter A. Elmore*
110. High-Voltage Engineering: Theory and Practice, Second Edition, Revised and Expanded, *Mazen Abdel-Salam, Hussein Anis, Ahdab El-Morshedy, and Roshdy Radwan*
111. EMI Filter Design: Second Edition, Revised and Expanded, *Richard Lee Ozenbaugh*
112. Electromagnetic Compatibility: Principles and Applications, Second Edition, Revised and Expanded, *David A. Weston*
113. Permanent Magnet Motor Technology: Design and Applications, Second Edition, Revised and Expanded, *Jacek F. Gieras and Mitchell Wing*

114. High Voltage Circuit Breakers: Design and Applications, Second Edition, Revised and Expanded, *Ruben D. Garzon*
115. High Reliability Magnetic Devices: Design and Fabrication, *Colonel Wm. T. McLyman*
116. Practical Reliability of Electronic Equipment and Products, *Eugene R. Hnatek*

Additional Volumes in Preparation

Battery Technology Handbook, Second Edition, *H. A. Kiehne*

PRACTICAL RELIABILITY OF ELECTRONIC EQUIPMENT AND PRODUCTS

EUGENE R. HNATEK
*Qualcomm Incorporated
San Diego, California*

MARCEL DEKKER, INC. NEW YORK • BASEL

ISBN: 0-8247-0832-6

This book is printed on acid-free paper.

Headquarters
Marcel Dekker, Inc.
270 Madison Avenue, New York, NY 10016
tel: 212-696-9000; fax: 212-685-4540

Eastern Hemisphere Distribution
Marcel Dekker AG
Hutgasse 4, Postfach 812, CH-4001 Basel, Switzerland
tel: 41-61-260-6300; fax: 41-61-260-6333

World Wide Web
http://www.dekker.com

The publisher offers discounts on this book when ordered in bulk quantities. For more information, write to Special Sales/Professional Marketing at the headquarters address above.

Copyright © 2003 by Marcel Dekker, Inc. All Rights Reserved.

Neither this book nor any part may be reproduced or transmitted in any form or by any means, electronic or mechanical, including photocopying, microfilming, and recording, or by any information storage and retrieval system, without permission in writing from the publisher.

Current printing (last digit):
10 9 8 7 6 5 4 3 2 1

PRINTED IN THE UNITED STATES OF AMERICA

To my mother Val, my wife Susan, and my Lord and Savior Jesus Christ
for their caring ways, encouragement, and unconditional love.
Words cannot adequately express my gratitude and feelings for them.
To my brother Richard for his loving care of our mother.

Preface

Reliability is important. Most organizations are concerned with fast time to market, competitive advantage, and improving costs. Customers want to be sure that the products and equipment they buy work as intended for the time specified. That's what reliability is: performance against requirements over time.

A number of excellent books have been written dealing with the topic of reliability—most from a theoretical and what I call a "rel math" perspective. This book is about electronic product and equipment reliability. It presents a practical "hands-on perspective" based on my personal experience in fielding a myriad of different systems, including military/aerospace systems, semiconductor devices (integrated circuits), measuring instruments, and computers.

The book is organized according to end-to-end reliability: from the customer to the customer. At the beginning customers set the overall product parameters and needs and in the end they determine whether the resultant product meets those needs. They basically do this with their wallets. Thus, it is imperative that manufacturers truly listen to what the customer is saying. In between these two bounds the hard work of reliability takes place: design practices and testing; selection and qualification of components, technology and suppliers; printed wiring assembly and systems manufacturing; and testing practices, including regulatory testing and failure analysis.

To meet any reliability objective requires a comprehensive knowledge of the interactions of the design, the components used, the manufacturing techniques employed, and the environmental stresses under which the product will operate. A reliable product is one that balances design-it-right and manufacture-it-correctly techniques with just the right amount of testing. For example, design verification testing is best accomplished using a logical method such as a Shewhart or Deming cycle (plan–do–check–act–repeat) in conjunction with accelerated stress and failure analysis. Only when used in this closed-feedback loop manner will testing help make a product more robust. Testing by itself adds nothing to the reliability of a product.

The purpose of this book is to give electronic circuit design engineers, system design engineers, product engineers, reliability engineers, and their managers this end-to-end view of reliability by sharing what is currently being done in each of the areas presented as well as what the future holds based on lessons-learned. It is important that lessons and methods learned be shared. This is the major goal of this book. If we are ignorant of the lessons of the past, we usually end up making the same mistakes as those before us did. The key is to never stop learning. The topics contained in this book are meant to foster and stimulate thinking and help readers extrapolate the methods and techniques to specific work situations.

The material is presented from a large-company, large-system/product perspective (in this text the words *product*, *equipment*, and *system* are interchangeable). My systems work experiences have been with large companies with the infrastructure and capital equipment resources to produce high-end products that demand the highest levels of reliability: satellites, measuring instruments (automatic test equipment for semiconductors), and high-end computers/servers for financial transaction processing. This book provides food for thought in that the methods and techniques used to produce highly reliable and robust products for these very complex electronic systems can be "cherry-picked" for use by smaller, resource-limited companies. The methods and techniques given can be tailored to a company's specific needs and corporate boundary conditions for an appropriate reliability plan.

My hope is that within this book readers will find some methods or ideas that they can take away and use to make their products more reliable. The methods and techniques are not applicable in total for everyone. Yet there are some ingredients for success provided here that can be applied regardless of the product being designed and manufactured. I have tried to provide some things to think about. There is no single step-by-step process that will ensure the production of a high-reliability product. Rather, there are a number of sound principles that have been found to work. What the reader ultimately decides to do depends on the product(s) being produced, the markets served, and the fundamental precepts under which the company is run. I hope that the material presented is of value.

ACKNOWLEDGMENTS

I want to acknowledge the professional contributions of my peers in their areas of expertise and technical disciplines to the electronics industry and to this book. This book would not have been possible without the technical contributions to the state of the art by the people listed below. I thank them for allowing me to use their material. I consider it an honor to have both known and worked with all of them. I deeply respect them personally and their abilities.

>David Christiansen, Compaq Computer Corporation, Tandem Division
>Noel Donlin, U.S. Army, retired
>Jon Elerath, Compaq Computer Corporation, Tandem Division; now with Network Appliance Inc.
>Charles Hawkins, University of New Mexico
>Michael Hursthopf, Compaq Computer Corporation, Tandem Division
>Andrew Kostic, IBM
>Edmond Kyser, Compaq Computer Corporation, Tandem Division; now with Cisco Systems
>Ken Long, Celestica
>Chan Moore, Compaq Computer Corporation, Tandem Division
>Joel Russeau, Compaq Computer Corporation, Tandem Division
>Richard Sevcik, Xilinx Inc.
>Ken Stork, Ken Stork and Associates
>Alan Wood, Compaq Computer Corporation, Tandem Division

David Christiansen, Michael Hursthopf, Chan Moore, Joel Russeau, and Alan Wood are now with Hewlett Packard Company. Thanks to Rich Sevcik for reviewing Chapter 8. His expert comments and suggestions were most helpful. Thanks to Joel Russeau and Ed Kyser for coauthoring many articles with me. Their contributions to these works were significant in the value that was provided to the readers. To my dear friend G. Laross Coggan, thanks for your untiring assistance in getting the artwork together; I couldn't have put this all together without you. To my production editor, Moraima Suarez, thanks for your editorial contribution. I appreciate your diligence in doing a professional job of editing the manuscript and your patience in dealing with the issues that came up during the production process.

Eugene R. Hnatek

Contents

Preface   v

1. **Introduction to Reliability**   **1**
   1.1 What Is Reliability?   1
   1.2 Discipline and Tasks Involved with Product Reliability   6
   1.3 The Bathtub Failure Rate Curve   6
   1.4 Reliability Goals and Metrics   12
   1.5 Reliability Prediction   14
   1.6 Reliability Risk   17
   1.7 Reliability Growth   23
   1.8 Reliability Degradation   24
   1.9 Reliability Challenges   26
   1.10 Reliability Trends   26
        References   27

2. **Basic Reliability Mathematics**   **29**
   2.1 Statistical Terms   29
   2.2 Statistical Distributions   34
   2.3 Plotting and Linearization of Data   45
   2.4 Confidence Limit and Intervals   45
   2.5 Failure Free Operation   52
   2.6 Reliability Modeling   52

	2.7	Practical MTBF and Warranty Cost Calculations	54
		Reference	58
		Appendix A	59
3.	**Robust Design Practices**		**61**
	3.1	Introduction	61
	3.2	Concurrent Engineering/Design Teams	63
	3.3	Product Design Process	65
	3.4	Translate Customer Requirements to Design Requirements	69
	3.5	Technology Assessment	69
	3.6	Circuit Design	74
	3.7	Power Supply Considerations	80
	3.8	Redundancy	82
	3.9	Component/Supplier Selection and Management	84
	3.10	Reliability Prediction	84
	3.11	Support Cost and Reliability Tradeoff Model	85
	3.12	Stress Analysis and Part Derating	87
	3.13	PCB Design, PWA Layout, and Design for Manufacture	95
	3.14	Thermal Management	114
	3.15	Signal Integrity and Design for Electromagnetic Compatibility	116
	3.16	Design for Test	121
	3.17	Sneak Circuit Analysis	138
	3.18	Bill of Material Reviews	138
	3.19	Design Reviews	139
	3.20	The Supply Chain and the Design Engineer	141
	3.21	Failure Modes and Effects Analysis	144
	3.22	Design for Environment	152
	3.23	Environmental Analysis	154
	3.24	Development and Design Testing	154
	3.25	Transportation and Shipping Testing	161
	3.26	Regulatory Testing	162
	3.27	Design Errors	164
		References	165
		Further Reading	166
4.	**Component and Supplier Selection, Qualification, Testing, and Management**		**167**
	4.1	Introduction	167
	4.2	The Physical Supply Chain	167
	4.3	Supplier Management in the Electronics Industry	170

	4.4	The Role of the Component Engineer	197
	4.5	Component Selection	202
	4.6	Integrated Circuit Reliability	204
	4.7	Component Qualification	217
		References	237
		Further Reading	238
		Appendix A: Supplier Scorecard Process Overview	238
		Appendix B: Self-Qualification Form for Programmable Logic ICs	242

5. Thermal Management — 247

- 5.1 Introduction — 247
- 5.2 Thermal Analysis Models and Tools — 248
- 5.3 Impact of High-Performance Integrated Circuits — 254
- 5.4 Material Considerations — 259
- 5.5 Effect of Heat on Components, Printed Circuit Boards, and Solder — 261
- 5.6 Cooling Solutions — 266
- 5.7 Other Considerations — 281
- References — 282
- Further Reading — 282

6. Electrostatic Discharge and Electromagnetic Compatibility — 283

- 6.1 Electrostatic Discharge — 283
- 6.2 Electromagnetic Compatibility — 292
- Further Reading — 306

7. Manufacturing/Production Practices — 309

- 7.1 Printed Wiring Assembly Manufacturing — 309
- 7.2 Printed Wiring Assembly Testing — 323
- 7.3 Environmental Stress Testing — 336
- 7.4 System Testing — 360
- 7.5 Field Data Collection and Analysis — 362
- 7.6 Failure Analysis — 368
- References — 375
- Further Reading — 375

8. Software — 377

- 8.1 Introduction — 377
- 8.2 Hardware/Software Development Comparison — 378
- 8.3 Software Availability — 379

8.4	Software Quality	380
	Reference	408

Appendix A: An Example of Part Derating Guidelines	*409*
Appendix B: FMEA Example for a Memory Module	*423*
Epilogue	*431*
Index	*435*

1
Introduction to Reliability

1.1 WHAT IS RELIABILITY?

To set the stage, this book deals with the topic of electronic product hardware reliability. Electronic products consist of individual components (such as integrated circuits, resistors, capacitors, transistors, diodes, crystals, and connectors) assembled on a printed circuit board; third party–provided hardware such as disk drives, power supplies, and various printed circuit card assemblies; and various mechanical fixtures, robotics, shielding, cables, etc., all integrated into an enclosure or case of some sort.

The term *reliability* is at the same time ambiguous in the general sense but very exacting in the practical and application sense when consideration is given to the techniques and methods used to ensure the production of reliable products. Reliability differs/varies based on the intended application, the product category, the product price, customer expectations, and the level of discomfort or repercussion caused by product malfunction. For example, products destined for consumer use have different reliability requirements and associated risk levels than do products destined for use in industrial, automotive, telecommunication, medical, military, or space applications.

Customer expectations and threshold of pain are important as well. What do I mean by this? Customers have an expectation and threshold of pain for the product they purchase based on the price paid and type of product. The designed-in reliability level should be just sufficient enough to meet that expectation and threshold of pain. Thus, reliability and customer expectations are closely tied to price. For example if a four- to five-function electronic calculator fails, the customer's level of irritation and dissatisfaction is low. This is so because both the purchase price and the original customer expectation for purchase are both low. The customer merely disposes of it and gets another one. However, if your Lexus engine ceases to function while you are driving on a busy freeway, your level of anxiety, irritation, frustration, and dissatisfaction are extremely high. This is because both the customer expectation upon purchase and the purchase price are high. A Lexus is not a disposable item.

Also, for a given product, reliability is a moving target. It varies with the maturity of the technology and from one product generation to the next. For example, when the electronic calculator and digital watch first appeared in the marketplace, they were state-of-the-art products and were extremely costly as well. The people who bought these products were early adopters of the technology and expected them to work. Each product cost in the neighborhood of several hundred dollars (on the order of $800–$900 for the first electronic calculator and $200–$400 for the first digital watches). As the technology was perfected (going from LED to LCD displays and lower-power CMOS integrated circuits) and matured and competition entered the marketplace, the price fell over the years to such a level that these products have both become disposable commodity items (except for high-end products). When these products were new, unique, and high priced, the customer's reliability expectations were high as well. As the products became mass-produced disposable commodity items, the reliability expectations became less and less important; so that today reliability is almost a "don't care" situation for these two products. The designed-in reliability has likewise decreased in response to market conditions.

Thus companies design in just enough reliability to meet the customer's expectations, i.e., consumer acceptance of the product price and level of discomfort that a malfunction would bring about. You don't want to design in more reliability than the application warrants or that the customer is willing to pay for. Table 1 lists the variables of price, customer discomfort, designed-in reliability, and customer expectations relative to product/application environment, from the simple to the complex.

Then, too, a particular product category may have a variety of reliability requirements. Take computers as an example. Personal computers for consumer and general business office use have one set of reliability requirements; computers destined for use in high-end server applications (CAD tool sets and the like) have another set of requirements. Computers serving the telecommunication industry must operate for 20-plus years; applications that require nonstop availability and 100% data integrity (for stock markets and other financial transaction applications, for example) have an even higher set of requirements. Each of these

TABLE 1 Key Customer Variables Versus Product Categories/Applications Environment

	Calculators	Personal computers	Pacemaker	Computers for banking applications	Auto	Airline	Satellite
Price	Low					→	Extremely high
Discomfort and repercussion caused by malfunction	Low					→	Extremely high
Designed-in reliability	Low					→	Extremely high
Customer expectations	Low					→	Extremely high

markets has different reliability requirements that must be addressed individually during the product concept and design phase and during the manufacturing and production phase.

Reliability cannot be an afterthought apart from the design phase, i.e., something that is considered only when manufacturing yield is low or when field failure rate and customer returns are experienced. Reliability must be designed and built (manufactured) in from the start, commensurate with market and customer needs. It requires a complete understanding of the customer requirements and an accurate translation of those requirements to the language of the system designer. This results in a design/manufacturing methodology that produces a reliable delivered product that meets customer needs. Electronic hardware reliability includes both circuit and system design reliability, manufacturing process reliability, and product reliability. It is strongly dependent on the reliability of the individual components that comprise the product design. Thus, reliability begins and ends with the customer. Figure 1 shows this end-to-end product reliability methodology diagrammatically.

Stated very simply, reliability is not about technology. It's about customer service and satisfaction and financial return. If a consumer product is reliable, customers will buy it and tell their friends about it, and repeat business will ensue. The same holds true for industrial products. The net result is less rework and low field return rate and thus increased revenue and gross margin. Everything done to improve a product's reliability is done with these thoughts in mind.

Now that I've danced around it, just what is this nebulous concept we are talking about? *Quality* and *reliability* are very similar terms, but they are not interchangeable. Both quality and reliability are related to variability in the electronic product manufacturing process and are interrelated, as will be shown by the bathtub failure rate curve that will be discussed in the next section.

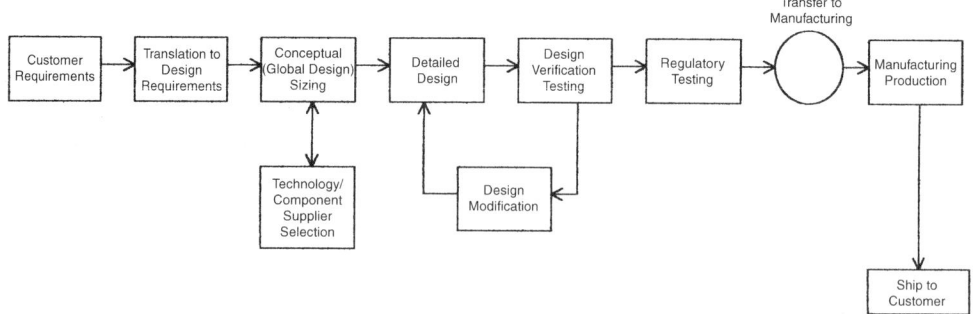

FIGURE 1 End-to-end product reliability.

Introduction to Reliability

Quality is defined as product performance against requirements at an instant in time. The metrics used to measure quality include

PPM: parts per million defective
AQL: acceptable quality level
LTPD: lot tolerance percent defective

Reliability is the performance against requirements over a period of time. Reliability measurements always have a time factor. IPC-SM-785 defines reliability as the ability of a product to function under given conditions and for a specified period of time without exceeding acceptable failure levels.

According to IPC standard J-STD-001B, which deals with solder joint reliability, electronic assemblies are categorized in three classes of products, with increasing reliability requirements.

Class 1, or general, electronic products, including consumer products. Reliability is desirable, but there is little physical threat if solder joints fail.

Class 2, or dedicated service, electronics products, including industrial and commercial products (computers, telecommunications, etc.). Reliability is important, and solder joint failures may impede operations and increase service costs.

Class 3, or high-performance, electronics products, including automotive, avionics, space, medical, military, or any other applications where reliability is critical and solder joint failures can be life/mission threatening.

Class 1 products typically have a short design life, e.g., 3 to 5 years, and may not experience a large number of stress cycles. Class 2 and 3 products have longer design lives and may experience larger temperature swings. For example, commercial aircraft may have to sustain over 20,000 takeoffs and landings over a 20-year life, with cargo bay electronics undergoing thermal cycles from ground level temperatures (perhaps as high as 50°C under desert conditions) to very low temperatures at high altitude (about -55°C at 35,000 feet). The metrics used to measure reliability include

Percent failure per thousand hours
MTBF: mean time between failure
MTTF: mean time to failure
FIT: failures in time, typically failures per billion hours of operation

Reliability is a hierarchical consideration at all levels of electronics, from materials to operating systems because

Materials are used to make components.
Components compose subassemblies.

Subassemblies compose assemblies.
Assemblies are combined into systems of ever-increasing complexity and sophistication.

1.2 DISCIPLINE AND TASKS INVOLVED WITH PRODUCT RELIABILITY

Electronic product reliability encompasses many disciplines, including component engineering, electrical engineering, mechanical engineering, materials science, manufacturing and process engineering, test engineering, reliability engineering, and failure analysis. Each of these brings a unique perspective and skill set to the task. All of these need to work together as a single unit (a team) to accomplish the desired product objectives based on customer requirements.

These disciplines are used to accomplish the myriad tasks required to develop a reliable product. A study of 72 nondefense corporations revealed that the product reliability techniques they preferred and felt to be important were the following (listed in ranked order) (1):

Supplier control	76%
Parts control	72%
Failure analysis and corrective action	65%
Environmental stress screening	55%
Test, analyze, fix	50%
Reliability qualification test	32%
Design reviews	24%
Failure modes, effects, and criticality analysis	20%

Each of these companies used several techniques to improve reliability. Most will be discussed in this book.

1.3 THE BATHTUB FAILURE RATE CURVE

Historically, the bathtub failure rate curve has been used to discuss electronic equipment (product) reliability. Some practitioners have questioned its accuracy and applicability as a model for reliability. Nonetheless, I use it for "talking purposes" to present and clarify various concepts. The bathtub curve, as shown in Figure 2, represents the instantaneous failure rate of a population of identical items at identical constant stress. The bathtub curve is a composite diagram that provides a framework for identifying and dealing with all phases of the lives of parts and equipment.

Observations and studies have shown that failures for a given part or piece of equipment consist of a composite of the following:

Introduction to Reliability

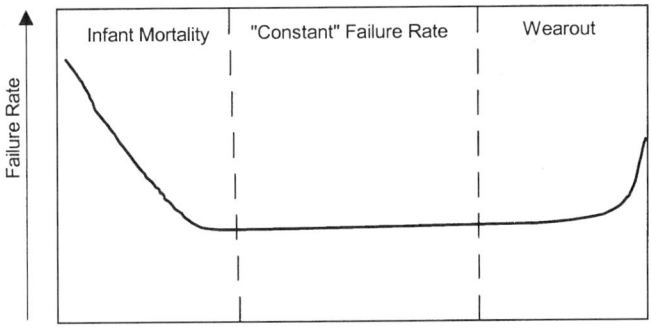

FIGURE 2   The bathtub curve.

Quality	Unrelated to stress Not time-dependent	Eliminated by inspection process and process improvements
Reliability	Stress-dependent	Eliminated by screening
Wearout	Time-dependent	Eliminated by replacement, part design, or new source
Design	May be stress- and/or time-dependent	Eliminated by proper application and derating

The bathtub curve is the sum of infant mortality, random failure, and wearout curves, as shown in Figure 3. Each of the regions is now discussed.

1.3.1   Region I—Infant Mortality/Early Life Failures

This region of the curve is depicted by a high failure rate and subsequent flattening (for some product types). Failures in this region are due to quality problems and are typically related to gross variations in processing and assembly. Stress screening has been shown to be very effective in reducing the failure (hazard) rate in this region.

1.3.2   Region II—Useful Life or Random Failures

Useful life failures are those that occur during the prolonged operating period of the product (equipment). For electronic products it can be much greater than 10 years but depends on the product and the stress level. Failures in this region are related to minor processing or assembly variations. The defects track with the defects found in Region I, but with less severity. Most products have acceptable

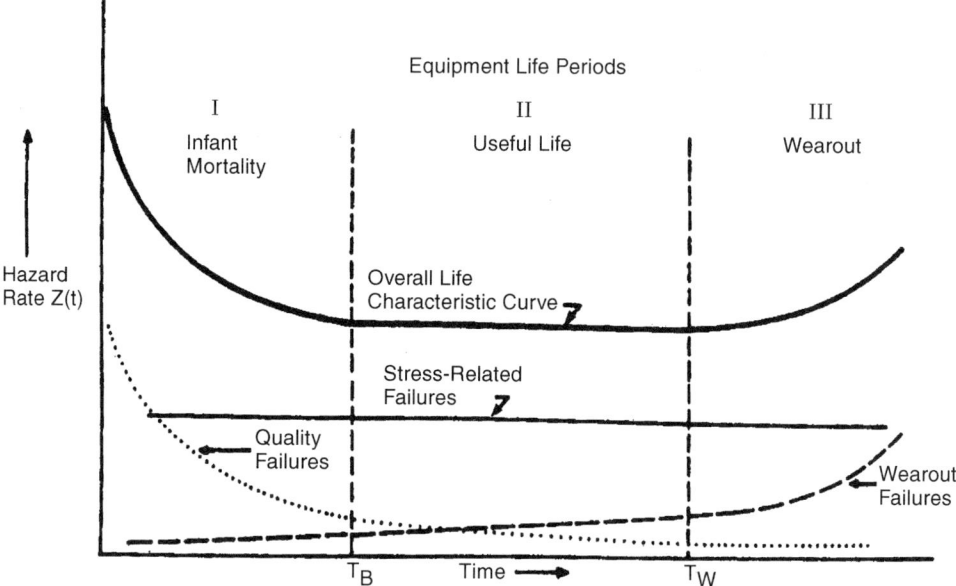

FIGURE 3 The bathtub curve showing how various failures combine to form the composite curve.

failure rates in this region. Field problems are due to "freak" or maverick lots. Stress screening cannot reduce this inherent failure rate, but a reduction in operating stresses and/or increase in design robustness (design margins) can reduce the inherent failure rate.

1.3.3 Region III—Aging and Wearout Failures

Failures in this region are due to aging (longevity exhausted) or wearout. All products will eventually fail. The failure mechanisms are different than those in regions I and II. It has been stated that electronic components typically wear out after 40 years. With the move to deep submicron ICs, this is dramatically reduced. Electronic equipment/products enter wearout in 20 years or so, and mechanical parts reach wearout during their operating life. Screening cannot improve reliability in this region, but may cause wearout to occur during the expected operating life. Wearout can perhaps be delayed through the implementation of stress-reducing designs.

Figures 4–8 depict the bathtub failure rate curves for human aging, a mechanical component, computers, transistors, and spacecraft, respectively. Note that since mechanical products physically wear out, their life cycle failure rate

Introduction to Reliability

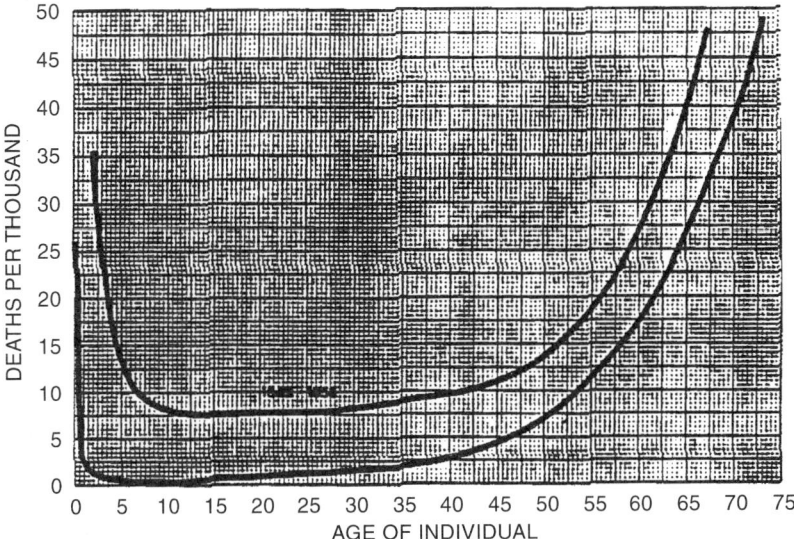

FIGURE 4 Human life cycle curve.

is very different from the electronic product life curve in the following ways: significantly shorter total life; steeper infant mortality; very small useful operating life; fast wearout.

Figure 9 shows that the life curve for software is essentially a flat straight line with no early life or wearout regions because all copies of a software program

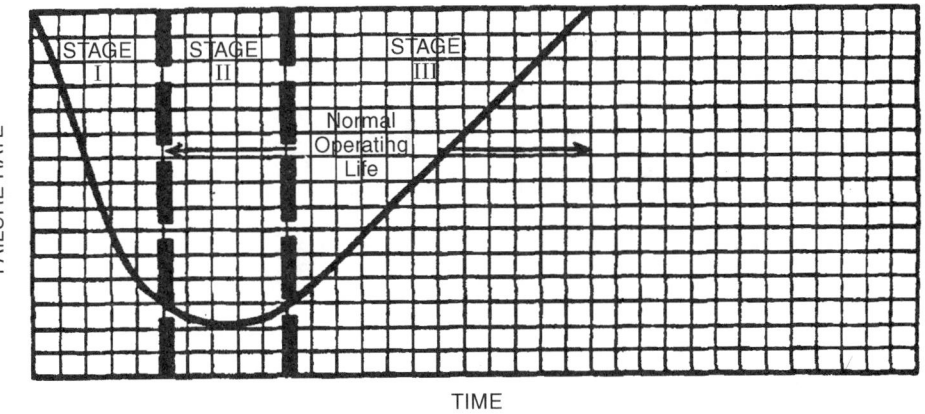

FIGURE 5 Mechanical component life cycle curve.

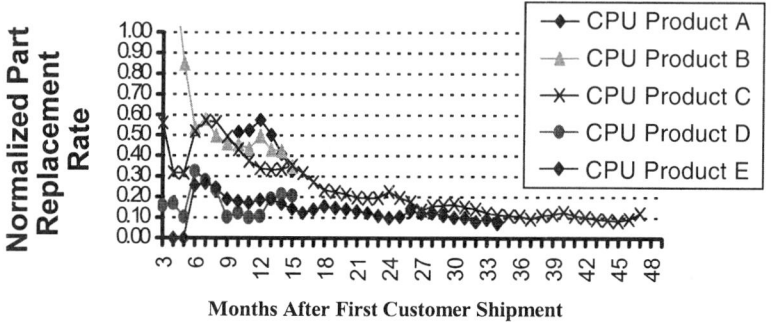

FIGURE 6 Computer failure rate curve.

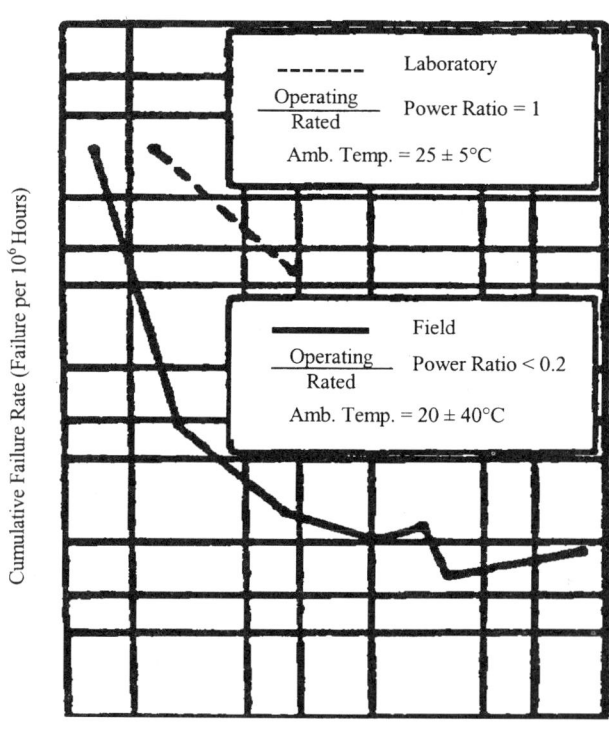

FIGURE 7 Failure rates for NPN silicon transistors (1 W or less) versus calendar operating time.

Introduction to Reliability

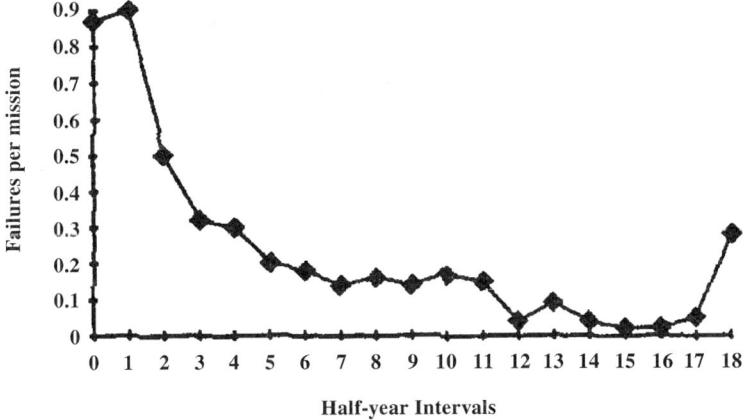

FIGURE 8 Failure rate for spacecraft in orbit.

are identical and software reliability is time-independent. Software has errors or defects just like hardware. Major errors show up quickly and frequently, while minor errors occur less frequently and take longer to occur and detect. There is no such thing as stress screening of software.

The goal is to identify and remove failures (infant mortalities, latent defects) at the earliest possible place (lowest cost point) before the product gets in the customer's hands. Historically, this has been at the individual component level but is moving to the printed wiring assembly (PWA) level. These points are covered in greater detail in Chapters 4 and 7.

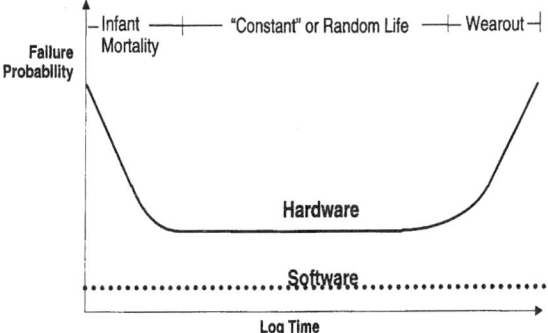

FIGURE 9 Typical software–hardware comparison life curve.

Let me express a note of caution. The bathtub failure rate curve is useful to explain the basic concepts, but for complete electronic products (equipment), the time-to-failure patterns are much more complex than the single graphical representation shown by this curve.

1.4 RELIABILITY GOALS AND METRICS

Most hardware manufacturers establish reliability goals for their products. Reliability goals constrain the design and prevent the fielding of products that cannot compete on a reliability basis. Reliability goals are based on customer expectations and demand, competitive analysis, comparisons with previous products, and an analysis of the technology capability. A combined top-down and bottom-up approach is used for goal setting and allocation. The top-down approach is based on market demand and competitive analysis. Market demand is measured by customer satisfaction surveys, feedback from specific customers, and the business impact of lost or gained sales in which hardware reliability was a factor. The top-down analysis provides reliability goals at a system level, which is the customer's perspective.

The bottom-up approach is based on comparing the current product to previous products in terms of complexity, technology capability, and design/manufacturing processes. Reliability predictions are created using those factors and discussions with component suppliers. These predictions are performed at the unit or board level, then rolled up to the system level to be compared with the top-down goals. If they do not meet the top-down goals, an improvement allocation is made to each of the bottom-up goals, and the process is iterated.

However, there is a wide gap between what is considered a failure by a customer and what is considered a failure by hardware engineering. Again, using computers as an example, the customer perceives any unscheduled corrective maintenance (CM) activity on a system, including component replacement, adjustment, alignment, and reboot as a failure. Hardware engineering, however, considers only returned components for which the failure can be replicated as a failure. The customer-perceived failure rate is significantly higher than engineering-perceived failure rate because customers consider no-trouble-found (NTF) component replacements and maintenance activity without component replacement as failures. This dichotomy makes it possible to have low customer satisfaction with regard to product reliability even though the design has met its failure rate goals. To accommodate these different viewpoints, multiple reliability metrics are specified and measured. The reliability goals are also translated based on customer expectations into hardware engineering goals such that meeting the hardware engineering goals allows the customer expectations to be met.

Typical reliability metrics for a high-reliability, high-availability, fault-

Introduction to Reliability

TABLE 2 Metric Definitions for a High-Reliability, High-Availability, Fault-Tolerant Computer

Metric	Definition
Corrective maintenance (CM) rate	A corrective maintenance activity such as a part replacement, adjustment, or reboot. CMs are maintenance activities done in a reactive mode and exclude proactive activity such as preventive maintenance.
Part replacement (PR) rate	A part replacement is any (possibly multiple) part replaced during a corrective maintenance activity. For almost all the parts we track, the parts are returned to the factory, so part replacement rate is equivalent to part return rate.
Failure rate	A returned part that fails a manufacturing or engineering test. Any parts that pass all tests are called no trouble found (NTF). NTFs are important because they indicate a problem with our test capabilities, diagnostics, or support process/training.

Note: All rates are annualized and based on installed part population.

tolerant computer are shown in Table 2. The CM rate is what customers see. The part (component) replacement (PR) rate is observed by the factory and logistics organization. The failure rate is the engineers' design objective. The difference between the failure rate and the PR rate is the NTF rate, based on returned components that pass all the manufacturing tests. The difference between the CM rate and PR rate is more complex.

If no components are replaced on a service call, the CM rate will be higher than the PR rate. However, if multiple components are replaced on a single service call, the CM rate will be lower than the PR rate. From the author's experience, the CM rate is higher than the PR rate early in the life of a product when inadequate diagnostics or training may lead to service calls for which no problem can be diagnosed. For mature products these problems have been solved, and the CM and PR rates are very similar.

Each of the stated reliability metrics takes one of three forms:

CM/PR/failure rate goal, based on market demand
Expected CM/PR/failure rate, based on predictions
Actual CM/PR/failure rate, based on measurement

The relationships among the various forms of the metrics are shown in Figure 10.

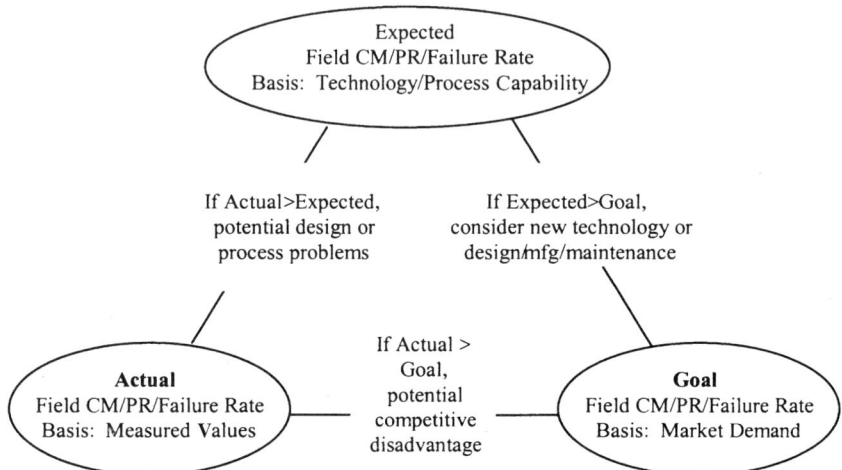

FIGURE 10 Reliability metric forms.

1.5 RELIABILITY PREDICTION

Customers specify a product's reliability requirements. The marketing/product development groups want an accurate quantitative ability to trade off reliability for performance and density. They also may require application-specific qualifications to meet the needs of different market segments. The designers want design for reliability requirements that will not impede their time to market. Manufacturing wants stable qualified processes and the ability to prevent reliability problems. And there is the continuous pressure to reduce the cost of operations.

Reliability modeling assists in calculating system-level reliability from subsystem data and depicts the interrelationship of the components used. Using reliability models, a designer can develop a system that will meet the reliability and system level requirements and can perform tradeoff studies to optimize performance, cost, or specific parameters.

Reliability prediction is performed to determine if the product design will meet its goals. If not, a set of quality initiatives or process improvements are identified and defined such that the goals will be met. Reliability process improvements are justified by relating them directly to improved field reliability predictions.

Reliability prediction is nothing more than a tool for getting a gross baseline understanding of what a product's potential reliability (failure rate) is. The number derived from the calculations is not to be an end-all panacea to the reliability issue. Rather it is the beginning, a call to understand what constitutes reliability for that product and what the factors are that detract from achieving higher reliability. This results in an action plan.

Initial reliability predictions are usually based on component failure rate models using either MIL-HDBK-217 or Bellcore Procedure TR-332. Typically one analyzes the product's bill of materials (BOM) for the part types used and plugs the appropriate numbers into a computer program that crunches the numbers. This gives a first "cut" prediction. However, the failure rates predicted are usually much higher than those observed in the field and are considered to be worst-case scenarios.

One of the criticisms of the probabilistic approach to reliability (such as that of MIL-HDBK-217) is that it does not account for interactions among components, materials, and processes. The failure rate for a component is considered to be the same for a given component regardless of the process used to assemble it into the final product. Even if the same process is used by two different assemblies, their methods of implementation can cause differences.

Furthermore, since reliability goals are based on competitive analysis and customer experience with field usage, handbook-based reliability predictions are unlikely to meet the product goals. In addition, these predictions do not take into account design or manufacturing process improvements possibly resulting from the use of highly accelerated life test (HALT) or environmental stress screening (ESS), respectively. Table 3 presents some of the limitations of reliability prediction.

Thus, reliability prediction is an iterative process that is performed throughout the design cycle. It is not a "once done, forever done" task. The initial reliability prediction is continually refined throughout the design cycle as the bill of materials gets solidified by factoring in test data, failure analysis results, and

TABLE 3 Limitations of Reliability Prediction

Simple techniques omit a great deal of distinguishing detail, and the very prediction suffers inaccuracy.
Detailed prediction techniques can become bogged down in detail and become very costly. The prediction will also lag far behind and may hinder timely hardware development.
Considerable effort is required to generate sufficient data on a part class/level to report statistically valid reliability figures for that class/level.
Component reliability in fielded equipment is very difficult to obtain due to lack of suitable and useful data acquisition.
Other variants that can affect the stated failure rate of a given system are uses, operator procedures, maintenance and rework practices, measurement techniques or definitions of failure, operating environments, and excess handling differing from those addressed by modeling techniques.

the degree to which planned reliability improvement activities are completed. Subsequent predictions take into account usage history with the component technology, suppliers, and specific component type (part number) as well as field data from previous products and the planned design and manufacturing activities. Field data at the Tandem Division of Compaq Computer Corporation has validated that the reliability projections are more accurate than handbook failure rate predictions.

1.5.1 Example of Bellcore Reliability Prediction

A calculated reliability prediction for a 56K modem printed wiring assembly was made using Bellcore Reliability Prediction procedure for Electronic Equipment, TR-332 Issue 5, December 1995. (Currently, Issue 6, December 1997, is the latest revision of Bellcore TR-332. The device quality level has been increased to four levels: 0, I, II, and III, with 0 being the new level. Table 4 describes these four levels.) Inherent in this calculation are the assumptions listed in Table 5.

Assuming component Quality Level I, the calculated reliability for the PWA is 3295 FITS (fails per 10^9 hr), which is equivalent to an MTBF of 303,481 hr. This failure rate is equivalent to an annual failure rate of 0.029 per unit, or 2.9 failures per hundred units per year. The assumption is made that

TABLE 4 Device Quality Level Description from Bellcore TR-332 (Issue 6, December 1997)

The device failure rates contained in this document reflect the expected field reliability performance of generic device types. The actual reliability of a specific device will vary as a function of the degree of effort and attention paid by an equipment manufacturer to factors such as device selection/application, supplier selection/control, electrical/mechanical design margins, equipment manufacturing process controls, and quality program requirements.

Quality Level 0 Commercial-grade, reengineered, remanufactured, reworked, salvaged, or gray-market components that are procured and used without device qualification, lot-to-lot controls, or an effective feedback and corrective action program by the equipment manufacturer.

Quality Level I Commercial-grade components that are procured and used without thorough device qualification or lot-to-lot controls by the equipment manufacturer.

Quality Level II Components that meet requirements of Quality Level I plus purchase specifications that explicitly identify important characteristics (electrical, mechanical, thermal, and environmental), lot control, and devices qualified and listed on approved parts/manufacturer's lists.

Quality Level III Components that meet requirements of Quality Levels I and II plus periodic device qualification and early life reliability control of 100% screening. Also an ongoing continuous reliability improvement program must be implemented.

Introduction to Reliability 17

TABLE 5 56K Modem Analysis Assumptions

An ambient air temperature of 40°C around the components (measured 0.5 in. above the component) is assumed.
Component Quality Level I is used in the prediction procedure. This assumes standard commercial, nonhermetic devices, without special screening or preconditioning. The exception is the Opto-couplers, which per Bellcore recommendation are assumed to be Level III.
Electrical stresses are assumed to be 50% of device ratings for all components.
Mechanical stress environment is assumed to be ground benign (GB).
Duty cycle is 100% (continuous operation).
A mature manufacturing and test process is assumed in the predicted failure rate (i.e., all processes under control).
The predicted failure rate assumes that there are no systemic design defects in the product.

---

there are no manufacturing test, or design problems that significantly affect field reliability. The results fall well within the normal range for similar hardware items used in similar applications. If quality Level II components are used the MTBF improves by a factor of about 2.5. One has to ask the following question: is the improved failure rate worth the added component cost? Only through a risk analysis and an understanding of customer requirements will one be able to answer this question.

	Failure rate (FITS)	MTBF (hr)	Annualized failure rate
Quality Level I	3295	303,481	0.029
Quality Level II	1170	854,433	0.010

The detailed bill-of-material failure rates for Quality Levels I and II are presented in Tables 6 and 7, respectively.

1.6 RELIABILITY RISK

It is important that the person or group of people who take action based on reliability prediction understand risk. Reliability predictions vary. Some of the source of risks include correct statistical distribution, statistical error (confidence limits), and uncertainty in models and parameters.

Reliability metrics revolve around minimizing costs and risks. Four cost elements to incorporate in metrics are

18   Chapter 1

TABLE 6   Reliability Calculation Assuming Quality Level I

ID	Generic name	Item code	Part name	QTY	FR
1.1.54	137240-007	IC-Memory	IC, SRAM,32Kx8,15ns,3.3V,SOJ-2	2	438.8
1.1.89	322078-001	IC-Memory	IC, FEPROM,256Kx8,3.3V,50ns,TSO	1	251.3
1.1.72	194774-001	IC-Analog	IC, SM,V/REG,3.3V,500MA,MAX604	1	155.3
1.1.96	009000-000	IC-Analog	IC, ANALOG, CODEC, MQFP44, LUCE	1	285.8
1.1.71	191952-001	IC-Digital	IC, EEPROM,512x8,24CO4,SOIC	1	198.2
1.1.80	322062-001	IC-Digital	IC, USB uCNTRLR,USS820,48TQFP	1	81.8
1.1.91	322081-001	IC-Digital	IC, DAT PMP,DSP1675T28,128TQFP	1	81.8
1.1.5	106146-099	Resistor	RES, SM,2.4K OHM,1/8W,5%	1	1.5
1.1.6	106146-118	Resistor	RES, SM,15K OHM,1/8W,5%	1	1.5
1.1.9	107263-100	Resistor	RES, SM,2.7K OHM,1/4W,5%	1	1.5
1.1.10	107263-111	Resistor	RES, SM,7.5K OHM,1/4W,5%	1	1.5
1.1.20	114740-238	Resistor	RES, SM,24.3 OHM,1%,1/10W,0805	2	3.0
1.1.23	119200-530	Resistor	RES, SM,20.0k OHM, 1/4W, 1%	2	3.0
1.1.29	119919-001	Resistor	RES, SM,0 OHM,1/16W,5%,0603	4	6.0
1.1.30	119919-034	Resistor	RES, SM 4.7 OHM,1/16W,5%,0603	1	1.5
1.1.31	119919-054	Resistor	RES, SM,33 OHM,1/16W,5%,0603	1	1.5
1.1.32	119919-070	Resistor	RES, SM,150 OHM,1/16W,5%,0603	4	6.0
1.1.34	119919-082	Resistor	RES, SM,470 OHM,1/16W,5%,0603	5	7.5
1.1.35	119919-086	Resistor	RES, SM,680 OHM,1/16W,5%,0603	1	1.5
1.1.36	119919-090	Resistor	RES, SM,1K OH,1/16W,5%,0603	1	1.5
1.1.37	119919-094	Resistor	RES, SM,1.5K OHM,1/16W,5%,0603	1	1.5
1.1.38	119919-097	Resistor	RES, SM,2K OHM,1/16W,5%,0603	1	1.5
1.1.39	119919-102	Resistor	RES, SM,3.3K OHM,1/16W,5%,0603	1	1.5
1.1.40	119919-112	Resistor	RES, SM,8.2K OHM,1/16W,5%,0603	1	1.5
1.1.41	119919-114	Resistor	RES, SM,10K OHM,1/16W,5%,0603	2	3.0
1.1.42	119919-134	Resistor	RES, SM,68KOHM,1/16W,5%,0603	1	1.5
1.1.43	119919-138	Resistor	RES, SM,100K OHM,1/16W,5%,0603	4	6.0
1.1.48	124637-013	Resistor	RES, SM,39 OHM,1W,5%	1	1.5
1.1.55	139708-001	Resistor	RES, SM,100 OHM,1%,1/16W,603	1	1.5
1.1.56	139708-006	Resistor	RES, SM,10K,1%,1/16W,603	18	27.0
1.1.57	139708-015	Resistor	RES, SM,191 OHM,1%,1/16W,603	2	3.0
1.1.58	139708-045	Resistor	RES, SM,34.0K,1%,1/16W,603	2	3.0
1.1.59	139708-096	Resistor	RES, SM,475 OHM,1%,1/16W,603	2	3.0
1.1.60	139708-133	Resistor	RES, SM,33.2k,1%,1/16W,603	2	3.0
1.1.61	139708-135	Resistor	RES, SM,16.2k,1%,1/16W,603	2	3.0
1.1.62	139708-170	Resistor	RES, SM,1.30K,1%,1/16W,0603	1	1.5
1.1.63	139708-194	Resistor	RES, SM,26.1K,1%,1/16W,0603	2	3.0
1.1.1	105077-157	Capacitor	CAP, SM,.047MFD,50V,5%,X7R	1	3.0
1.1.2	105077-163	Capacitor	CAP, SM,50V,X7R,5%,.15uF,1812	2	6.0
1.1.3	105079-236	Capacitor	CAP, SM,820PF,50V,10%,NPO	2	6.0
1.1.14	109764-013	Capacitor	CAP, SM,1MFD,20V,20%,TANT	1	3.0
1.1.15	109764-017	Capacitor	CAP, SM,4.7MFD,20V,20%,TANT	1	3.0
1.1.22	117467-726	Capacitor	CAP, 22MFD,35V ALEL	2	246.0
1.1.25	119917-115	Capacitor	CAP, SM,15pF,5%,50V,COG,603	2	6.0
1.1.26	119917-118	Capacitor	CAP, SM,27pF,5%,50V,COG,603	4	12.0
1.1.27	119917-121	Capacitor	CAP, SM,47pF,5%,50V,COG,603	2	6.0
1.1.45	119949-001	Capacitor	CAP, SM,22uf,35V,20%	3	9.0

Introduction to Reliability

TABLE 6 Continued

ID	Generic name	Item code	Part name	QTY	FR
1.1.46	119949-003	Capacitor	CAP, SM,10uF,16V,20%,ALEL	5	225.0
1.1.50	129621-012	Capacitor	CAP, SM,CER,Y5V,0.1uF,16v,0603	32	96.0
1.1.51	129621-021	Capacitor	CAP, SM,CER,Y5V,10uF,35V,1210	1	3.0
1.1.52	129633-201	Capacitor	CAP, SM,25V,10%,0.015,0603	1	3.0
1.1.73	198183-002	Capacitor	CAP, SM,.47UF,X7R,250V,1825	1	3.0
1.1.83	322066-001	Switch	SWITCH, PUSH-PUSH,PB,THRU/HOLE	1	45.0
1.1.85	322068-001	Relay	IC, SM,SLD ST RLY,2FORMA,400V	1	75.0
1.1.86	322070-001	Relay	IC, SM,PWR SW,1.2A,TPS2041D, SO	1	75.0
1.1.8	106899-016	Connector	CONN, PCB,T/JK,6P,LOW PRO	2	7.2
1.1.82	322065-001	Connector	CONN, USB,4P,T/H,TYPE-B	1	2.4
1.1.4	106125-001	LF Diode	DIODE, SM,GNL PRP...PIN2NC	3	54.0
1.1.17	110118-001	LF Diode	DIODE, SM,ZENER,10V	2	36.0
1.1.18	110118-012	LF Diode	DIODE, SM,ZENER,18V	2	36.0
1.1.24	119606-001	LF Diode	DIODE, SM,DUAL,SWITCHING	2	72.0
1.1.67	187108-002	LF Diode	VSTR, SM,275V,250A,2215	1	30.0
1.1.68	187108-005	LF Diode	VSTR, SM,100V,250A,2215	1	30.0
1.1.93	353914-001	LF Diode	DIODE, SM,TVSARRAY,LOWCAP,500W	1	18.0
1.1.11	107269-002	LF Transistor	XSTR, SM,NPN.........MMBTA42	1	18.0
1.1.16	110098-001	LF Transistor	XSTR, SM,PNP,MED PWR......2907	1	18.0
1.1.49	128920-001	LF Transistor	XSTR, SM,NPN,HGH GAIN,MMBT6429	1	18.0
1.1.74	204109-002	LF Transistor	XSTR, SM,DGTL FET,N-CH,25V,.2A	1	60.0
1.1.90	322080-001	LF Transistor	XSTR, NPN,PWR,80V,1.5A,BD139	1	18.0
1.1.92	342615-001	LF Transistor	IC, SM,LD SW,8V,6323L,SSOT-6	1	120.0
1.1.19	110204-005	Optoelectronic	DIODE, SM,LED,PURE GREEN,XPRNT	4	36.0
1.1.77	298956-001	Optoelectronic	IC, SM,OPTOCPLR,300% CTR	2	54.0
1.1.78	298957-001	Optoelectronic	IC, SM,OPTOCPLR,DUAL,300% CTR	1	54.0
1.1.12	107352-009	Inductive	FB, SM,80 OHM,500mA,1806	4	6.0
1.1.13	107352-013	Inductive	FB, SM,600 OHM,200mA,805	2	3.0
1.1.64	141639-001	Inductive	SPKR, XDCR,MINI PCB MMT	1	21.0
1.1.66	176560-001	Inductive	XFMR, MINI,V.32bis	1	12.0
1.1.65	160642-011	Crystal	XTAL, 12.00MHZ,20PF,30PPM	1	75.0
1.1.69	187131-008	Crystal	XTAL, SM,29.4912MHz,20PF,20PPM	1	75.0

1. The cost of a failure in the field
2. The cost of lost business due to unacceptable field failures
3. Loss of revenue due to reliability qualification delaying time to market
4. The cost of the lost opportunity to trade off "excess" reliability safety margins for increased performance/density

Note that the first two items represent a cost associated with failures that occur in a *small subpopulation* of the devices produced. In contrast the last two terms represent an opportunity to increase the profits on *every part* produced. Economic

TABLE 7 Reliability Calculation Assuming Quality Level II

ID	Generic name	Item code	Part name	QTY	FR
1.1.54	137240-007	IC-Memory	IC, SRAM,32Kx8,15ns,3.3V,SOJ-2	2	146.3
1.1.89	322078-001	IC-Memory	IC, FEPROM,256Kx8,3.3V,50ns,TSO	1	83.8
1.1.72	194774-001	IC-Analog	IC, SM,V/REG,3.3V,500MA,MAX604	1	51.8
1.1.96	009000-000	IC-Analog	IC, ANALOG, CODEC, MQFP44, LUCE	1	95.3
1.1.71	191952-001	IC-Digital	IC, EEPROM,512x8,24CO4,SOIC	1	66.1
1.1.80	322062-001	IC-Digital	IC, USB uCNTRLR,USS820,48TQFP	1	27.3
1.1.91	322081-001	IC-Digital	IC, DAT PMP,DSP1675T28,128TQFP	1	27.3
1.1.5	106146-099	Resistor	RES, SM,2.4K OHM,1/8W,5%	1	0.5
1.1.6	106146-118	Resistor	RES, SM,15K OHM,1/8W,5%	1	0.5
1.1.9	107263-100	Resistor	RES, SM,2.7K OHM,1/4W,5%	1	0.5
1.1.10	107263-111	Resistor	RES, SM,7.5K OHM,1/4W,5%	1	0.5
1.1.20	114740-238	Resistor	RES, SM,24.3 OHM,1%,1/10W,0805	2	1.0
1.1.23	119200-530	Resistor	RES, SM,20.0k OHM, 1/4W, 1%	2	1.0
1.1.29	119919-001	Resistor	RES, SM,0 OHM,1/16W,5%,0603	4	2.0
1.1.30	119919-034	Resistor	RES, SM 4.7 OHM,1/16W,5%,0603	1	0.5
1.1.31	119919-054	Resistor	RES, SM,33 OHM,1/16W,5%,0603	1	0.5
1.1.32	119919-070	Resistor	RES, SM,150 OHM,1/16W,5%,0603	4	2.0
1.1.34	119919-082	Resistor	RES, SM,470 OHM,1/16W,5%,0603	5	2.5
1.1.35	119919-086	Resistor	RES, SM,680 OHM,1/16W,5%,0603	1	0.5
1.1.36	119919-090	Resistor	RES, SM,1K OH,1/16W,5%,0603	1	0.5
1.1.37	119919-094	Resistor	RES, SM,1.5K OHM,1/16W,5%,0603	1	0.5
1.1.38	119919-097	Resistor	RES, SM,2K OHM,1/16W,5%,0603	1	0.5
1.1.39	119919-102	Resistor	RES, SM,3.3K OHM,1/16W,5%,0603	1	0.5
1.1.40	119919-112	Resistor	RES, SM,8.2K OHM,1/16W,5%,0603	1	0.5
1.1.41	119919-114	Resistor	RES, SM,10K OHM,1/16W,5%,0603	2	1.0
1.1.42	119919-134	Resistor	RES, SM,68KOHM,1/16W,5%,0603	1	0.5
1.1.43	119919-138	Resistor	RES, SM,100K OHM,1/16W,5%,0603	4	2.0
1.1.48	124637-013	Resistor	RES, SM,39 OHM,1W,5%	1	0.5
1.1.55	139708-001	Resistor	RES, SM,100 OHM,1%,1/16W,603	1	0.5
1.1.56	139708-006	Resistor	RES, SM,10K,1%,1/16W,603	18	9.0
1.1.57	139708-015	Resistor	RES, SM,191 OHM,1%,1/16W,603	2	1.0
1.1.58	139708-045	Resistor	RES, SM,34.0K,1%,1/16W,603	2	1.0
1.1.59	139708-096	Resistor	RES, SM,475 OHM,1%,1/16W,603	2	1.0
1.1.60	139708-133	Resistor	RES, SM,33.2k,1%,1/16W,603	2	1.0
1.1.61	139708-135	Resistor	RES, SM,16.2k,1%,1/16W,603	2	1.0
1.1.62	139708-170	Resistor	RES, SM,1.30K,1%,1/16W,0603	1	0.5
1.1.63	139708-194	Resistor	RES, SM,26.1K,1%,1/16W,0603	2	1.0
1.1.1	105077-157	Capacitor	CAP, SM,.047MFD,50V,5%,X7R	1	1.0
1.1.2	105077-163	Capacitor	CAP, SM,50V,X7R,5%,.15uF,1812	2	2.0
1.1.3	105079-236	Capacitor	CAP, SM,820PF,50V,10%,NPO	2	2.0
1.1.14	109764-013	Capacitor	CAP, SM,1MFD,20V,20%,TANT	1	1.0
1.1.15	109764-017	Capacitor	CAP, SM,4.7MFD,20V,20%,TANT	1	1.0
1.1.22	117467-726	Capacitor	CAP, 22MFD,35V ALEL	2	82.0
1.1.25	119917-115	Capacitor	CAP, SM,15pF,5%,50V,COG,603	2	2.0
1.1.26	119917-118	Capacitor	CAP, SM,27pF,5%,50V,COG,603	4	4.0
1.1.27	119917-121	Capacitor	CAP, SM,47pF,5%,50V,COG,603	2	2.0
1.1.45	119949-001	Capacitor	CAP, SM,22uf,35V,20%	3	3.0

Introduction to Reliability

TABLE 7 Continued

1.1.46	119949-003	Capacitor	CAP, SM,10uF,16V,20%,ALEL	5	75.0
1.1.50	129621-012	Capacitor	CAP, SM,CER,Y5V,0.1uF,16v,0603	32	32.0
1.1.51	129621-021	Capacitor	CAP, SM,CER,Y5V,10uF,35V,1210	1	1.0
1.1.52	129633-201	Capacitor	CAP, SM,25V,10%,0.015,0603	1	1.0
1.1.73	198183-002	Capacitor	CAP, SM,.47UF,X7R,250V,1825	1	1.0
1.1.83	322066-001	Switch	SWITCH, PUSH-PUSH,PB,THRU/HOLE	1	15.0
1.1.85	322068-001	Relay	IC, SM,SLD ST RLY,2FORMA,400V	1	25.0
1.1.86	322070-001	Relay	IC, SM,PWR SW,1.2A,TPS2041D, SO	1	25.0
1.1.8	106899-016	Connector	CONN, PCB,T/JK,6P,LOW PRO	2	2.4
1.1.82	322065-001	Connector	CONN, USB,4P,T/H,TYPE-B	1	0.8
1.1.4	106125-001	LF Diode	DIODE, SM,GNL PRP...PIN2NC	3	18.0
1.1.17	110118-001	LF Diode	DIODE, SM,ZENER,10V	2	12.0
1.1.18	110118-012	LF Diode	DIODE, SM,ZENER,18V	2	12.0
1.1.24	119606-001	LF Diode	DIODE, SM,DUAL,SWITCHING	2	24.0
1.1.67	187108-002	LF Diode	VSTR, SM,275V,250A,2215	1	10.0
1.1.68	187108-005	LF Diode	VSTR, SM,100V,250A,2215	1	10.0
1.1.93	353914-001	LF Diode	DIODE, SM,TVSARRAY,LOWCAP,500W	1	6.0
1.1.11	107269-002	LF Transistor	XSTR, SM,NPN.........MMBTA42	1	6.0
1.1.16	110098-001	LF Transistor	XSTR, SM,PNP,MED PWR......2907	1	6.0
1.1.49	128920-001	LF Transistor	XSTR, SM,NPN,HGH GAIN,MMBT6429	1	6.0
1.1.74	204109-002	LF Transistor	XSTR, SM,DGTL FET,N-CH,25V,.2A	1	20.0
1.1.90	322080-001	LF Transistor	XSTR, NPN,PWR,80V,1.5A,BD139	1	6.0
1.1.92	342615-001	LF Transistor	IC, SM,LD SW,8V,6323L,SSOT-6	1	40.0
1.1.19	110204-005	Optoelectronic	DIODE, SM,LED,PURE GREEN,XPRNT	4	12.0
1.1.77	298956-001	Optoelectronic	IC, SM,OPTOCPLR,300% CTR	2	54.0
1.1.78	298957-001	Optoelectronic	IC, SM,OPTOCPLR,DUAL,300% CTR	1	54.0
1.1.12	107352-009	Inductive	FB, SM,80 OHM,500mA,1806	4	2.0
1.1.13	107352-013	Inductive	FB, SM,600 OHM,200mA,805	2	1.0
1.1.64	141639-001	Inductive	SPKR, XDCR,MINI PCB MMT	1	7.0
1.1.66	176560-001	Inductive	XFMR, MINI,V.32bis	1	4.0
1.1.65	160642-011	Crystal	XTAL, 12.00MHZ,20PF,30PPM	1	25.0
1.1.69	187131-008	Crystal	XTAL, SM,29.4912MHz,20PF,20PPM	1	25.0

pressures are going to force increased attention on reliability's role in improving time to market and enhancing performance.

There are two ways to increase the value of reliability predictions. First, rather than a point prediction, the capability is needed to develop curves of reliability levels versus design, manufacturing, and end use variables (Fig. 11). This will allow optimization of the reliability given the economics of a particular marketplace. Second, risk needs to be quantified so it can be factored into technology decisions.

Let's use the bathtub curve to try to answer this question. As mentioned before, the bathtub curve depicts a product's reliability (i.e., failure rate) through-

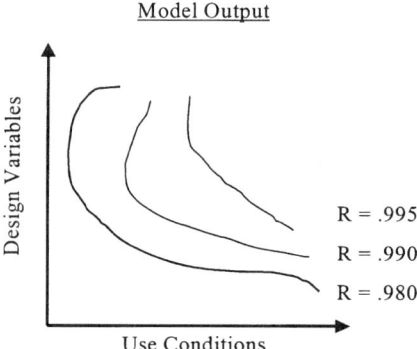

FIGURE 11 Curves of reliability levels as a function of design and use conditions.

out its life. Figure 12 shows the bathtub curve with a vertical line placed at the product's design life requirements. If a high margin exists between the lifetime requirement and the wearout time, a high cost is incurred for having this design margin (overdesign for customer requirements), but there is a low reliability risk. If the wearout portion of the curve is moved closer to the lifetime requirement (less design margin), then a lower cost is incurred but a greater reliability risk presents itself. Thus, moving the onset of wearout closer to the lifetime expected by the customer increases the ability to enhance the performance of all products, is riskier, and is strongly dependent on the accuracy of reliability wearout models. Thus, one must trade off (balance) the high design margin versus cost. Several prerequisite questions are (1) why do we need this design margin and (2) if I

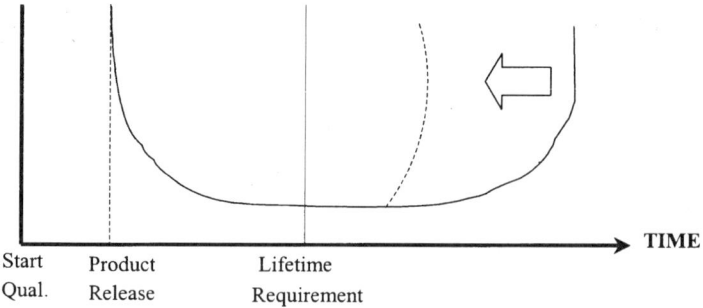

FIGURE 12 Bathtub curve depicting impact of short versus long time duration between product lifetime requirement specifications and wearout.

didn't need to design my product with a larger margin, could I get my product to market faster?

This begs the question what level of reliability does the customer for a given product really need. It is important to understand that customers will ask for very high levels of reliability. They do this for two reasons: (1) they don't know what they need and (2) as a safety net so that if the predictions fall short they will still be okay. This requires that the designer/manufacturer work with the customer to find out the true need. Then the question must be asked, is the customer willing to pay for this high level of reliability? Even though the customer's goal is overall system reliability, more value is often placed on performance, cost, and time to market. For integrated circuits, for example, it is more important for customers to get enhanced performance, and suppliers may not need to fix or improve reliability. Here it's okay to hold reliability levels constant while aggressively scaling and making other changes.

1.7 RELIABILITY GROWTH

Reliability growth is a term used to describe the increase in equipment mean time to failure that comes about due to improvements in design and manufacturing throughout the development, preproduction, and early production phases. The model originally proposed by Duane (3) is probably the most well known for forecasting reliability growth. Since the burn-in process also in effect enhances the reliability, there has been some confusion regarding growth due to corrective actions in design and production, and growth due to burn-in.

Figures 13a and 13b illustrate the separate effects of burn-in and MTTF

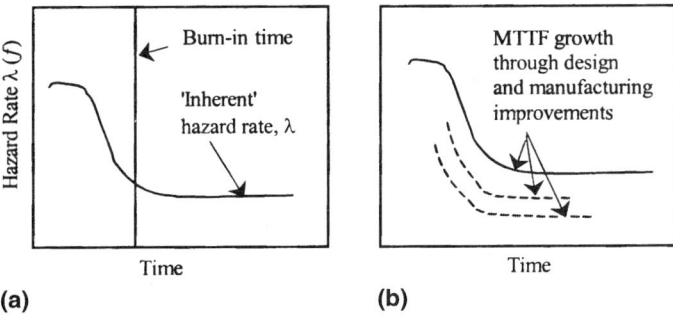

FIGURE 13 (a) Reliability enhancement through burn-in. (b) Reliability enhancement through MTTF growth.

growth. Burn-in removes the weak components and in this way brings the equipment into its useful life period with a (supposedly) constant hazard rate λ (see Fig. 13a). Reliability growth through design and manufacturing improvements, on the other hand, steadily reduces the inherent hazard rate in the useful life period of the product, i.e. it increases the MTTF. The corrective actions we speak of when discussing burn-in are primarily directed toward reducing the number of infant mortality failures. Some of these improvements may also enhance the MTTF in the useful life period, providing an added bonus. The efforts expended in improving the MTTF may very well reflect back on early failures as well. Nonetheless, the two reliability enhancement techniques are independent.

1.8 RELIABILITY DEGRADATION

Degradation can be defined as the wearing down of the equipment through unwanted actions occurring in items in the equipment. An example would be component degradation. Degradation over time slowly erodes or diminishes the item's effectiveness until an eventual failure occurs. The cause of the failure is called the failure mechanism. A graphic example of degradation is the wearing of land by unwanted action of water, wind, or ice, i.e., soil erosion.

Product or equipment reliability degradation can occur due to process-induced manufacturing defects and assembly errors, the variable efficiency of conventional manufacturing and quality control inspection processes, and the latent defects attributable to purchased components and materials. The last has historically caused irreparable problems in the electronics industry and requires that strict process control techniques be used in component manufacturing.

The problem here is the unknown number of latent defects in marginal or weakened components which can fail under proper conditions of stress, usually during field operation.

Some of the things that can be done to prevent reliability degradation are the following:

1. "Walk the talk" as regards quality. This requires a dedication to quality as a way of life from the company president down to the line worker.
2. Implement an effective quality control program at the component, PWA, module, subsystem, and system levels.
3. Design for manufacturing, testability, and reliability.
4. Use effective statistical quality control techniques to remove variability.
5. Implement manufacturing stress screens.
6. Improve manufacturing and test equipment preventative maintenance actions and eliminate poorly executed maintenance.

Introduction to Reliability 25

7. Train the work and maintenance forces at all levels and provide essential job performance skills.
8. Include built-in test equipment and use of fault-tolerant circuitry.

1.8.1 Component Degradation

Component degradation is typically a change which occurs with time that causes the component's operational characteristics to change such that the component may no longer perform within its specification parameters. Operation degradation will occur through the accumulation of thousands of hours of component operation. The component may eventually fail due to wearout. If a component such as a semiconductor device is used within its design constraints and properly manufactured, it will provide decades of trouble-free operation.

Component Degradation Mechanisms

Typical IC degradation mechanisms include

1. Electrical overstress
2. Operation outside of a component's design parameters
3. Environmental overstress
4. Operational voltage transients
5. Test equipment overstress (exceeding the component's parameter ratings during test)
6. Excessive shock (e.g., from dropping component on hard surface)
7. Excessive lead bending
8. Leaking hermetically sealed packages
9. High internal moisture entrapment (hermetic and plastic packages)
10. Microcracks in the substrate
11. Chemical contamination and redistribution internal to the device
12. Poor wire bonds
13. Poor substrate and chip bonding
14. Poor wafer processing
15. Lead corrosion due to improperly coated leads
16. Improper component handling in manufacturing and testing
17. Use of excessive heat during soldering operations
18. Use of poor rework or repair procedures
19. Cracked packages due to shock or vibration
20. Component inappropriate for design requirements

Looking through the list of degradation mechanisms indicates, it is clear that they can be eliminated as potential failure mechanisms resulting in high cost savings. These mechanisms can be eliminated by use of proper component design, manufacturing, and derating processes and by ensuring that the correct component is used in the application.

It is difficult to detect component degradation in a product until the product ceases functioning as intended. Degradation is very subtle in that it is typically a slowly worsening condition.

1.9 RELIABILITY CHALLENGES

Electronics is in a constant state of evolution and innovation, especially for complex products. This results in some level of uncertainty as regards reliability and thus poses challenges. Two of these are as follows:

1. The ratio of new to tried and true portions of electronic systems is relatively high, therefore, reliability information may be largely unknown.
2. There is basically no statistically valid database for new technology that is in a constant state of evolution. Predictions cannot be validated until an accepted database is available.

1.10 RELIABILITY TRENDS

The integration of technology into every dimension of our lives has allowed customers to choose among many options/possibilities to meet their needs. This has led to a raised expectation of customized products. We have come from a one-size-fits-all product and reliability mindset to one of customized products and reliability of manufacturing batches/lots of a single unit. This has increased the number of possible solutions and product offerings.

Product designers and manufacturers have been driven to cut development times; product lifetimes have decreased due to the pace of innovation, and shorter times to market and times to revenue have resulted. Concurrently, the time between early adoption and mass adoption phases for new electronic and digital products has been compressed.

In the past the physical life environment was important in reliability prediction. Today's shortened product life has caused one to question the underlying methodology of the past. What are the concerns of a product that will be produced with an expected life of 18 months or less? A product that could be considered to be disposable? Do we simply toss out the methods that worked in the past, or do we step back and decide which of our tools are the most appropriate and applicable to today's product development and life cycles and customer expectations? In this book various tools and methods are presented that do work for high-end electronic products. It is up to the readers to decide which of the methods presented make sense and should be used in the changing conditions facing them in their own company in their chosen marketplace.

ACKNOWLEDGMENT

Portions of Section 1.4 are extracted from Ref. 4, courtesy of the Tandem Division of Compaq Computer Corporation Reliability Engineering Department.

REFERENCES

1. Criscimagna NH. Benchmarking Commercial Reliability Practices. IITRI, 1997.
2. Reliability Physics Symposium, 1981.
3. Duane JT. Learning curve approach to reliability monitoring. IEEE Transactions on Aerospace. Vol. 2, No. 2, pp. 563–566, 1964.
4. Elerath JG et al. Reliability management and engineering in a commercial computer environment. 1999. International Symposium on Product Quality and Integrity. Courtesy of the Tandem Division of Compaq Computer Corporation Reliability Engineering Dept.

2
Basic Reliability Mathematics

Detailed discussions of the commonly used measures of reliability, statistical terms, and distributions can be found in any basic reliability textbook. This chapter provides a review of basic reliability mathematics.

2.1 STATISTICAL TERMS

Each individual failure mechanism of an item or set of items has its own distribution in time, which may or may not begin at time zero. For example, if a group of products has a subset with cracks of a certain size, the cracks will grow during the service life of the product and cause failures of the subset according to their own distribution before wearout of the entire lot. If the same lot has a different subset with somewhat smaller cracks, they will fail later than the first, according to a different distribution but still before wearout. Another subset with a different defect will fail according to another distribution, etc.

These individual distributions may also represent the probability of random events which can overstress an item and cause immediate failure. Examples of such events are electrical surges, electrical overstress, dropping or damaging an item, rapid temperature changes, and the like. Usually the probability of such random events is considered constant throughout the useful life region.

It is often impractical to use an entire distribution as a variable. Instead, parameters of distributions that are important to the task at hand are used. These parameters could be almost any feature of a life distribution, but some parameters have become common, such as the mean, median, mode, and standard deviation. We are most familiar with these terms as they relate to the normal distribution. They are calculated in different ways but have the same meaning for other distributions.

A common terminology in reliability work is the time at which a certain percentage of items fail: t_{50}, t_{16}, and t_{75}, for instance, are the times by which 50%, 16%, and 75% of the items in a sample have failed. The median of a normal distribution is t_{50}.

The location parameter of a distribution locates it in time. For a normal distribution, the location parameter is the mean. Location parameters are also called measures of central tendencies, measures of central values, and measures of location.

The shape parameter provides a quantitative measure of the shape, or spread, of a distribution. For a normal distribution, the shape parameter is the standard deviation. Shape parameters are also called measures of variation.

Variability exists in all design and manufacturing processes. The goal is to reduce this variability to a very narrow distribution, thus reducing costs.

Variables may be either discrete or continuous. A variable that is described by a probabilistic law is called a random variable. The properties of a random variable are specified by the set of possible values it may take, together with associated probabilities of occurrence of these values. Reliability studies deal with both discrete and continuous random variables. The number of failures in a given interval of time is an example of a discrete variable. The time from part installation to part failure and the time between successive equipment failures are examples of continuous random variables. The distinction between discrete and continuous variables (or functions) depends upon how the problem is treated and not necessarily on the basic physical or chemical processes involved.

The commonly used measures of reliability are defined and summarized in Table 1.

2.1.1 Part Replacement Rate for Computer Applications

The part replacement rate (PRR) is the number of times a part is removed from a system. The PRR is calculated as follows:

$$\text{PRR/year} \cong \frac{\text{PRR}}{\text{year}} \, 8760 \left(\frac{\text{Number of removals}}{\text{Run hours}} \right) = \frac{8760}{\text{MTBPR}} \quad (2.11)$$

Note: 8760 hr/year = 365 days × 24 hr.

Basic Reliability Mathematics

TABLE 1 Basic Reliability Functions and Their Definitions

1. Cumulative distribution function (CDF), F(t).

   $$F(t) = P\{T \leq t\} \tag{2.1}$$

   This is the probability that failure takes place at a time T less than or equal to t.

   $$F(t) = \int f(t)dt \tag{2.2}$$

2. Probability density function (PDF), f(t), is the probability that a system survives for a specified period of time.

   $$f(t)\,\Delta t = P\{t \leq T \leq t + \Delta t\} \tag{2.3}$$

   This is the probability that failure takes place at a time T between t and t + Δt.

   $$f(t) = F'(t) = \int f(t)dt \tag{2.4}$$

3. Reliability (survivor) function, R(t), represents the probability that an item will not have failed by a given time t, that is, that the item will be reliable.

   $$R(t) = P\{T > t\} \tag{2.5}$$

   This is the probability that a system or unit or module or component operates without failure for length of time t. In its general form it is equal to 1 minus the CDF:

   $$R(t) = 1 - F(t) \tag{2.6}$$

4. Hazard rate function is the probability that an item will fail during the next time interval, Δt, given that it is functioning at the beginning of that interval.

   $$\lambda t = \frac{f(t)}{R(t)} = \frac{f(t)}{1 - F(t)} \tag{2.7}$$

   For ICs this is the instantaneous rate of failure of a population of ICs that have survived to time t.

5. Mean time to failure (MTTF).

   $$\text{MTTF} = \int_0^\infty t f(t)dt \tag{2.8}$$

   Substituting for f(t) and integrating by parts gives

   $$\text{MTTF} = \int_0^\infty t\left[\frac{-dR(t)}{dt}\right]dt = -tR(t)\bigg|_0^\infty + \int_0^\infty R(t)dt \tag{2.9}$$

   The term tR(t) = 0 at t = 0, since R(t) $\leq$ 1.

   R(t) $\to$ 0 exponentially as t $\to$ ∞, so that $tR(t)\bigg|_\infty = 0$

   Therefore,

   $$\text{MTTF} = \int_0^\infty R(t)dt \tag{2.10}$$

2.1.2 Mean Time Between Failure

Mean time between failure (MTBF) is the average time that a unit operates between failures. It is the actual field MTBF and is calculated as follows and measured in hours:

$$\text{MTBF} = \frac{\text{Run hours}}{\text{Number of failures}} \quad (2.12)$$

2.1.3 Mean Time to Failure

Mean time to failure (MTTF) was defined in Table 2.1. Oftentimes the terms *mean time between failure* and *mean time to repair* (MTTR) are also used as measures of failure. These are graphically depicted and defined in Figure 1.

Since MTTF is usually several years, while MTTR is at most a few hours, we can normally make the following approximation:

$$\text{MTBF} \cong \text{MTTF}$$

$$\text{Availability} = \frac{\text{MTTF}}{\text{MTTF} + \text{MTTR}} \quad (2.13)$$

We often need to distinguish between component-level and system-level MTBF. Note that MTBSF is mean time between *system* failures. Often this is hundreds of years.

MTTF Calculation Examples

Example 1

a. Calculate the MTTF from the following data:

10 failures
Times to failure (in hours): 65,000, 56,000, 86,000, 72,000, 48,000, 92,000, 66,000, 71,000, 78,000, 66,000

$$\text{MTBF} = \frac{\Sigma \text{ (times)}}{10} = \frac{700,000}{10} = 70,000 \text{ hr}$$

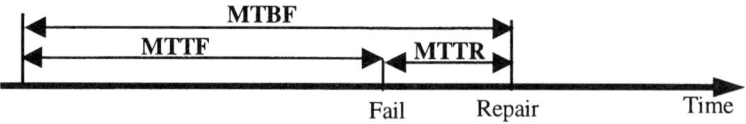

FIGURE 1 Relationship between MTTF, MTBF, and MTTR. *Note*: if MTBF is defined as mean time *before* failure, then MTBF = MTTF.

Basic Reliability Mathematics

b. Each repair took 1.25 hr. Compute the availability. Show this as outage minutes per year.

$$A = \frac{MTTF}{MTTF + MTTR} = \frac{70{,}000}{70{,}000 + 1.25}$$
$$= .999982 \text{ (unitless, measure of uptime)}$$

To scale this to something more meaningful, we convert this to minutes per year:

Minutes per year = 60 min/hr × 24 hr/day × 365 day/year = 525,600
Outage = 525,600 (1 − .999982) = 9.46 min/year

This is easier to understand than .999982.

So, how does this translate to cost? Large computer system business application outages can cost a customer between $1000 and $5000 per outage minute.

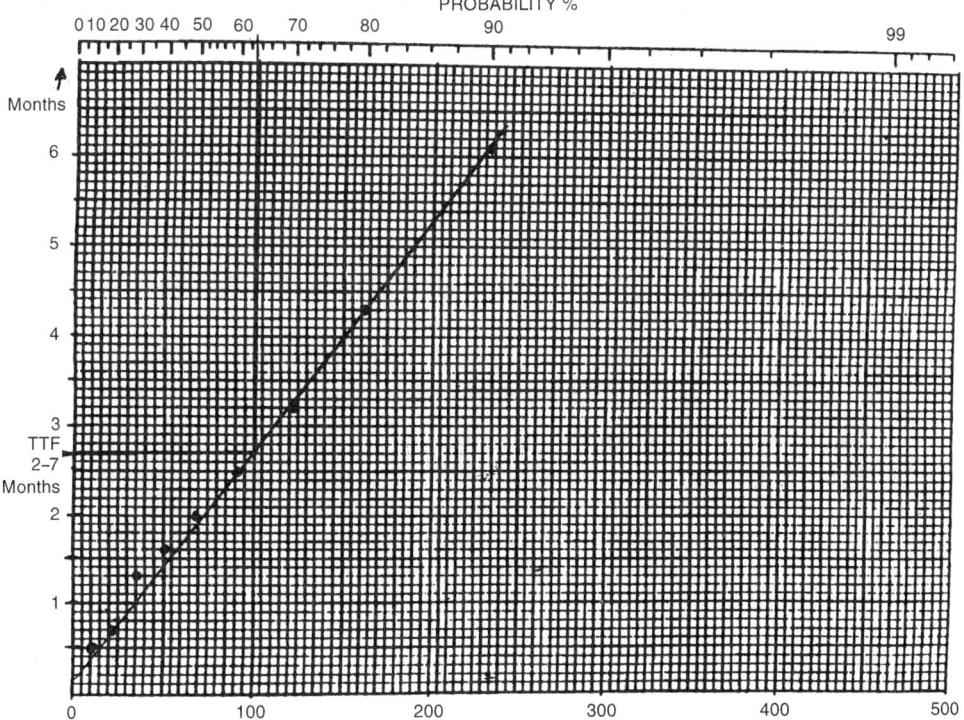

FIGURE 2 Determining MTTF.

Example 2. A particular software application running on a mainframe computer experiences failures at the following times (in months):

1. 3.2   6. 9.3
2. 4.8   7. 15.4
3. 5.3   8. 19.7
4. 5.5   9. 22.2
5. 7.3

What is the MTTF, as determined from the plot of time versus probability? If the failures seem to represent a constant failure rate, then what is that rate, λ?

Here is a schematic look at where the failures occur on the time scale:

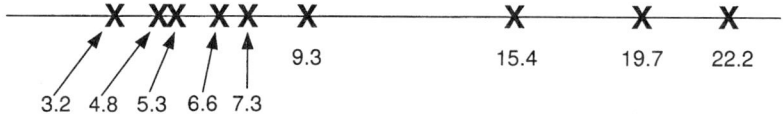

1. First calculate intervals (delta-t's): 3.2, 1.6, 0.5, 1.3, 0.7, 2.0, 6.1, 4.3, 2.5.
2. Put these in ascending order: 0.5, 0.7, 1.3, 1.6, 2.0, 2.5, 3.2, 4.3, 6.1.
3. Divide the interval of 100% by N + 1 = 9 + 1 = 10: 10%, 20%, 30%, etc.
4. Plot on exponential paper time values on the y axis, percentage on the x axis (Fig. 2).
5. Find the 63% point, drop down the 63% line, and find the y-axis intercept. It's about 2.7 months. This is the MTTF. If this is a constant failure rate (it appears to be), then the failure rate λ is the inverse of the MTTF: $\lambda = 1/2.8 = .37$ per month.

2.2 STATISTICAL DISTRIBUTIONS

The basic tool of the reliability engineer is the life distribution (bathtub curve or failure rate curve), which may also be called the failure distribution. Life distributions are a method of describing failures in time mathematically. They can be either probabilistic (a combination of smaller distributions of different failure mechanisms or deterministic (a single distribution representing a single failure mechanism). The ordinate of a life distribution is usually some function of time, and the abscissa is usually some measure of failure probability or failure rate.

Many standard statistical distributions may be used to model various reliability parameters. However, a relatively small number of statistical distributions satisfies most reliability needs. The particular distribution used depends upon the nature of the data. The most appropriate model for a particular application may

Basic Reliability Mathematics

be decided either empirically or theoretically or by a combination of both approaches.

Life distributions can have any shape, but some standard forms have become commonly used over the years, and reliability engineers generally try to fit their data to one of them so that conclusions can be drawn. The normal, lognormal, Weilbull, and exponential distributions will be presented due to their widespread use in reliability. Full descriptions and examples of their use are given in any good introductory statistical text.

Figures 3 and 4 summarize the shape of common failure density, reliability, and hazard rate functions for each of the distributions.

2.2.1 Continuous Distributions

Normal (or Gaussian) Distribution

There are two principal reliability applications of the normal distribution. These are (1) The analysis of items which exhibit failure due to wear, such as mechanical devices, and (2) the analysis of manufactured items and their ability to meet specifications. No two parts made to the same specifications are exactly alike. The variability of parts leads to a variability in hardware products incorporating those parts. The designer must take this part variability into account; otherwise the equipment may not meet the specification requirements due to the combined effects of part variability.

Use of the normal distribution in this application is based upon the central limit theorem. It states that the sum of a large number of identically distributed random variables, each with a finite mean μ and a standard deviation σ, is normally distributed. Thus, the variations in parameters of electronic component parts due to manufacturing are considered to be normally distributed.

The normal distribution is not as common in reliability work as the other three distributions in this section, but it is included here because of its familiarity.

Lognormal Distribution

If the natural logarithm of a function is found to be distributed normally, then the function is said to be lognormal. As shown in Figure 3, μ defines the mean of the distribution and σ defines its standard deviation. A third parameter, t, representing the minimum life, may also be incorporated in the lognormal distribution. Physical examples of the lognormal distribution are the fatigue life of certain types of mechanical components and incandescent light bulbs. Light bulbs eventually suffer filament deterioration and empirically follow a lognormal distribution. Semiconductor failures also frequently follow a lognormal distribution.

The lognormal distribution is very common in reliability work. It is used to model and describe failures in any system which is subject to stresses (such as accelerated life failures in time) which are multiplicative, and in which the

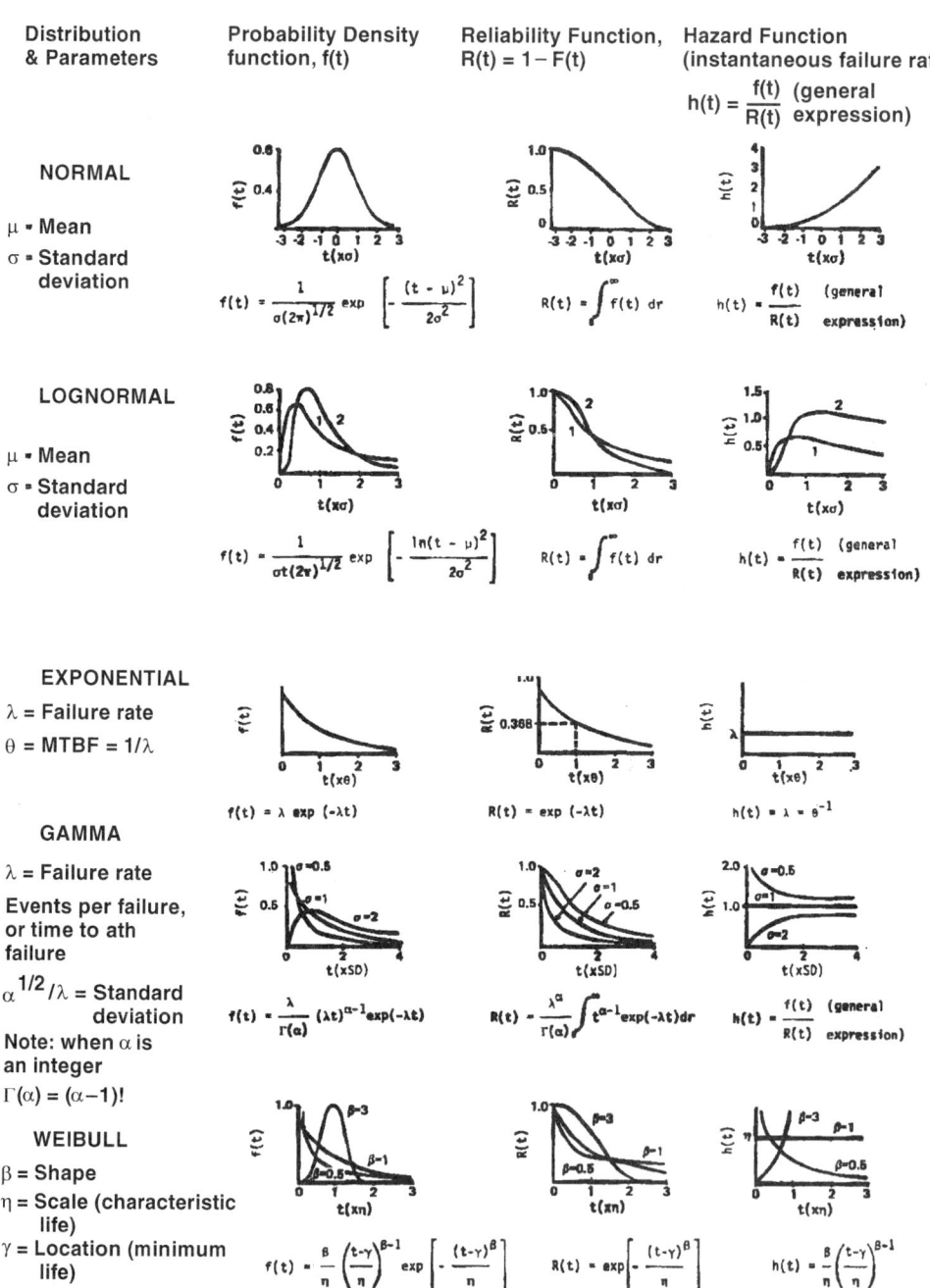

FIGURE 3 Commonly used continuous reliability distribution.

Basic Reliability Mathematics

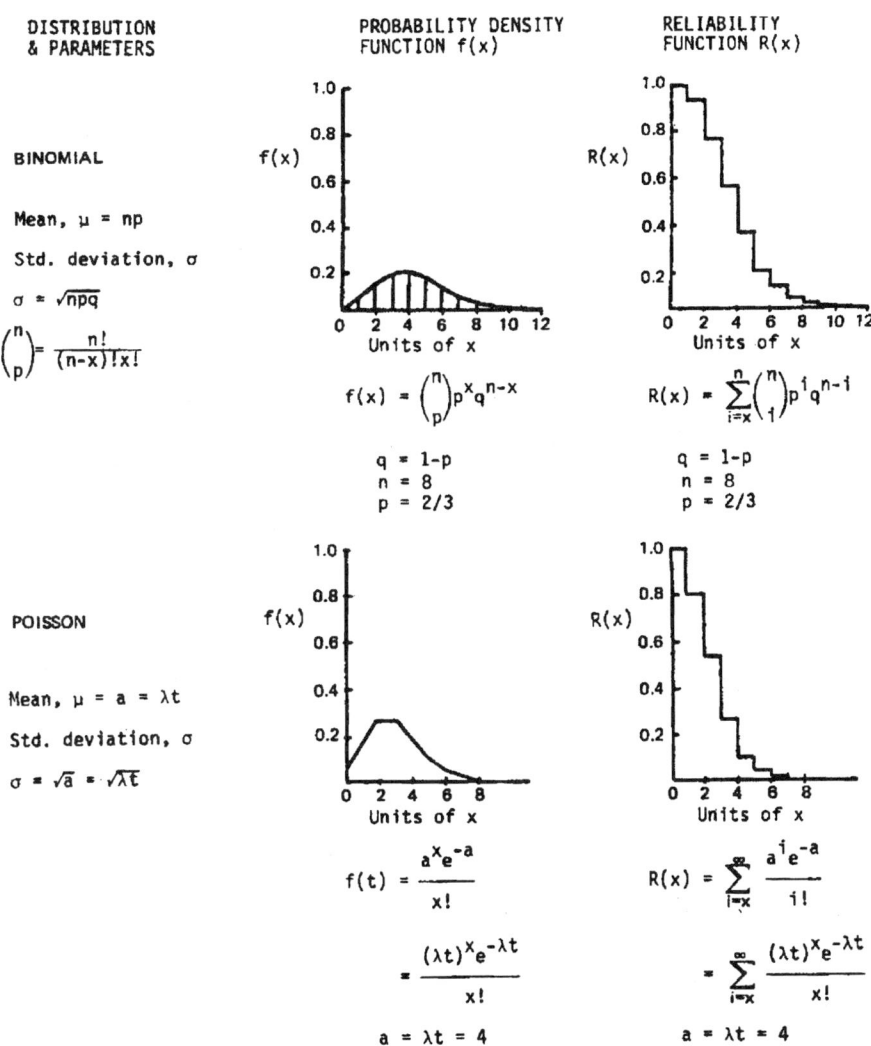

FIGURE 4 Commonly used discrete distributions.

susceptibility to failure of an item is dependent on its age. It is usually assumed that most semiconductor failures follow a log normal distribution. For integrated circuits, diffusion, temperature, and high voltage have linked relations to failures due to chemical reactions such as contamination. Crack propagation, oxide growth, and wear are examples of this.

The Exponential Distribution

The exponential distribution is one of the most important distributions in reliability work. It is used almost exclusively for reliability prediction of electronic equipment. The exponential model is useful and simple to understand and use; it is good for introductory concepts and calculations, has a constant hazard rate, and is a special case of the Weibull distribution, occurring when the Wiebull slope β is equal to 1. An example of an exponential distribution for failure in time of a given product is depicted by the probability density function of Figure 5. Figure 6 shows graphical plots of the probability density, cumulative distribution, reliability, and hazard rate function for the exponential model. From a statistical point of view, the exponential distribution is only applicable to failures which are totally random. If the failure rate is constant, its probability of failure does not change with time, that is, it is independent of past history. A component or product that fails according to the exponential failure rate is as good the moment before failing as it was at the time it was put in service. The cause of failure is external, such as an electrical surge or sudden mechanical load. A common cause of such failures in electronic products is handling damage.

From a reliability point of view, the exponential distribution is usually not representative of a single mechanism. Instead, it is the sum of several other distributions, which may be both random and phenomenological. Since most products are subject to these mixed-mode failures, the exponential distribution is often used to empirically describe and predict their reliability.

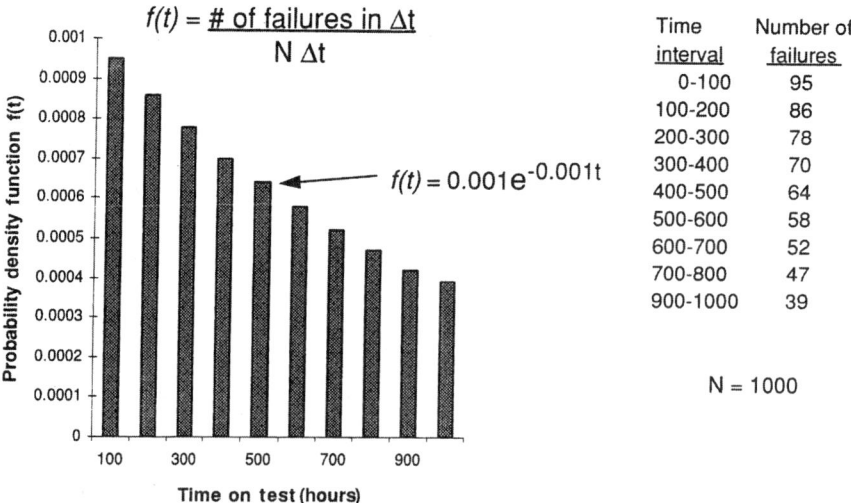

Time interval	Number of failures
0-100	95
100-200	86
200-300	78
300-400	70
400-500	64
500-600	58
600-700	52
700-800	47
900-1000	39

N = 1000

$$f(t) = \frac{\text{\# of failures in } \Delta t}{N \Delta t}$$

$$f(t) = 0.001 e^{-0.001t}$$

FIGURE 5   Example of exponential distribution.

Basic Reliability Mathematics

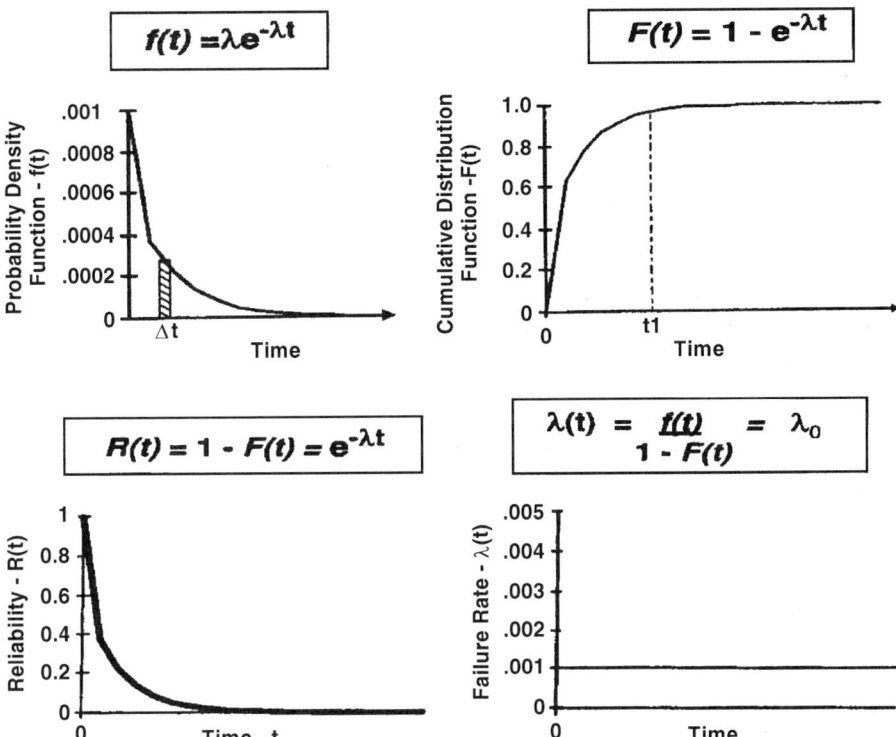

FIGURE 6 Reliability measures for the exponential model.

Some major advantages of the exponential distribution are:

There is a single, easily estimated parameter, λ.
It is mathematically very tractable.
It has wide applicability.
It is additive; for example, the sum of a number of independent exponentially distributed variables is exponentially distributed.

Specific applications include

1. Items whose failure rate does not change significantly with age.
2. Complex repairable equipment without excessive amounts of redundancy.
3. Equipments from which the infant mortality or early failures have been eliminated. This is done by "burning-in" the equipment for some time period.

For the exponential distribution, replacing the time-varying failure rate $\lambda(t)$ with a constant λ gives

$\lambda(t)$	= constant failure rate	(2.14)
PDF: $f(t)$	= $\lambda e^{-\lambda t}$	(2.15)
CDF: $f(t)$	= $1 - e^{-\lambda t}$	(2.16)
Reliability: $R(t)$	= $e^{-\lambda t}$	(2.17)
Mean time between failure: MTBF	= $1/\lambda$	(2.18)

If a component, like an integrated circuit, fails purely at random and is not subject to burn-in or wearout (at least for a specified portion of its lifetime), then the exponential model is applicable.

Reliability Calculation Example: Disk Drive Using the Exponential Distribution

If the probability density function of the lifetime of a disk drive is given by $f(t) = 0.1e^{-0.1t}$ (t is in years; $a = -0.1$ in the integration formula),

a. What is the probability the disk will fail in the first year?
b. What is the probability the disk will operate without failure for at least 5 years?

There are two ways to solve each part of the problem. Both are presented.

a. P(fail in first year): one can use the basic formula

$$F(1) = \int_0^1 0.1e^{-0.1t}\, dt = 0.1 \left(\frac{e^{-0.1t}}{-0.1}\right)\Big|_0^1 = -e^{-.1} + 1$$

$$= -.905 + 1 = .095$$

or one can use the relation $F(t) = 1 - e^{-\lambda t} = 1 - e^{-0.1(1)} = 1 - .905 = .095$.

b. Chance of failure in 5 years:

$$F(1) = \int_0^5 0.1e^{-0.1t}\, dt = 0.1 \left(\frac{e^{-0.1t}}{-0.1}\right)\Big|_0^5 = -e^{-.5} + 1$$

$$= -.607 + 1 = .393$$

So success is $1 - .393 = .607$.

Or, this could be calculated using $R(t)$ to begin with: $R(5) = e^{-0.1(5)} - e^{-0.1\infty} = .607$.

Reliability Calculations Example: Integrated Circuits Using the Exponential Distribution

Integrated Circuit Example 1. Assume an exponential distribution of failures in time; $\lambda = 5 \times 10^{-5}$ failures per hour and 100,000 parts.

Basic Reliability Mathematics 41

What is the probability that an IC will fail within its first 400 hr?

Using the cumulative function, $F(t) = 1 - e^{-(5 \times 10^{-5})(400)} = 0.0198$.

What fraction of parts are functioning after 1000 hr?

Using the reliability function, $R(t) = 1 - F(t) = 1 - [1 - e^{-(5 \times 10^{-5})(1000)}]$ = 0.9512, or 95,123 parts.

Integrated Circuit Example 2. For the conditions stated in Example 1, what fallout of parts do you expect between 2000 and 2500 hr?

Using the cumulative function and taking the difference at 2000 and 2500 hr, respectively:

$F(t) = 1 - e^{-(5 \times 10^{-5}(2500)} = 0.1175$, or 11,750 failed parts

$F(t) = 1 - e^{-(5 \times 10^{-5}(2000)} = 0.09516$, or 9,516 failed parts

The fallout is $11{,}750 - 9516 = 2234$ ICs between 2000 and 2500 hr.

What is the expected mean time to failure for a single part?

$\text{MTTF} = \lambda^{-1} = 1/(0.5 \times 10^{-5}) = 20{,}000$ hr.

Integrated Circuit Example 3. Let's convert to FITs, or failures in 10^9 hours. In Examples 1 and 2, $\lambda = 5 \times 10^{-5}$, or 50,000 FITs, which is terrible. Assume 200 FITs and 1 million parts shipped.

How many of these 1 million ICs will fail in 1 year?

Using the cumulative function, $F(t) = 1 - e^{-\lambda T}$, where $\lambda T = 200$ (FITs) $\times 10^{-9} \times 8760$ (hr), so, $F(t) = 1 - e^{-200} = 0.00175$. Thus 1751 ICs are expected to fail in the first year. This example assumes an accurate match of data to correct PDF.

Integrated Circuit Example 4. An IC is going on a 2-year mission to Jupiter. What FIT rate is needed to ensure a probability of failure <0.001?

2 years = 17,520 hr

$F(t) = 17{,}520 = 1 - e^{-\lambda(17{,}520)} = 10^{-3}$

$\lambda = \ln\left(\dfrac{1}{1 - 10^{-3}}\right)\left(\dfrac{1}{17{,}520}\right) = 57$ FITs

2.2.2 Weibull Distribution

The Weibull distribution has a wide range of applications in reliability analysis. It is a very general distribution which, by adjustment of the distribution parameters, can be made to model a wide range of life distribution characteristics. The formulas for calculating failure rate, PDF, CDF, and reliability are as follows:

$$\text{Failure rate: } \lambda(t) = \frac{\beta}{\eta}\left(\frac{t}{\eta}\right)^{\beta-1} \tag{2.19}$$

$$\text{PDF: } f(t) = \frac{\beta}{\eta}\left(\frac{t}{\eta}\right)^{\beta-1} \exp\left(\frac{-t}{\eta}\right)^{\beta} \tag{2.20}$$

$$\text{CDF: } F(t) = 1 - \exp\left(\frac{-t}{\eta}\right)^{\beta} \tag{2.21}$$

$$\text{Reliability: } R(t) = \exp\left(\frac{-t}{\eta}\right)^{\beta} \tag{2.22}$$

where
  β = shape parameter
  η = scale parameter or characteristic life (at which 63.2% of the population will have failed)

The Weibull distribution failure rate is plotted in Figure 7.

Depending upon the value of β, the Weibull distribution function can also take the form of the following distributions:

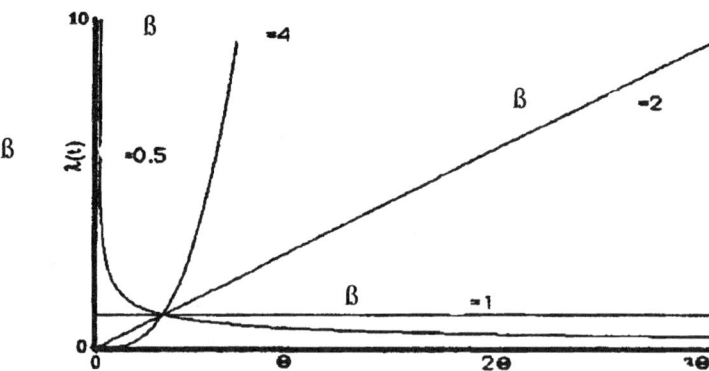

FIGURE 7 Weibull distribution failure rate curve.

Basic Reliability Mathematics

Beta value	Distribution type	Hazard rate
<1	Gamma	Decreasing
1	Exponential	Constant
2	Lognormal	Increasing/decreasing
3.5	Normal (approximately)	Increasing

The Weibull distribution is one of the most common and most powerful in reliability because of its flexibility in taking on many different shapes for various values of β. For example, it can be used to describe each of the three regions of the bathtub curve as follows and as shown in Figure 8.

Infant mortality (decreasing failure rate): β < 1
Useful life (constant failure rate): β = 1
Wearout (increasing failure rate): β > 1

The Weibull plot is a graphical data analysis technique to establish the failure distribution for a component or product with incomplete failure data. Incomplete means that failure data for a power supply or disk drive, for example, does not include both running and failure times because the units are put into service at different times. The Weibull hazard plot provides estimates of the distribution parameters, the proportion of units failing by a given age, and the behavior of the failure rates of the units as a function of their age. Also, Weibull hazard plotting answers the reliability engineering question, does the data support the engineering conjecture that the failure rate of the power supply or disk drive increases with their age? If so, there is a potential power supply/disk drive wear-out problem which needs to be investigated.

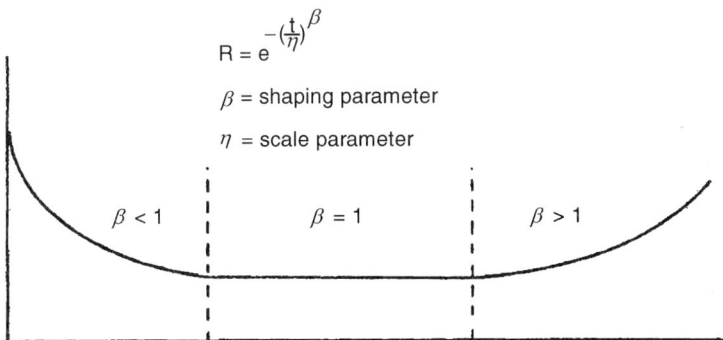

$$R = e^{-(\frac{t}{\eta})^{\beta}}$$

β = shaping parameter

η = scale parameter

FIGURE 8 The Weibull distribution and the bathtub curve.

Both the lognormal and Weibull models are widely used to make predictions for components, disk drives, power supplies, and electronic products/equipment.

2.2.3 Discrete Distributions

Binomial Distribution

The binomial distribution is very useful in both reliability and quality assurance, it is used when there are only two possible outcomes, such as success or failure, and probability remains the same for all trials. The probability density function (PDF) of the binomial distribution and the cumulative distribution function (CDF) are shown in Figure 4.

The probability density function f(x) is the probability of obtaining exactly x good items and (n − x) bad items in a sample of n items, where p is the probability of obtaining a good item (success) and q [or (1 − p)] is the probability of obtaining a bad item (failure).

The cumulative distribution function is the probability of obtaining r or fewer successes in n trials.

Computations involving the binomial distribution become rather unwieldly for even small sample sizes. However, complete tables of the binomial PDF and CDF are available in many statistics texts.

Poisson Distribution

This distribution is used quite frequently in reliability analysis. It can be considered an extension of the binomial distribution when n is infinite. In fact, it is used to approximate the binomial distribution when n > 20 and p < .05.

If events are Poisson distributed, they occur at a constant average rate, and the number of events occurring in any other time interval are independent of the number of events occurring in any other time interval. For example, the number of failures in a given time would be given by

$$f(x) = \frac{a^x e^{-a}}{x!} \qquad (2.23)$$

where x is the number of failures and a is the expected number of failures. For the purpose of reliability analysis, this becomes

$$f(x; \lambda, t) = \frac{(\lambda t)^x e^{-\lambda t}}{x!} \qquad (2.24)$$

where
λ = failure rate
t = length of time being considered
x = number of failures

Basic Reliability Mathematics

The reliability function R(t), or the probability of zero failures in time t, is given by

$$R(t) = \frac{(\lambda t)^0 e^{-\lambda t}}{0!} = e^{-\lambda t} \qquad (2.25)$$

that is, simply the exponential distribution.

2.3 PLOTTING AND LINEARIZATION OF DATA

Data of any distribution are often manipulated to achieve straight lines, as shown in Figure 9. This allows easy identification of constraints from slopes and intercepts.

Figure 10 and Table 2 show the linear transformation of the lognormal and Weibull distributions, respectively.

2.4 CONFIDENCE LIMIT AND INTERVALS

The use of confidence intervals is a technique to quantify uncertainty. Confidence in an estimate increases as a higher percentage of the population is sampled.

Failure rate can be expressed as a point estimate. The point estimate is a single number—the best guess as to the value of a random number.

$$\text{Point estimate of MTTF} = \frac{\text{Number hours}}{\text{number failures}}$$

A confidence level is the probability that a given parameter lies between two limits or is above a lower limit or is below an upper limit (see Fig. 11). Statistical estimates of parameters such as failure rates are usually expressed in

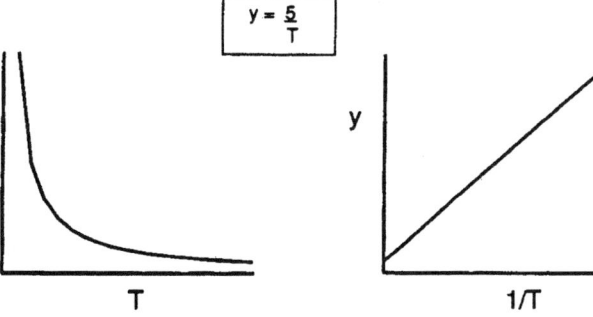

FIGURE 9 Linearizing data. Useful plot is natural log of time to failure (y axis) versus cumulative percent fails (x axis).

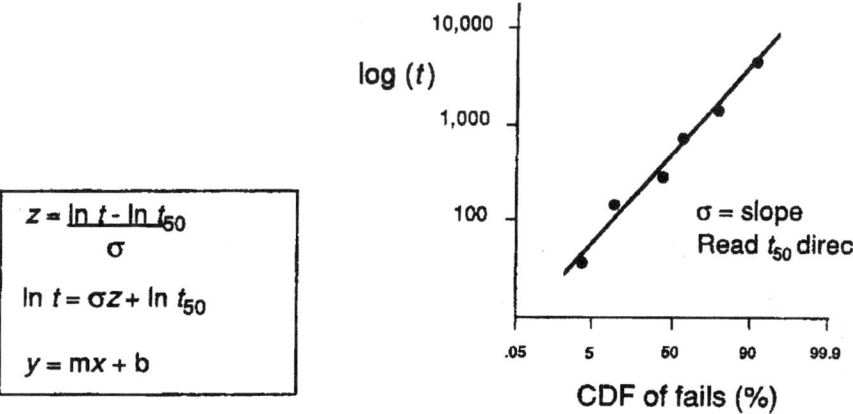

FIGURE 10 Lognormal plot linearization.

terms of intervals, with an associated probability, or confidence that the true value lies within such intervals. The end points of the intervals are called *confidence limits* and are calculated at a given confidence level (probability) using measured data to estimate the parameter of interest.

The upper plus lower confidence limits (UCL + LCL) for a confidence level must always total 100%. The greater the number of failures, the closer the agreement between the failure rate point estimate and upper confidence limit of the failure rate. Numerical values of point estimates become close when dealing with cases of 50 or more failures.

Here is how confidence limits work. A 90% upper confidence limit means that there is a 90% probability that the true failure rate will be less than the rate computed. A 60% upper confidence limit means that there is a 60% probability

TABLE 2 Weibull Plot Linear Transformation Equation

Manipulate Weibull into a straight equation so that constants m and c can be observed directly from the plot.

$F(t) = 1 - e^{-(t/c)^m}$	(2.26)
$1 - F(t) = e^{-(t/c)^m}$	(2.27)
$-\ln[1 - F(t)] = (t/c)^m$	(2.28)
$\ln\{-\ln[1 - F(t)]\} = m \ln(t) - m \ln(c)$ (plot)	(2.29)
$y = mx + b$	(2.30)

Basic Reliability Mathematics

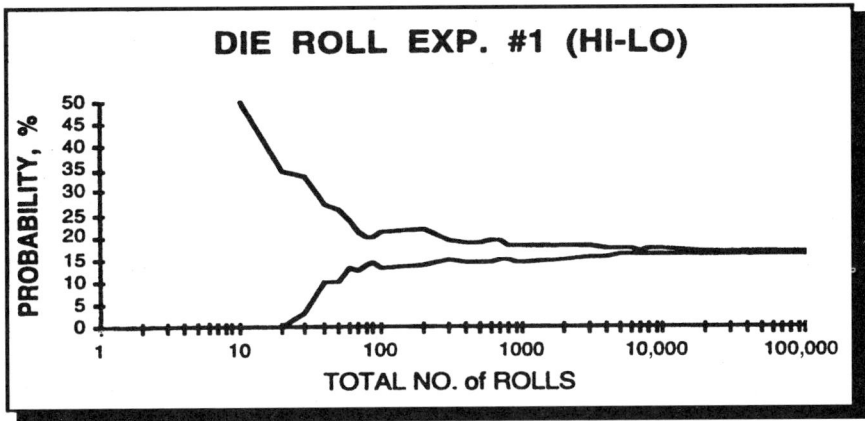

FIGURE 11 Confidence levels showing the probability that a parameter lies between two lines.

that the true failure rate will be less than the rate computed. Both of these are diagrammed in Figure 12. Conversely, the true failure rate will be greater than the computed value in 10% and 40% of the cases, respectively. The higher the confidence level, the higher the computed confidence limit (or failure rate) for a given set of data.

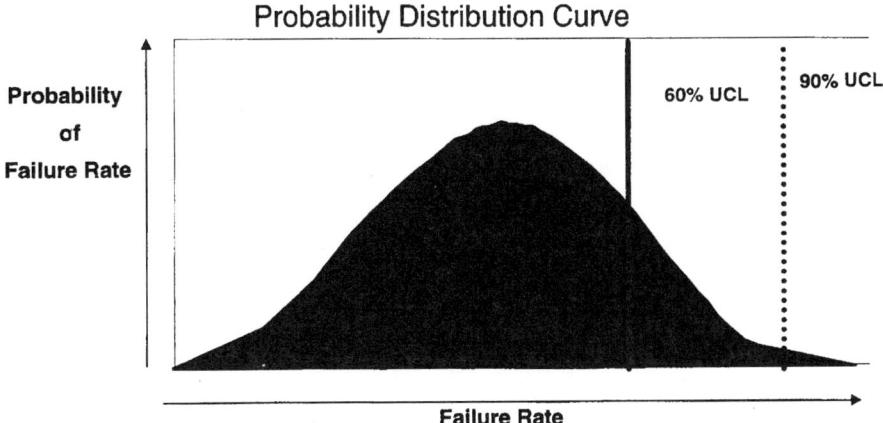

FIGURE 12 Graphical depiction of upper confidence level.

A confidence interval is a range that contains the value of a random number with a specified probability. A 90% confidence interval for the MTTF means that there is a 90% probability that the "true" value of the MTTF is within the interval (Fig. 13). We try to center the confidence interval around the point estimate. As we get more data, the interval decreases or the confidence increases, as shown in Figure 14. Confidence intervals are often called two-sided confidence limits (Fig. 15).

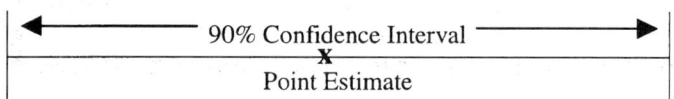

FIGURE 13 Definition of confidence interval. The confidence intervals consist of the interval and the confidence level for the interval.

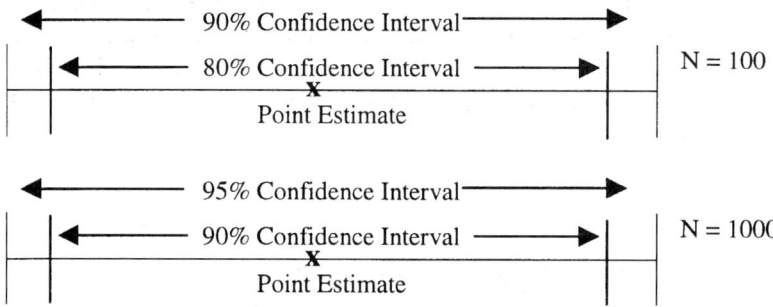

FIGURE 14 Impact of data on confidence limits and intervals. The 90% lower confidence limit is less than true MTTF with 90% confidence. The 90% upper confidence limit is greater than true MTTF with 90% confidence (90% LCL, 90% UCL) and has an 80% confidence interval.

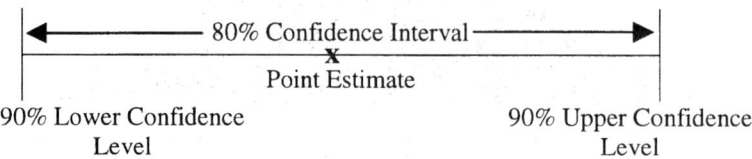

FIGURE 15 Two-sided confidence limits.

Basic Reliability Mathematics

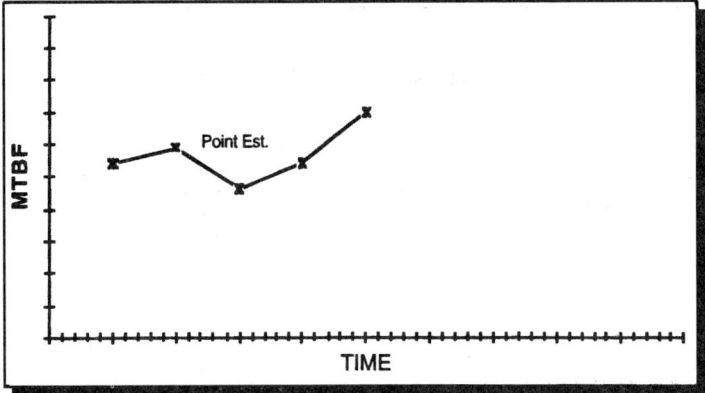

FIGURE 16 Point estimate varies over time.

Use of Confidence Levels

Figure 16 graphically shows how confidence limits are developed when starting with a MTBF point estimate. Figure 17 adds upper and lower confidence limits to the graph, and Figure 18 has 80% two-sided confidence limits installed. Figure 19 shows the MTBF and confidence limits for a power supply plotted for 11 calendar months.

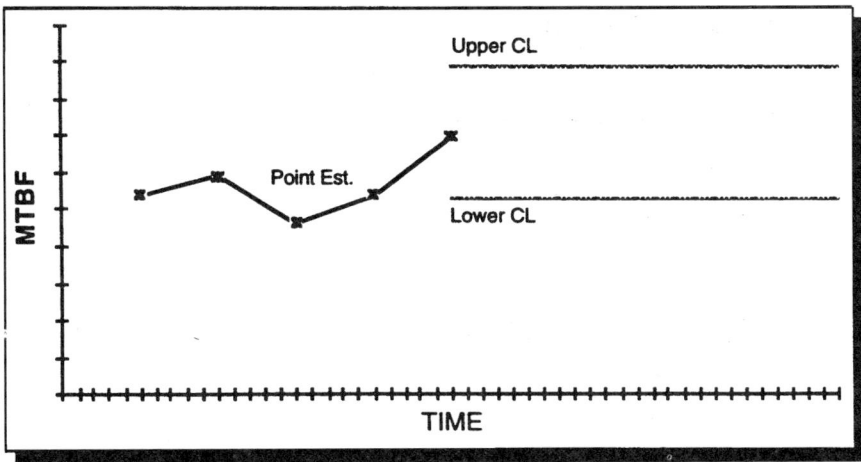

FIGURE 17 LCL and UCL added to Figure 16. LCL < CL% probability that true (future) MTBF < UCL.

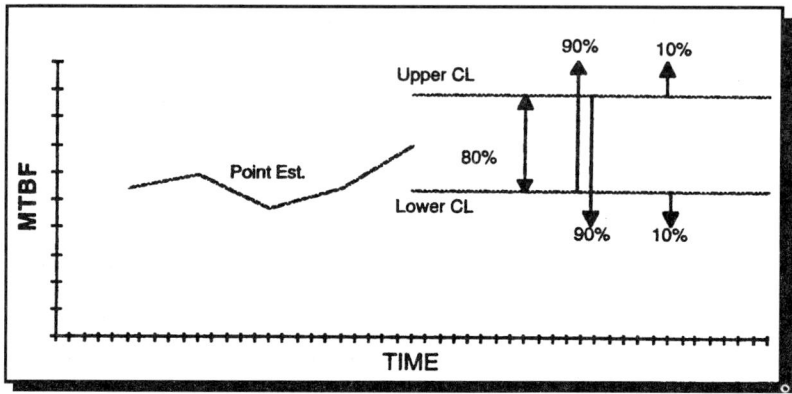

FIGURE 18 80% two-sided confidence limits.

The equations to be used for MTBF confidence interval calculations are as follows:

$$\text{MTBF}_{\text{UCL}} = \frac{2T}{X^2_{\text{UCL}, Y}} \quad \text{UCL} = 100 - \frac{(100\text{-CI})}{2 \text{ (in percent)}} \quad (2.31)$$

$$\text{MTBF}_{\text{LCL}} = \frac{2T}{X^2_{\text{LCL}, Y}} \quad \text{LCL} = 100 - \frac{(100\text{-CI})}{2} \quad (2.32)$$

where
  Y = degrees of freedom
  T = number of unit hours

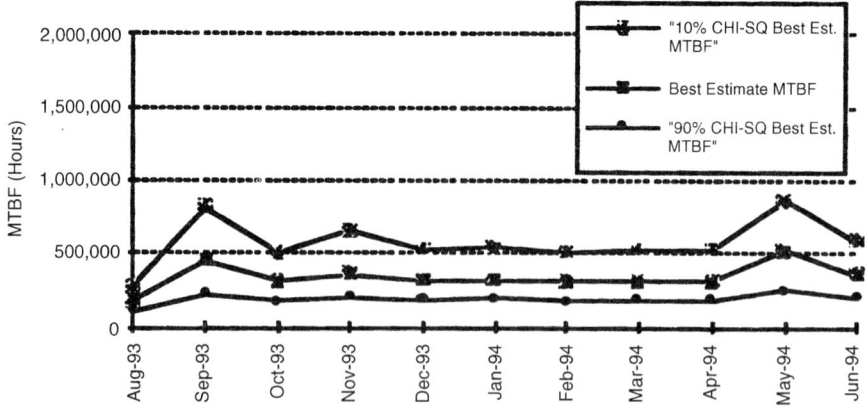

FIGURE 19 Confidence limit example: power supply (3-month rolling average).

Basic Reliability Mathematics

For time terminated testing,

$$Y = 2n + 2 \tag{2.33}$$

For failure terminated testing,

$$Y = 2n \tag{2.34}$$

where n = number of failures.

The sequence of steps used for determining lower and upper confidence limits using a chi-square table is as follows:

1. Determine desired level of confidence, LCL and UCL (determined from the CI).
2. Count number of failures, n.
3. Determine total unit-hours T.
4. Calculate the degrees of freedom, Y (it will be either 2n or 2n + 2).
5. Look up chi-squared values for LCL and UCL.
6. Calculate limits.

Example: UCL, LCL, and MTBF Determination. In a large computer system, ten disk drives have been tested to failure, and the MTTF is calculated at 145,000 hr. What are the LCL and UCL for an 80% confidence interval?

$T = 145{,}000 \times 10 = 1{,}450{,}000$ hr (total testing time)

$Y = 2n = 20$ (failure-terminated testing)

CI = 80% (LCL = 10%; UCL = 90%)

$\chi^2_{LCL, \gamma} = \chi^2_{.1, 20} =$ _____ (look up in table, Appendix A of Chapter 2)

$MTBF_{LCL} = \dfrac{2(1{,}450{,}000)}{28.4} =$ _____ hr

$\chi^2_{UCL, \gamma} = \chi^2_{.9, 20} =$ _____ (look up in table, Appendix A of Chapter 2)

$MTBF_{ULCL} = \dfrac{2(1{,}450{,}000)}{12.4} =$ _____ hr

Solution

$LCL = \dfrac{100 - 80}{2} = 10\%$ $\qquad UCL \dfrac{100 + 80}{2} = 90\%$

$\chi^2_{LCL, 2n} = \chi^2_{.1, 20} = 28.412$ $\qquad \chi^2_{UCL, 2n} = \chi^2_{.9, 20} = 12.443$

$MTBF_{LCL} = \dfrac{2(1{,}450{,}000)}{28.412} = 102{,}070$ $\quad MTBF_{UCL} = \dfrac{2(1{,}450{,}000)}{12.443} = 233{,}062$

Note also that the MTBF is not in the center of the confidence interval.

2.5 FAILURE FREE OPERATION

The probability of failure free operation is calculated using the following formula, which is based on the exponential distribution:

$$Ps = EXP\left(\frac{-T}{MTBF}\right) \qquad (2.35)$$

where
Ps = probability of failure free operation for time T
T = operating period of interest
MTBF = MTBF of product in same units as T = $1/\lambda$
EXP = exponential of the natural logarithm

Let's use this equation to calculate the probability of failure free operation for a number of different operating times all with an MTBF of 4000 hr.
If T = 8000 hr,

$$Ps = EXP\left(\frac{-8000}{4000}\right) = .1353 = 13.53\%$$

If T = 4000 hr,

$$Ps = EXP\left(\frac{-4000}{4000}\right) = .3679 = 36.79\%$$

If T = 168 hr,

$$Ps = EXP\left(\frac{-168}{4000}\right) = .9588 = 95.88\%$$

If T = 8 hr,

$$Ps = EXP\left(\frac{-8}{4000}\right) = .9980 = 99.8\%$$

2.6 RELIABILITY MODELING

A system or product contains a number of different subassemblies. A block diagram of a series-parallel combination of a power supply and two central processing units (CPUs) is shown in Figure 20.

The probability that a system successfully completes its intended operation life is given by

Basic Reliability Mathematics 53

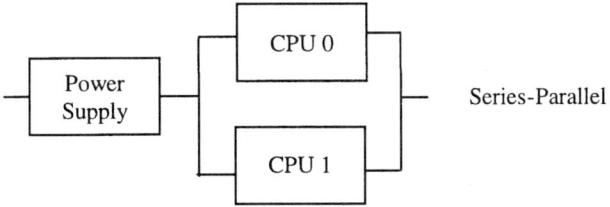

FIGURE 20 Series-parallel connection of power supply and computer CPUs.

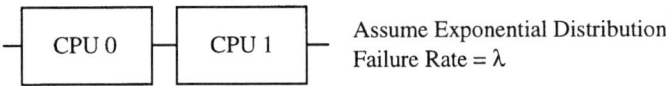

FIGURE 21 Series connection of two computer central processing units.

Reliability = Probability that power supply *and* CPU0 or CPU1 works

For a series system (Fig. 21) the reliability is given as

$$\text{Reliability} = R(A)R(B) = e^{-\lambda t}e^{-\lambda t} = e^{-2\lambda t} \quad (2.36)$$

Note, this is the same as a single unit with failure rate 2λ.

For a larger series system (Fig. 22) the total failure rate of the system is the sum of the failure rates of the individual units:

$$\text{System failure rate } \lambda = \lambda_{CPU} + \lambda_{disk} + \lambda_{power} + \lambda_{keyboard} + \lambda_{monitor} \quad (2.37)$$

$$\text{Reliability} = e^{-\lambda t} \quad (2.38)$$

A parallel combination of CPUs and the possible four system states is shown in Figure 23. Here,

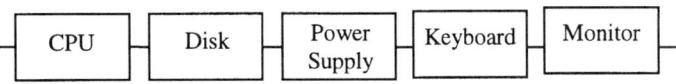

FIGURE 22 Complete system series network connection.

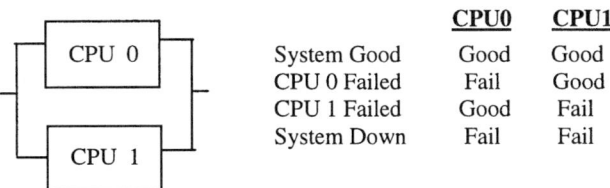

	CPU0	CPU1
System Good	Good	Good
CPU 0 Failed	Fail	Good
CPU 1 Failed	Good	Fail
System Down	Fail	Fail

FIGURE 23 Series-parallel connection of power supply and computer CPUs.

Failure rate $= \lambda$

$$\text{Reliability} = R(A) + R(B) - R(A)R(B) = e^{-\lambda t} + e^{-\lambda t} - e^{-\lambda t} e^{-\lambda t} \quad (2.39)$$
$$= 2e^{-\lambda t} + e^{-2\lambda t}$$

Let's work a practical example using the series-parallel network shown in Figure 20. Assume that power supply and CPU reliabilities have exponential distributions with failure rates of λ_{PS} and λ_{CPU}, respectively.

1. What is the system reliability? Leave it in the form of an exponential expression.
2. What is the system failure rate? (Trick question.)

We can start by calculating the CPU reliability:

$$R_{CPU0,CPU1} = 2e^{-\lambda \, CPU \, t} - e^{2\lambda \, CPU \, t} \quad (2.40)$$

Note that we cannot combine the exponentials. Now multiply by the term for the power supply:

$$R_{PS + CPUs} = e^{-\lambda \, PS \, t}(2e^{-\lambda CPU \, t} - e^{2\lambda \, CPU \, t}) \quad (2.41)$$

This is a complex expression.

The system failure rate Question 2 cannot be easily computed because of the time dependence; we can't add the exponentials.

This illustrates a basic fact about reliability calculations: unless the model is reasonably easy to use mathematically (as the exponential certainly is), it takes some powerful computing resources to perform the reliability calculations.

2.7 PRACTICAL MTBF AND WARRANTY COST CALCULATIONS

Let's put a lot of this into practice by looking at a detailed example to determine the MTBF impact on warranty cost. From the reliability distribution function,

$$R(t) = Ne^{-\lambda t} \text{ the number still surviving without failure} \quad (2.42)$$

where
- N = the number of units shipped. We will use $N = 100$.
- λ = the constant failure rate (in failures per million hours). (FIT is defined as failures per billion hours of operation, sometimes used in place of λ, but is smaller by three orders of magnitude. So 1000 λ is 1 FIT.)

$$\lambda = \frac{1}{\text{MTBF}} \quad (2.43)$$

Basic Reliability Mathematics

where MTBF is mean time between failure. So Eq. (2.42) becomes

$$R(t) = Ne^{-(t)/(MTBF)} \quad (2.44)$$

For example, let, t = 1000 hr, MTBF = 1000 hr, and N = 100 new VCRs or TV sets or ATE systems. Then,

$$R(t) = 100e^{-(1000 \text{ hr})/(1000 \text{ hr between failures})}$$
$$= 100 \, (2.71^{-1000/1000}) = 100 \, (2.71^{-1}) = 100 \, (.37)$$

R(t) = 37 units "still working without a failure"

This also means that 63 units *had* failures. But how many failures were there? With a 1000-hr MTBF there will be 1 per 1000, or .001 λ per million, failure rate. This means that every 1000 hr, a system will experience a failure. With 100 systems there will be 100 failures in those 1000 hr. But these failures will show up in only 63 units; the other 37 units will exhibit no failures during this time period. This means that of the 63 units that had failures, some had more than one failure.

Please note that there were *63 units that exhibited 100 total failures in one MTBF period*, and there were 37 units that went through this period without a failure. So how many of the 63 units had 1, 2, 3, ..., n failures?

$$P(n) = \left(\frac{\lambda^n t^n}{n!}\right) e^{-\lambda t} \quad (2.45)$$

where
 P(n) = percent of units exhibiting failures
 t = time duration
 n = number of failures in a single system (e.g., 1, 2, 3, ..., n)

Let's learn how many units will have 1, then 2, then 3, etc., failures per unit in the group of 63 units that will exhibit these 100 failures.

For zero failures (note, 1! is defined as equaling 1),

$$P(0) = \left[\frac{0.001^0 (1000^0)}{1}\right] 2.171^{-.001(1000)}$$

$$= \left[\frac{1(1)}{1}\right] 2.71^{-1}$$

$$= 1 \, (.37)$$

P(0) = .37, or 37%

So with 100 units there will be 37 units exhibiting zero failures in one MTBF time period.

For one failure,

$$P(1) = \left[\frac{0.001^1(1000^1)}{1}\right] 2.71^{-1}$$

$$P(1) = \left(\frac{1}{1}\right).37$$

P(1) = .37, or 37% will exhibit one failure.

So with 100 units there will be 37 units exhibiting one failure in one MTBF time period.

For two failures,

$$P(2) = \left[\frac{0.001^2(1000^2)}{2}\right] 0.37$$

$$P(2) = \left(\frac{1}{2}\right).37$$

P(2) = 18%

So with 100 units there will be 18 or 19 units exhibiting two failures in one MTBF time period.

P(3) = 6 units exhibiting three failures in one MTBF.

P(4) = 1 or 2 units exhibiting four failures in one MTBF.

P(5) = maybe 1 unit exhibiting five failures in one MTBF.

A simpler way of finding the percentage of failures encountered in some time period is

$$P(f) = \lambda t \tag{2.46}$$

Find how many will fail in one-hundredth of an MTBF time period.

$$P(f) = \frac{0.001(1000)}{100 \text{ hr}} \simeq 0.001(10) = 0.01, \text{ or } 1\%$$

Using 100 units this means that one unit exhibits the very first failure in 10 hr. So the time to first failure is 10 hr. Which one out of the 100 units will fail is a mystery, however.

Now let's move to the warranty issue. A system has an MTBF of 4000 hr. An engineer makes a recommendation for a hardware change that costs $400.00 per unit to install and raises the system's MTBF to 6500 hr. What is known:

The average cost of a field failure is $1,800.00/failure.

Basic Reliability Mathematics 57

The system operates 16 hr/day, 6 days/week. This roughly equals 16 hr (6 days)(52 weeks/year) = 5000 hr/year.

The warranty period is 1 year.

So the number of failures, λt, is $(1/4000)(5000) = 125\%$/year, or 125 annual failures per 100 systems shipped (assuming the 100 units were all shipped at the same time). At $1,800.00/failure, then $1800 (125) = $225,000 annual warranty cost.

If the MTBF is increased to 6500 hr, the number of failures, λt, is $5000/6500 = 77\%$ of the 100 units shipped, or 77 failures. At $1,800.00/failure then $1,800 (*77) = $138,600 annual warranty cost.

The net savings is thus

Original warranty cost	$225,000
New warranty cost	−138,600
Improvement cost	−40,000
Savings	$46,600

The return on investment is $46,400/$40,000 = 1.16.

The good news is that as the MTBF increases, the number of failures that occur per system decreases. The following table shows how many units had 1, 2, 3, etc., failures in both MTBF cases.

Number of failures	If the MTBF = 4000 hr	If the MTBF = 6500 hr
0	28–29	46–47
1	35–36	35–36
2	22–23	13–14
3	9–10	3–4
4	2–3	0–1
Total range	114–124	70–80

The MTBF of assemblies operating in parallel, or as a system, is defined as

$$\text{MTBF} = \frac{1}{1/\text{MTBF1} + 1/\text{MTBF2} + 1/\text{MTBF3} + \cdots + 1/\text{MTBFn}} \quad (2.47)$$

or, more clearly,

$$\text{MTBF} = \frac{1}{1/\text{MTBF1} + 1/\text{MTBF2} + 1/\text{MTBF3} + 1/\text{MTBF4} + 1/\text{MTBF5}}$$

Note: when a system consists of subassemblies, each subassembly has an MTBF significantly greater than the desired MTBF of the end system.

If the goal MTBF of the system is specified, the MTBF of each subassembly must be allocated so that the combination of the subassemblies meets the desired MTBF of the whole system.

For example, a system is composed of five subassemblies each with the following MTBFs. When these are assembled into the system will the system MTBF goal of 4000 hr be met?

Subassembly 1 MTBF = 20,000 hr
Subassembly 2 MTBF = 14,000 hr
Subassembly 3 MTBF = 33,000 hr
Subassembly 4 MTBF = 24,000 hr
Subassembly 5 MTBF = 18,000 hr

The system MTBF is derived using the relationship

$$\text{MTBF Total} = \frac{1}{1/\text{MTBF1} + \text{MTBF2} + \text{MTBF3} + \text{MTBF4} + \text{MTBF5}}$$

or

$$\text{MTBF Total} = \frac{1}{\lambda 1 + \lambda 2 + \lambda 3 + \lambda 4 + \lambda 5}$$

$$\text{MTBF} = \frac{1}{.000041 + .000071 + .000030 + .000042 + .000056}$$

$$= \frac{1}{0.000240} = 4167 \text{ hrs}$$

Thus, the system consisting of these five subassemblies has a system MTBF of 4167 hr. This is an acceptable number because it exceeds the specified 4000 hour goal.

ACKNOWLEDGMENT

Portions of this chapter have been excerpted from the reference. Section 2.6 was generated by Ted Kalal, and is used with permission.

REFERENCE

Wood A. Reliability Concepts Training Course, Tandem Computer Corporation, Cupertino, California.

APPENDIX A

Percentage Points of the V Chi-Square Distribution

χ^2_α

ν	$\chi^2_{.995}$	$\chi^2_{.99}$	$\chi^2_{.975}$	$\chi^2_{.95}$	$\chi^2_{.90}$	$\chi^2_{.80}$	$\chi^2_{.75}$	$\chi^2_{.70}$
1	.0000393	.000157	.000982	.00393	.0158	.0642	.102	.148
2	.0100	.0201	.0506	.103	.211	.446	.575	.713
3	.0717	.115	.216	.352	.584	1.005	1.213	1.424
4	.207	.297	.484	.711	1.064	1.649	1.923	2.195
5	.412	.554	.831	1.145	1.610	2.343	2.675	3.000
6	.676	.872	1.237	1.635	2.204	3.070	3.455	3.828
7	.989	1.239	1.690	2.167	2.833	3.822	4.255	4.671
8	1.344	1.646	2.180	2.733	3.490	4.594	5.071	5.527
9	1.735	2.088	2.700	3.325	4.168	5.380	5.899	6.393
10	2.156	2.558	3.247	3.940	4.865	6.179	6.737	7.267
11	2.603	3.053	3.816	4.575	5.578	6.989	7.584	8.148
12	3.074	3.571	4.404	5.226	6.304	7.807	8.438	9.034
13	3.565	4.107	5.009	5.892	7.042	8.634	9.299	9.926
14	4.075	4.660	5.629	6.571	7.790	9.467	10.165	10.821
15	4.601	5.229	6.262	7.261	8.574	10.307	11.306	11.721
16	5.142	5.812	6.908	7.962	9.312	11.152	11.192	12.624
17	5.697	6.408	7.564	8.672	10.085	12.002	12.792	13.531
18	6.265	7.015	8.231	9.390	10.865	12.857	13.675	14.440
19	6.844	7.633	8.907	10.117	11.651	13.716	14.562	15.352
20	7.434	8.260	9.591	10.851	12.443	14.578	15.452	16.266
21	8.034	8.897	10.283	11.591	13.240	15.445	16.344	17.182
22	8.643	9.542	10.982	12.338	14.041	16.314	17.240	18.101
23	9.260	10.196	11.688	13.091	14.848	17.187	18.137	19.021
24	9.886	10.856	12.401	13.848	15.659	18.062	19.037	19.943
25	10.520	11.524	13.120	14.611	16.473	18.940	19.939	20.867

APPENDIX A (Continued)

χ^2_α

ν	$\chi^2_{.50}$	$\chi^2_{.30}$	$\chi^2_{.25}$	$\chi^2_{.20}$	$\chi^2_{.10}$	$\chi^2_{.05}$	$\chi^2_{.025}$	$\chi^2_{.01}$	$\chi^2_{.005}$
1	.455	1.074	1.323	1.642	2.706	3.841	5.024	6.635	7.879
2	1.386	2.408	2.773	3.219	4.605	5.991	7.378	9.210	10.597
3	2.366	3.665	4.108	4.642	6.251	7.815	9.348	11.345	12.838
4	3.357	4.878	5.385	5.989	7.779	9.488	11.143	13.277	14.860
5	4.351	6.064	6.626	7.289	9.236	11.070	12.832	15.086	16.750
6	5.348	7.231	7.841	8.558	10.645	12.592	14.449	16.812	18.548
7	6.346	8.383	9.037	9.803	12.017	14.067	16.013	18.475	20.278
8	7.344	9.524	10.219	11.030	13.362	15.507	17.535	20.090	21.955
9	8.343	10.656	11.389	12.242	14.684	16.919	19.023	21.666	23.589
10	9.342	11.781	12.549	13.442	15.987	18.307	20.483	23.209	25.188
11	10.341	12.899	13.701	14.631	17.275	19.675	21.920	24.725	26.757
12	11.340	14.011	14.845	15.812	18.549	21.920	23.337	26.217	28.300
13	12.340	15.119	15.984	16.985	19.812	22.362	24.736	27.688	29.819
14	13.339	16.222	17.117	18.151	21.064	23.685	26.119	29.141	31.319
15	14.339	17.322	18.245	19.311	22.307	24.996	27.488	30.578	32.801
16	15.338	18.418	19.369	20.465	23.542	26.296	28.845	32.000	34.267
17	16.338	19.511	20.489	21.615	24.769	27.587	30.191	33.409	35.718
18	17.338	20.601	21.605	22.760	25.989	28.869	31.526	34.805	37.156
19	18.338	21.689	22.718	23.900	27.204	30.144	32.852	36.191	38.582
20	19.337	22.775	23.828	25.038	28.412	31.410	34.170	37.566	39.997
21	20.337	23.858	24.935	26.171	29.615	32.671	35.479	38.932	41.401
22	21.337	24.939	26.039	27.301	30.813	33.924	36.781	40.289	42.796
23	22.337	26.018	27.141	28.429	32.007	35.172	38.076	41.638	44.181
24	23.337	27.096	28.241	29.553	33.196	36.415	39.364	42.980	45.558
25	24.337	28.172	29.339	30.675	34.382	37.652	40.646	44.314	46.928

3
Robust Design Practices

3.1 INTRODUCTION

Technology has created an increasingly complex problem for delivering reliability. More systems are designed today faster than ever before under shrinking cost margins and using more electronics with ever more complex devices that no one has the time to test thoroughly. Concurrently, the tolerance for poor reliability is shrinking, even while expectations are rising for rapid technology changes and shorter engineering cycles.

Grueling project schedules, thirst for performance and cost competitiveness result in corners being cut in design, with as little verification and testing being done as possible. In all of this what is a company to do to deliver high-reliability products to its customers? Here are some hints. Use common platforms, product architectures, and mainstream software and keep new engineering content down as much as possible to have evolutionary rather than revolutionary improvements. Use proven or preferred components. Perform margin analysis to ensure performance and reliability beyond the stated specifications. Conduct accelerated stress screening to expose and correct defects in the engineering phase before shipping any products. Conduct detailed design reviews throughout the design process composed of multidisciplinary participants.

This chapter discusses the importance of the design stage and specifically the elements of design that are vital and necessary to producing a reliable product.

Design has the most impact on the reliability outcome of a product. 80% of the reliability and production cost of a product is fixed during its design. Reliability must be designed in a product. This requires a lot of forethought and a conscious formal effort to provide just the required margin needed for customer application. A given product's projected life in its intended application determines the amount of margin (robustness) that will be designed in and for which the customer is willing to pay. Figure 1 shows the product life cycles of various categories of computers. This figure shows that not all products require high reliability and that not all segments of a given product category have the same reliability requirements, i.e., a one size fits all mind set. For example, personal computers (PCs) are commodity items that are becoming disposable (much as calculators and cell phones are). Computers used in financial transaction applications (stock markets, banks, etc.), automobiles, telecommunication equipment, and satellites have much more stringent reliability requirements.

How product reliability has been accomplished has differed greatly between U.S. and Japanese companies. The Japanese companies generate many change notices during product design, continually fine tuning and improving the design in bite-sized pieces until the design is frozen upon release to production. Companies in the United States, on the other hand, are quick to release a product to production even though it contains known "bugs" or deficiencies. Geoffrey Moore, in *Living on the Fault Line,* calls this "going ugly early" to capture market share. After being released to production a number of changes are made to correct these deficiencies. But in the meantime, plan on customer returns and complaints.

Reliability begins at the global system concept design phase; moves to the detailed product design phase that encompasses circuit design, application-

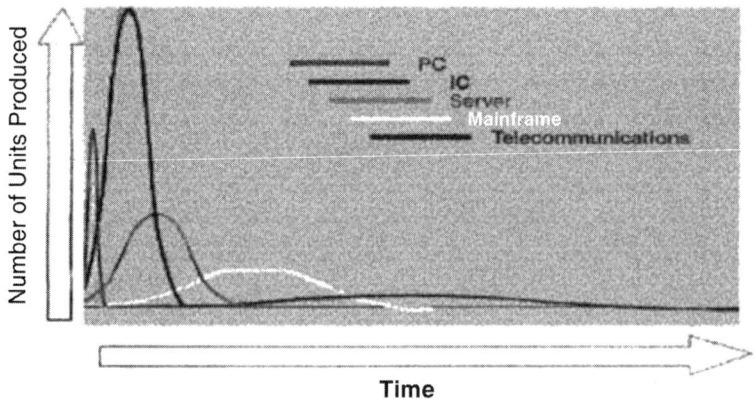

FIGURE 1 Product life cycles for various categories of computers. (Courtesy of Andrew Kostic, IBM.)

Robust Design Practices

TABLE 1 Detailed Design Tasks

Product design, including design for manufacture, design for test, and design for electromagnetic compatibility
Logic circuit design
Design verification, simulation, and emulation of application-specific ICs
PWA design/layout and design for manufacture
System software development
Software reviews
Enclosure design (mechanical and thermal)
Reliability production and update (throughout the design phase)
Part selection/finalization (throughout the design phase)
BOM reviews (throughout the design phase)
Failure modes and effects analyses
Hardware design reviews (throughout the design phase)
PWA test vectors and system test programs development
Manufacturing qualification testing of new packaging and technology
Diagnostic test development
Signal integrity and timing analysis tests
Thermal analysis and testing
PWA outsource provider qualification
Appropriate hardware design verification testing

specific integrated circuit (ASIC) design, mechanical interconnect design, printed wire assembly (PWA) design (including design for manufacturability), thermal design, and industrial design (case, cover, enclosure, etc.); and concludes with design for reliability, testability, and electromagnetic compatibility and design verification testing. Reliability is strongly dependent on the individual components (piece parts) used, suppliers selected, and the manufacturing and test processes developed.

Table 1 takes the detailed design block of Figure 1 and expands it to show the typical tasks involved in producing a high-end computer design. These tasks, some of which can occur in parallel, are applicable to the design of all products, but the choice as to which of these tasks and how many are used for a particular situation depends on the amount of robustness (design margin) needed for the intended customer application.

3.2 CONCURRENT ENGINEERING/DESIGN TEAMS

The quality of a product is highly dependent on the human organization that turns it out. Today's designs require close cooperation, alignment, and integration among all appropriate technical disciplines to fulfill the design objectives, beginning at the earliest place in the design cycle. This method of operation has been given several names, including *concurrent engineering*. Concurrent engineering can be defined as (IDA Report R-338, 1988)

> A systematic approach to the concurrent design of products and their related processes, including manufacturing and support. This approach is intended to cause developers, from the outset, to consider all elements of the product life cycle from conception through disposal, including quality, cost, schedule, and user requirements.

Thus, concurrent engineering is the cooperation of multiple engineering disciplines early in and throughout the design cycle. It is important because a team organization can have the most immediate effect on quality and reliability. Experience shows that concurrent engineering leads to

> Shorter product design and manufacturing
> Lower field returns
> Significant cost savings

Table 2 provides three examples of the benefits of the application of concurrent engineering.

A typical concurrent engineering product design team includes system designers; logic designers; ASIC designers; PWA designers; mechanical, thermal, and structural designers; reliability engineers; test engineers; component engineers; manufacturing engineers; industrial designers; and regulatory (electromagnetic compatibility and safety) engineers. Reliability engineers can be considered the glue that keeps the design project together since they are concerned with

> Time degradation of materials
> Physical measurements
> Electrical and electronic measurements
> Equipment design
> Processes and controls
> System performance simulation and analysis

TABLE 2 Benefits of Concurrent Engineering

Case study	Cost	Schedule	Quality
AT&T	Circuit pack repair cut 40%	Total process time cut to 46% of baseline	Defects down 30%
Hewlett-Packard	Manual costs down 42%	Development cycle time cut 35%	Field failure rate cut 60%
IBM	Labor hours cut 45%	Design cycle cut 40%	Fewer ECOs

Synthesis
Piece parts and suppliers selected for the design

However, a startling change is taking place. Design (concurrent engineering) teams are shrinking in size. This is due to the increased integration possible with integrated circuit technology and the available design tools.

3.3 PRODUCT DESIGN PROCESS

A development process needs to be both flexible and grounded on certain basic fundamentals and processes. One fundamental concept is that work proceeds sequentially through certain phases of a product's development. Although the process is iterative, much of the work in each of the phases is performed concurrently, and earlier phases are often revisited as the work is modified, the sequence remains much the same throughout and phases complete sequentially. For example, though the product design may be modified as implementation proceeds, major design always precedes major implementation. Movement of a development project from one phase to the next represents an increased level of commitment to the work by a company. Exit from a phase is a result of interorganizational concurrence with the phase review deliverables required in that phase. Concurrence implies commitment by each organization to the activities required in the next phase. Each phase has a specific objective. During each phase, major functions perform certain activities to achieve the objective. Typical product life cycle phases for a complex electrical equipment development project are

> Phase 0: Concept. Identification of a market need
> Phase 1: Investigation and Requirements. Response to the need with a product description
> Phase 2: Specification and Design. Design of the product(s) that constitute the development effort
> Phase 3: Implementation and Verification. Implementation and testing of the product(s)
> Phase 4: Preproduction and Introduction. Preparation of the product for general availability
> Phase 5: Production and Support. Review of the performance of the product(s) in the field to determine future related work and to codify lessons learned for next generation development.

Let's look at each phase in greater detail.

3.3.1 Phase 0: Concept

Phase 0 answers the question is this the right product for the market. The objective of Phase 0 is to identify, as accurately as possible, the market need or opportunity for a development effort and to communicate the need or opportunity to Development (a.k.a. Engineering) so that they can perform a feasibility study.

Marketing identifies the need for the development effort/project and communicates the need to Product Management. If a project begins with a proposal of a market opportunity from Development, Marketing reviews the proposal, identifies the market for the program, and then communicates the information to Product Management, in much the same way as if the idea had originated in Marketing. Product Management produces a high-level product definition and market requirements document to define the market need or opportunity and communicate it to Development.

3.3.2 Phase 1: Investigation and Requirements

Phase 1 answers the question will we commit to develop and build this product. The objective of Phase 1 review is to reach a decision to commit funds necessary to design and specify a development project (in Phase 2) and to enter it into a company's strategic plans.

Development (with the support of other organizations such as Manufacturing, Industrial Design, Product Assurance, and Reliability Engineering, for example) responds to the requirements document created in Phase 0 with a detailed Product Description. The Product Description addresses each of the requirement document's product objectives with a corresponding statement of planned product function, performance, documentation, and other product attributes.

All organizations (whether involved in the project at this phase or not) prepare a preliminary Operating Plan with estimated dates to support the Product Description. A Program Manager is assigned to lead the effort and forms the core team, with assistance from the heads of the major functions.

3.3.3 Phase 2: Specification and Design

Phase 2 answers the question can we approve this project for Marketing with committed Beta (external partner evaluation units) and first-customer-ship (FCS) dates. The objective of Phase 2 Review is to approve the completion of the product design and to commit both the ship dates and the funds necessary for its implementation.

Development completes the specification and design of the product, including creation of project level deliverables. Product Assurance works with Development to create a test plan for the products in the project (test libraries for software; test suites, facilities, and equipment for hardware test). Reliability Engineering reviews the piece part components and suppliers that have been selected, generates reliability prediction calculations, and compares the results with the product/market goals. Support defines the field product support strategy.

All functions commit to their efforts by producing "final" operating plans. *Final*, in this sense, means that a plan is complete, contains committed dates, and emphasizes the activities that each function will perform during Phase 3. Operating plans are reviewed at each phase to ensure that they contain the information needed during the next phase of the development effort.

3.3.4 Phase 3: Implementation and Verification

Phase 3 answers the question can we authorize shipments to our Beta partners. The object of Phase 3 review is to release the product(s) for Beta testing.

Phase 3 typically includes several subphases. After hardware design and development, the preproduction product typically goes through both engineering and manufacturing detailed verification testing. In software, Development completes the design and coding, inspects the code, and performs unit and product testing; Product Assurance performs system testing of the integrated product(s) as a whole, and Program Management coordinates all Alpha (internal engineering and manufacturing) unit testing. Since many products combine both software and hardware, the subphases must overlap, i.e., a hardware unit must be available for software testing. So for life cycle purposes, a single Phase 3 is recognized.

Before completing Phase 3, Product Assurance verifies that Beta criteria have been met. The goal is to thoroughly validate the functionality and quality of the product so that the product that goes to Beta partners is the same product that goes to FCS.

3.3.5 Phase 4: Preproduction and Introduction

Phase 4 answers the question should we announce and authorize FCS of this product. The objective of Phase 4 review is to announce and authorize general availability of the product, i.e., transfer it to production.

Development produces all project-specific release-related material. Marketing produces data sheets, collateral support material, and announcement material for the product. Support manages Beta testing to ensure that the product is ready for general availability (the necesssary support structure is in place). A postpartum review of Phase 4 should include a review and analysis of the entire development effort and how it could have been executed better.

3.3.6 Phase 5: Production and Support

Phase 5 answers the question is this product successful. The objective of Phase 5 review is to ensure that the product is and continues to be successful in the field by defining (if necessary) maintenance plans, product upgrade and enhancement plans, and new marketing strategies or retirement plans.

TABLE 3 Component Engineering Product Design Phase Deliverables

Phase 0	Phase 1	Phase 2	Phase 3	Phase 4	Phase 5
Technology roadmap	Spicer model and library support Limited special studies experiments Circuit simulation performance model	Spicer model and library support Special studies experiments Circuit simulation performance model	Spicer model and library support Special studies experiments Manufacturing test support	Spicer model and library support Special studies experiments Manufacturing test support	Spicer model and library support Special studies experiments Manufacturing test support
Concept BOM review and needs assessment	BOM review and risk assessment	BOM review support changes and new qualification plans	BOM review support	BOM review support	Emergency product support
Quality key technologies/suppliers before use (DRAM, SRAM, microprocessor)	Component qualification plan and matrix	Potential suppliers list complete Qualify components to plan	Supplier qualification complete Component qualification complete for first source	Component qualification complete for first source	Approved vendor list management, process and product changes, requalification as needed
	Electrical testing and support for failure analysis	Electrical testing and support for failure analysis	Electrical testing and support for failure analysis Supplier defect and corrective action support	Electrical testing and support for failure analysis Supplier defect and correction action support Manufacturing support	Electrical testing and support for failure analysis Supplier defect and corrective action support Manufacturing support

Robust Design Practices 69

All involved functions meet periodically to review the performance of the product in the field. The first such Phase 5 review meeting should take place approximately 6 months after the product has become generally available (a shorter time for products with an extremely short lifetime, such as disk drives or consumer products), unless critical problems arise that require convening the meeting sooner. At the first Phase 5 review subsequent review dates are selected.

A Phase 5 review ends by recommending any of the following product actions:

Continue to manufacture and maintain the product.
Replace the product with the next generation.
Enhance the product.
Retire/make obsolete the product.

Table 3 is an example of the typical phase deliverables for a component engineering organization at a high-end computer server manufacturer.

3.4 TRANSLATE CUSTOMER REQUIREMENTS TO DESIGN REQUIREMENTS

Marketing typically describes, develops, and documents the market need or opportunity for a product based on customer requests and inputs. This document contains such general information as product function, weight, color, performance (sensitivity, selectivity, power output, and frequency response, for example), fit with earlier versions/generations of the product, product migration, price, availability date (market window of opportunity), some measure of reliability or warranty allowance, and so on (this is the Phase 0 input). From this, a product description (Phase 1) is developed that addresses specific objectives and translates the general requirements to a high-level engineering requirements document that is used as the basis for the technology assessment and detailed design activities that follow.

3.5 TECHNOLOGY ASSESSMENT

Early on in the conceptual design stage it is important to understand what technology needs are required to fulfill the documented product description as listed in the engineering requirements document. For example, for a new computer server development effort, such issues as system speed, microprocessor type, bus architecture, memory requirements (DRAM, SRAM, size, speed, etc.), application-specific integrated circuit requirements, on-board power requirements, and interconnect needs (how? mechanical? electrical? size? bus interface?) must be addressed. The process starts with a questionnaire, similar to that shown in Table 4, sent from Component Engineering to the design organization. Component

TABLE 4 Component and Technology Questionnaire

---

1. What interconnect schemes do you plan to use? Do they involve using only existing qualified hardware, or are there new interconnect requirements? (If new, we would like to work with you on requirements definition, application verification, supplier selection, and supplier and part qualification.)

2. What basic technologies are you planning to use? What will the basic voltages for your circuits be (e.g., 1.8, 2.5, 3.3, 5, 12, or 15 V? Other?) Please identify the component types that will be used. How many total new components do you anticipate using that will need to be qualified? How many component part numbers do you anticipate using in your BOM (all will need source checks and may need approved vendor list/source control drawing generation/updating)? Specify the following:

    DRAM (speed/type/size/package)
        SIMM style
    SRAM (speed/type/size/package)
        SIMM style
    PLD/FPGA
        Tools
        Prototype only or production too?
    ASICs (speed/size/package)
        Tools
    Microprocessor
        Support tools
        Bus structures
    Microperipherals
        SCSI
        Special communication circuits
        Special functions
    Special modules/custom circuits
        DC/DC converters
        Delay lines
        Oscillators
        Special functions (i.e., line interface modules)
    Analog ICs
    Digital logic ICs
    Delay lines (standard or custom/active or passive/single edge or both)
    Fiber optic/optic interface
    Passive components
        Terminators
        Filters
        Other
    Discrete Semiconductors
        LED
        FET
        Other

Robust Design Practices 71

TABLE 4   Continued

> PON circuit (please identify which one)
> Nonvolatile memory (type/size/package/speed)
>   EPROM
>   EEPROM
>   Flash
>   SEEPROM
> Fuses
> Switches
> Voltage regulators
> Mechanical fasteners
> Connectors
> Cabling
> Clock-driving circuits and phase-locked loops

Of these technologies, which do you anticipate will be custom or will need to be procured to nonstandard requirements? Which are new to your group? Which do you believe will require special studies (for ground bounce, SSO/SSI, edge rate, jitter, timing, load sensitivity, symmetry, power)? Which do you believe need special characterization or qualification analysis?

    3.   What kind of PWA or like assemblies will you be using? What kind of material?

    4.   What manufacturing processes do you anticipate using (MCM, TAB, BGA, CSP, flip chip, SMT, two-sided SMT, SMT-PTH, PTH)?

    5.   Are you planning to use any of the following categories of devices:

> ECL
> TTL/bipolar (AS/ALS/LS/S/F)
> <16-MB or smaller DRAMs?
> <4-MB or smaller SRAMs?
> Bipolar PLDs?
> 8- or 16-bit microprocessors?
> EEPROMs (PLDs slower than 10 nsec?)

If so, you will be facing severe sourcing risks and may face end-of-life issues.

    6.   Which devices will need SPICE model support? How will you verify timing? Signal integrity? Loading? Component derating?

    What derating guidelines will be used? Identify the following:

> Clock design margin?
> Analog margin, signal integrity?
> Thermal/junction temperatures of components?
> Power plane noise/decoupling?
> Power of IPS/UPS?

TABLE 4 Continued

7. Regarding testing:

   Do you plan to test each supplier's part in your application?
   Do you plan to measure all signals for timing, noise margin, and signal integrity?
   Do you plan to test power fail throughout the four corners of your product specification?
   Do you plan to perform HALT evaluation of the product (test to destruction–analyze–fix–repeat)? If so, which assemblies?
   Will all suppliers' parts be used in the HALT evaluation?
   Will you be testing worst/best case parts (which ones)?
   Will you HALT these?

8. What kind of manufacturability verification/process verification do you plan to do?

9. What special equipment/software will be needed to program the programmable logic?

10. For new suppliers, what discussions and design work and supplier qualification/component qualification has been done to date?

    Who has been the principal contact or interface between the circuit designer and the supplier?
    Are these discussions/efforts documented?

11. What are the key project assumptions?

    Major design assemblies
    Design reviews planned (which/when/who)
    List of key designers and managers (and their responsibilities today)
    Schedule:
      Design specifications
      Prototypes
      Alpha
      Beta
      FCS
    Number of alpha tests (by whom?)
    Number of prototypes (built by whom?)
    Product support life
    Product sales life
    Total product quantity during product life (per year if estimated)
    Design and phase review dates (Phases 1,2,3,4,5)

Source: Tandem Computer Division, Compaq Computer Corporation.

Engineering works with Design Engineering at the concept phase to determine the type of components that will be needed in the circuit design, outlining the scope of work regarding components for both parties. Typically, the component engineer sits down with the design team members and works through the myriad technical issues listed in the questionnaire, starting at a general level and then digging down to more specific component issues. This helps determine if new technologies are needed to meet the product requirements. If new technologies are required, then the component engineer needs to find answers to the following questions:

> Does the technology exist?
> Is the technology currently in development? When will it be available?
> Will the technology development need to be funded?
> Will the technology need to be acquired?
> To use this product/technology do I need to have a contingency plan developed to ensure a continuous souce of supply for manufacturing?

There is also the issue of old products and technologies, those products and technologies that the designers have used before in previous designs and are comfortable with. In some cases the designers simply copy a portion of a circuit that contains these soon-to-be obsolete components. How long will the technologies continue to be manufactured (1.0-μm CMOS in a 0.25-μm CMOS world, for example)? How long will the specific product be available? Where is it in its life cycle?

This process helps to direct and focus the design to the use of approved and acceptable components and suppliers. It also identifies those components that are used for design leverage/market advantage that need further investigations by the component engineer; those suppliers and components that need to be investigated for acceptability for use and availability for production; and those components and suppliers requiring qualification.

The answers allow the component engineer and the designers to develop a technology readiness and risk assessment and answer the question is it possible to design the product per the stated requirements. A flow diagram showing user technology needs and supplier technology availability and the matching of these is shown in Figure 2.

As the keeper of the technology, the component engineer

> Is responsible for keeping abreast of the technology road maps for assigned circuit functions (connectors, memory, microprocessor, for example)
> Conducts technology competitiveness analyses
> Works with the design organization to develop a project technology sizing
> Understands the technology, data, and specifications

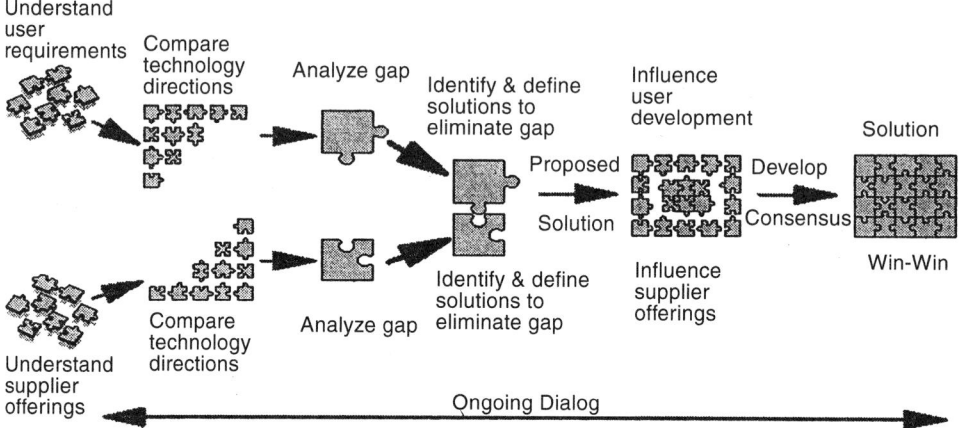

FIGURE 2 Matching user technology requirements and supplier offerings. (Courtesy of Andrew Kostic, IBM.)

Ensures that the design guidelines and computer-aided design (CAD) methods and libraries are available and meet the designer's requirements

All of this activity results in developing a work plan, helps focus the design effort, and facilitates in developing a "first cut" bill of materials (BOM).

3.6 CIRCUIT DESIGN

The design of today's electronic products and equipment is a complex task. This is primarily due to the steady increase in integrated circuit density, functional capability, and performance, for both digital and analog ICs. Concomitant improvements have been made in interconnect devices (printed circuit boards, connectors, and back-planes) and the materials used to make these devices; discrete semiconductors (e.g., power transistors); power supplies; disk drives; and the like.

A second reason for increased design complexity is that the electronics industry is evolving into a new horizontal structure around the four levels of packaging: chip-level integration (ICs), board-level miniaturization surface mount technology, productization (PCs, peripherals, communication sets and instrumentation), and system integration. Each of these requires distinctly different design, manufacturing, and management skills and techniques.

3.6.1 Circuit Types and Characteristics

A given electronic circuit often contains both analog and digital sections. There are significant technical differences between these sections and careful attention

Robust Design Practices

must thus be paid as to how they are electrically interconnected and physically located on the printed circuit board (PCB). As digital IC processes migrate to below 0.25-μm (the deep submicron realm), the resultant ICs become much noisier and much more noise sensitive. This is because in deep submicron technology interconnect wires (on-chip metallization) are jammed close together, threshold voltages drop in the quest for higher speed and lower operating power, and more aggressive and more noise-sensitive circuit topologies are used to achieve even greater IC performance. Severe chip-level noise can affect timing (both delay and skew) and can cause functional design failures.

The analog circuitry needs to be electrically isolated from the digital circuitry. This may require the addition of more logic circuitry, taking up board space and increasing cost. Noise can be easily coupled to analog circuitry, resulting in signal degradation and reduced equipment performance. Also, interfacing analog and digital ICs (logic, memory, microprocessors, phase-locked loops, digital-to-analog and analog-to-digital converters, voltage regulators, etc.) is not well defined. Improper interfacing and termination can cause unwanted interactions and crosstalk to occur. Then, too, the test philosophy (i.e., design for test, which will be discussed in a later section) must be decided early on. For example, bringing out analog signals to make a design testable can degrade product performance. Some of the pertinent differences between analog and digital ICS are listed in Table 5.

3.6.2 Design Disciplines and Interactions

Design of electronic products involves dealing with myriad issues. These include component (part) tolerances, selection/restriction of part suppliers, component/part selection and qualification, design for test (in-circuit test, self-test), understanding failure modes and mechanisms, understanding the relationship in component volume and price cycles, and understanding component life cycles. Many assemblies have design elements that are not optimized, resulting in increased

TABLE 5 Analog and Digital IC Techology Comparison

Analog	Digital
Transistors full on	Transistors either on or off
Large feature size ICs	Cutting edge feature size ICs
High quiescent current	Low quiescent current
Sensitive to noise	Sensitive to signal edge rates
Design tools not as refined/sophisticated	Sophisticated design tools
Tool incompatibility/variability	Standard tool sets
Simulation lags that of digital ICs	Sophisticated simulation process

labor time, increased production and rework costs, as well as yield and quality issues.

Knowledgeable engineers with a breadth of experience in designing different types of products/systems/platforms and who are crosstrained in other disciplines form the invaluable backbone of the design team. All of the various design disciplines (each with their own experts)—circuit design (analog and digital), printed circuit board design and layout (i.e., design for manufacturability), system design (thermal, mechanical and enclosure), design for test, design for electromagnetic compatibility, design for diagnosability (to facilitate troubleshooting), design for reliability, and now design for environment (DFE)—are intertwined and are all required to develop a working and producible design. Each one impacts all of the others and the cost of the product. Thus, a high level of interaction is required among these disciplines. The various design tasks cannot be separated or conducted in a vacuum.

The design team needs to look at the design from several levels: component, module, printed wiring assembly (PWA)—a PCB populated with all the components and soldered—and system. As a result, the design team is faced with a multitude of conflicting issues that require numerous tradeoffs and compromises to be made to effect a working and manufacturable design, leading to an iterative design process. The availability and use of sophisticated computer-aided design tools facilitate the design process.

Tools

Circuit design has changed dramatically over the past decade. For the most part, gone are the days of paper and pencil design followed by many prototype breadboard iterations. The sophistication of today's computer-aided design tools for digital integrated circuit designs allows simulations to be run on virtual breadboard designs, compared with the desired (calculated) results, debugged, and corrected—and then the process repeated, resulting in a robust design. This is all done before committing to an actual prototype hardware build. But there are two big disconnects here. First, digital IC design tools and methods are extremely effective, refined, and available, whereas analog IC and mixed-signal IC tools are not as well defined or available. The analog tools need to be improved and refined to bring them on a par with their digital equivalents. Second, there tends to be a preoccupation with reliance on simulation rather than actually testing a product. There needs to be a balance between simulation before prototyping and testing after hardware has been produced.

There are many CAD and electronic design automation (EDA) tools available for the circuit designer's toolbox. A reasonably comprehensive list is provided in Table 6. Electronic systems have become so large and complex that simulation alone is not always sufficient. Companies that develop and manufacture large digital systems use both simulation and hardware emulation. The rea-

Robust Design Practices

TABLE 6 Examples of CAD and EDA Design Tools

Concept and entry	Verification	Design reuse
HDL	Simulation	Intellectual property
Schematic capture	Verilog simulation	Libraries
Behavioral synthesis	VHDL simulation	Memories/macros
RTL synthesis	Cycle-based simulation	
	Acceleration and emulation	
	Fault simulation	
	SPICE simulation	
	HW/SW coverification	
Design for test		*Virtual prototype*
DFT tools		Hardware/software
Automatic test pattern generation		Silicon
PLD design	*Special purpose tools*	*Miscellaneous*
FPGA place and route	Design for manufacture	Analog design
FPGA tool set	Mechanical design	DSP tool set
	Wire harness design	Mixed signal design
IC design		
Acceleration/emulation	Extractors	Physical verification
CBIC layout	Floor planning	Process migration
Custom layout	Gate array layout	Reliability analysis
Delay calculator	Metal migration analysis	Signal integrity analysis
EMI analysis	Power analysis	Thermal analysis
SPICE	Timing analysis	
PCB design		
EMI analysis	Physical verification	Thermal analysis
MCM/hybrid design	Power analysis	Timing analysis
PCB design	Signal integrity analysis	Virtual prototype evaluation
Autorouter		

sons are twofold and both deal with time to market: (1) Simulation cycle time is several orders of magnitude slower than emulation. Faster simulations result in quicker design verification. (2) Since today's systems are software intensive and software is often the show stopper in releasing a new product to market, system designers cannot wait for the availability of complete hardware platforms (i.e., ASICs) to begin software bring-up.

A methodology called the *electronic test bench* aids the design verification task. The "intelligent test bench" or "intelligent verification environment" has been developed to further ease the designer's task. The intelligent test bench is a seamless, coherent, and integrated EDA linkage of the myriad different kinds of verification tools available and is controlled by a single, high-level test bench. The smart test bench is driven by the increasing variety of point tools needed for complex IC logic verification. It includes simulation, hardware/software co-verification, emulation, formal model checking, formal equivalency checking, and static analysis tools, including lint checkers, code coverage tools and "white box" verification tools that generate monitors or checkers. It makes choices about which tool to use on which portion of the design and derives a test plan. The value of the intelligent test bench is that it eliminates the need for engineers to spend time writing test benches, and it is transparent in that it utilizes familiar software languages.

Prior to use, the various models and libraries need to be qualified. In the case of the SPICE models for the various electronic components used in the design, several questions need to be asked, including:

Is there a viable model for a given functional component?
What is the accuracy of the model?
Is it usable in our system?

The qualification of component SPICE models can be segmented into four levels:

Level 0: model not provided by component supplier; substitute model.
Level 1: validate model to functional specifications.
Level 2: validate DC paramters and limited AC/timing parameters.
Level 3: compare to driving transmission line on PCB; compare quantitatively.

to unspecified parameters (simulation to measurement) with limited timing.

3.6.3 Understanding the Components

Here's the situation: a given circuit design by itself is okay, and the component by itself is okay, but the design/component implementation doesn't work. The following example illustrates the point. The characteristic of a circuit design is that it generates low frequency noise, which by itself is not a problem. The design uses phase-locked loop ICs (PLLS), which are sensitive to low frequency noise to achieve the performance required. The PLL by itself is also okay, but the design doesn't work. A designer must understand the circuit design and the characteristics of the components used.

Experience is showing that the circuit designer needs to understand both the circuit design application and the characteristics of the components. Does the component function match the application need? In many instances the designer is not exactly sure what components or component characteristics are needed or even what a given component does (i.e., how it performs). Examples of questions

Robust Design Practices

facing the designer include the following: How does one use a simple buffer? With or without pull-up resistors? With or without pull-down resistors? No connection? How is the value of the resistors chosen (assuming resistors are used)? It has been found that oftentimes the wrong resistor values are chosen for the pull-up/pull-down resistors. Or how do you interface devices that operate at different voltage levels with characterization data at different voltages (these ICs have different noise margins)? This also raises the question what is the definition of a logic 1 and a logic 0. This is both a mixed voltage and mixed technology issue.

Additionally, the characteristics of specific components are not understood. Many of the important use parameters are not specified on the supplier's data sheets. One needs to ask several questions: What parameters are important (specified and unspecified)? How is the component used in my design/application? How does it interface with other components that are used in the design?

Examples of important parameters, specified or not specified, include the following:

1. Input/output characteristics for digital circuit design.

    Bus hold maximum current is rarely specified. The maximum bus hold current defines the highest pull-up/pull-down resistors for a design.

    It is important to understand transition thresholds, especially when interfacing with different voltage devices. Designers assume 1.5-V transition levels, but the actual range (i.e., 1.3 V to 1.7 V) is useful for signal quality analysis.

    Simultaneous switching effect characterization data with 1, 8, 16, 32, and more outputs switching at the same time allows a designer to manage signal quality, timing edges, edge rates, and timing delay as well as current surges in the design,

    Pin-to-pin skew defines the variance in simultaneously launched output signals from package extremes.

    Group launch delay is the additional delay associated with simultaneous switching of multiple outputs.

2. Functional digital design characteristic.

    Determinism is the characteristic of being predictable. A complex component such as a microprocessor should provide the same output in the same cycle for the same instructions, consistently.

3. Required package considerations.

    Data on the thermal characteristics (thermal resistance and conductance), with and without the use of a heat sink, in still air and with various air flow rates are needed by the design team.

    Package capacitance and inductance values for each pin must be provided. Oftentimes a single value of each parameter is shown for the entire package. Package differences should be specified.

Power dissipation when a device is not at maximum ratings is needed. Power curves would be helpful. What does the power and current look like when a component switches from and to the quiescent state?

The maximum junction temperature and the conditions under which it occurs must be specified. When the IC is powered to its nominal ambient operating condition, what is its junction temperature? Are there any restrictions on the component's use when power supply current, frequency of operation, power dissipation, and ambient temperature are considered simultaneously?

Oftentimes the designer is not sure which supplier's component will work in the design. This is too much information for a designer to know about components. What is required is the support and collaboration of a component engineer at the beginning and throughout the design cycle who is an expert knowledge source of the functional components/suppliers the designer is using or contemplating using. Conducting regularly scheduled bill-of-material reviews is a good way to ensure that the right component for the application, a currently produced component (one that is not being made obsolete), and the right supplier are chosen. A later section discusses the concept of BOM reviews in greater detail.

3.7 POWER SUPPLY CONSIDERATIONS

Power supply voltages and requirements have changed over the past 10 years, being driven by the Semiconductor Industry Association (SIA) IC technology road map. The industry has migrated from 5-V requirements to 3.3 V, then to 2.5 V, and now to 1.8 V and below. This presents some significant system design issues such as dealing with ICs that have mixed voltage (VCC) levels, noise margins, crosstalk, and the like.

This change in supply voltage has driven the PWA and system designers to consider the issue of using centralized versus decentralized (on PWA) or distributed power supplies. Supply voltages less than 3.3 V have caused problems when trying to use centralized power architectures. Lower voltages drive current higher, causing resistive drops in the back-planes of large systems. This makes it difficult to distribute the required power with efficiency and safety. This also raises the questions: Do all PWAs get the same voltage and do all components on a given PWA get the same and correct voltage? To distribute the higher currents without significant voltage drops in these systems requires the use of large and expensive conductors. Other problems with centralized power include greater inductance and noise issues. As voltage and current pass through the wire or PCB trace, there is a greater loss of voltage.

Historically, power distribution in communications and large computer systems (base stations, switches, routers, and servers) has been accomplished using back-plane mounted DC/DC converters to convert a 24–48 V distribution bus to usable voltage rails for the analog and digital processing functions in a system. Higher power density requirements, standard packaging footprints, increasing reliability requirements (in MTBF), and cost drive these designs.

More and more PWA and system designers are turning to the use of DC/DC conversion at the point of use (on the PWA) to help remedy the situation. The DC/DC converters for distributed power applications come in a standard footprint called a brick. Brick, half-brick, and quarter-brick form factors cover power levels from 500 to 5 W, respectively. Actual power density depends on such system factors as heat sinking, ambient temperature, and air flow as well as the efficiency and packaging detail of the converter. The use of distributed or on-board DC/DC converters is not without its issues. These converters must deal with unique load characteristics such as high di/dt (rate of change of current with respect to time), precision voltage tolerance, or multiple-output sequencing. Their use is driving thermal design and packaging innovations, PCB materials, and layout issues to achieve proper heat sinking/cooling.

Portable applications require minimal DC/DC converter power dissipation, size, and weight. Converters for these applications use special control techniques and higher frequency operation. Analog IC suppliers have developed specialized product families that power supply and product/system designers use to meet the needs of these specific converter applications. However, this can present issues for the systems engineer such as noise, voltage, and current spikes and other electromagnetic interference (EMI) circuit interfering issues. Another issue in portable designs is power management under wide load conditions (switching from the quiescent, or sleep, operating mode to a fully active mode). To prolong battery life, DC/DC conversion must be efficient at both heavy and light loads.

When regulation is performed on a circuit card, a small switching regulator may provide single or multiple output voltages. Low dropout (LDO) voltage regulators or charge pumps are used to service additional loads. Power control ICs for portable applications at lower current levels often include integrated power switches for optimal size and cost, while at higher current levels pulse width modulated (PWM) control is provided to external FET switches. Robust control and protection features are needed not only to protect the regulator/converter itself, but to protect the loads as well. To maintain efficiency during extended, lightly loaded conditions portable power ICs must be able to change to a low-frequency mode of operation to reduce gate charge loss in the power switches (transistors).

Thus, the design and selection of the power distribution system is a critical systems design issue that permeates circuit design; system design; component selection; supplier selection; PCB design and layout; thermal, mechanical, and enclosure design (including cooling); reliability; and Electromagnetic Compatibility (EMC). See also later sections on Thermal Management, Signal Integrity, and Design for EMC for further power supply consideration details.

3.8 REDUNDANCY

Redundancy is often employed when a design must be fail safe, or when the consequences of failure are unacceptable, resulting in designs of extremely high reliability. Redundancy provides more than one functional path or operating element where it is critical to maintain system operability (The word *element* is used interchangeable with *component, subassembly,* and *circuit path*). Redundancy can be accomplished by means of hardware or software or a combination of the two. I will focus here on the hardware aspects of redundancy. The use of redundancy is not a panacea to solve all reliability problems, nor is it a substitute for a good initial design. By its very nature, redundancy implies increased complexity and cost, increased weight and space, increased power consumption, and usually a more complicated system checkout and monitoring procedure. On the other hand, redundancy may be the only solution to the constraints confronting the designer of a complex electronic system. The designer must evaluate both the advantages and disadvantages of redundancy prior to its incorporation in a design.

Depending on the specific application, numerous different approaches are available to improve reliability with a redundant design. These approaches are normally classified on the basis of how the redundant elements are introduced into the circuit to provide an alternative signal path. In general, there are two major classes of redundancy:

1. *Active* (or *fully on*) *redundancy*, where external components are not required to perform a detection, decision, or switching function when an element or path in the structure fails
2. Standby redundancy, were external components *are* required to detect, make a decision, and then to switch to another element or path as a replacement for the failed element or path

Redundancy can consist of simple parallel redundancy (the most commonly used form of redundancy), where the system will function if one or both of the subsystems is functional, or more complex methods—such as N-out-of-K arrangements, where only N of a total of K subsystems must function for system operation—and can include multiple parallel redundancies, series parallel redundancies, voting logic, and the like.

For simple parallel redundancy, the greatest gain is achieved through the addition of the first redundant element; it is equivalent to a 50% increase in the system life. In general, the reliability gain for additional redundant elements decreases rapidly for additions beyond a few parallel elements. Figure 3 shows

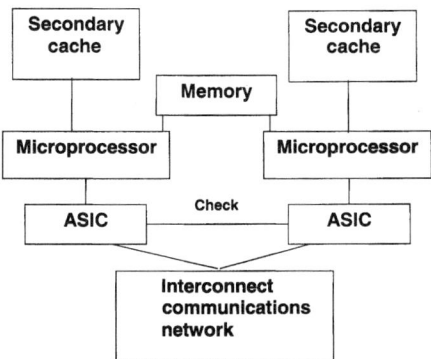

FIGURE 3 CPU with parallel redundancy.

an example of a parallel redundant circuit. This is a block diagram of a computer central processing unit with parallel secondary cache memory, microprocessors, and ASICs.

Adding redundant elements (additional circuitry) may have the effect of reducing rather than improving reliability. This is due to the serial reliability of the switching or other peripheral devices needed to implement the particular redundancy configuration. Care must also be exercised in applying redundancy to insure that reliability gains are not offset by increased failure rates due to switching devices, error detectors, and other peripheral devices needed to implement the redundancy configurations. An example of this is the use of standby or switching redundancy. This occurs when redundant elements are energized (i.e., in their quiescent states) but do not become part of the circuit until they are switched in and only become part of the circuit after the primary element fails. Thus, the redundancy gain is limited by the failure mode or modes of the switching device, and the complexity increases due to switching. In many applications redundancy provides reliability improvement with cost reduction. However, simple backup redundancy is not necessarily the most cost effective way to compensate for inadequate reliability. The circuit designer has the responsibility to determine what balance of redundancy alternatives is most effective, if any. This is a significant factor in total life cycle cost considerations. Redundancy may be easy and cost effective to incorporate if a circuit block or assembly is available off the shelf in comparison to starting a design from scratch or conducting a redesign. Redundancy may be too expensive if the item is costly or too heavy if the weight limitations are exceeded and so on.

These are some of the factors which the electronic circuit designer must consider. In any event, the designer should consider redundancy for reliability improvement of critical items (of low reliability) for which a single failure could cause loss of system or of one of its major functions, loss of control, unintentional actuation of a function, or a safety hazard. Redundancy is commonly used in the aerospace industry. Take two examples. The Apollo spacecraft had redundant on-board computers (more than two, and it often landed with only one computer operational), and launch vehicles and deep space probes have built-in redundancy to prevent inadvertent firing of pyrotechnic devices.

3.9 COMPONENT/SUPPLIER SELECTION AND MANAGEMENT

With the dependence of today's products on high-performance integrated circuits, disk drives, and power supplies, it is important that thorough and accurate processes are in place to ensure that the right technology and functional components are selected from the right suppliers. This means that components with the required performance characteristics and reliability have been selected for the design and that appropriate steps have been taken to assure that these same components are procured and delivered for use in production.

The selection process begins with the dialog between component engineering and design engineering in completing the technology assessment questionnaire discussed earlier in conjunction with technology and functional road maps. Obsolete or soon-to-be obsolete components and or technologies need to be identified and avoided. Also, the use of single- or sole-sourced components need to be identified, the risks assessed, and a proactive contingency plan developed during the detailed design phase to ensure a continuous source of supply during manufacturing. Depending on customer requirements and corporate culture, involved actions can include supplier design and manufacturing process audits and verifications; verification of supplier conducted reliability tests; analysis of the use of single- or sole-sourced components; exhaustive component qualification and new package evaluation and testing; accelerated stress testing during design to determine operating margins and device sensitivities; device characterization testing focusing on four-corner testing and testing of unspecified parameters; in-application (i.e., product) testing; and supplier performance monitoring. Chapter 4 discusses component and supplier selection and qualification in detail.

3.10 RELIABILITY PREDICTION

Reliability prediction is the process of developing a model or assigning/calculating reliability values or failure rates to each component and subsystem. MIL-HDBK-217 and Bellcore Procedure TR-332 are two documents that list component failure rate data. Even though of questionable accuracy, these models provide a starting point from which to proceed based on practical experience

Robust Design Practices

with various components and subassemblies. Reliability prediction is used and iteratively refined throughout the design cycle for comparing designs, estimating warranty costs, and predicting reliability. However, predictions are most useful when comparing two or more design alternatives. Although prediction metrics are notoriously inaccurate, the factors that cause the inaccuracies (such as manufacturing, quality, end-use environments, operator problems, mishandling, etc.) are usually the same for each competing design. Therefore, although the absolute values may be incorrect, the relative values and rankings tend to be valid. A detailed discussion and illustrative examples can be found in Chapter 1.

3.11 SUPPORT COST AND RELIABILITY TRADEOFF MODEL

A support cost model is used to assess cost effectiveness of tradeoffs, such as determining whether the additional cost of a higher reliability unit will pay for itself in terms of reduced service calls.

A complete product life cycle cost model includes development, production, and support costs. Experience has shown that the financial focus of most projects is on development and production costs. Support costs are often overlooked because they are harder to quantify and do not directly relate to financial metrics such as percent R&D expenditure or gross margins. They do, however, relate to profit, and a product reliability problem can affect the bottom line for many years.

Figure 4 shows the structure of a typical support cost model. This model helps evaluate total product cost and ensure that reliability is appropriately considered in design tradeoffs. Support cost is a function of many factors, such as reliability, support strategies, warranty policies, stocking locations, repair depots, and restoration times. Support costs include the number of service calls multiplied by the cost of a service call, highlighting the dependence of support cost on reliability. Other important factors, such as months of spares inventory, have also been included in the cost of a service call. The support cost model is implemented in a spreadsheet to make it easy for anyone to use. Costs are calculated for the expected product life and discounted back to present value. The model can be used for both developed and purchased products and is flexible enough to use with standard reliability metrics. For example, inventory, repair, and shipping costs are removed for service calls that do not result in a part replacement, and material repair cost is removed from returned units that do not fail in test.

The life cycle cost model is often used in the supplier selection process. An example of its use in power supply supplier selection is presented. Supplier A offered a low-cost, low-reliability power supply, while Supplier B offered a higher-cost, high-reliability power supply. Purchasing wanted to use supplier A and find another low-cost second source. The life cycle cost analysis shown in Table 7 was performed. Development costs were minimal since development was

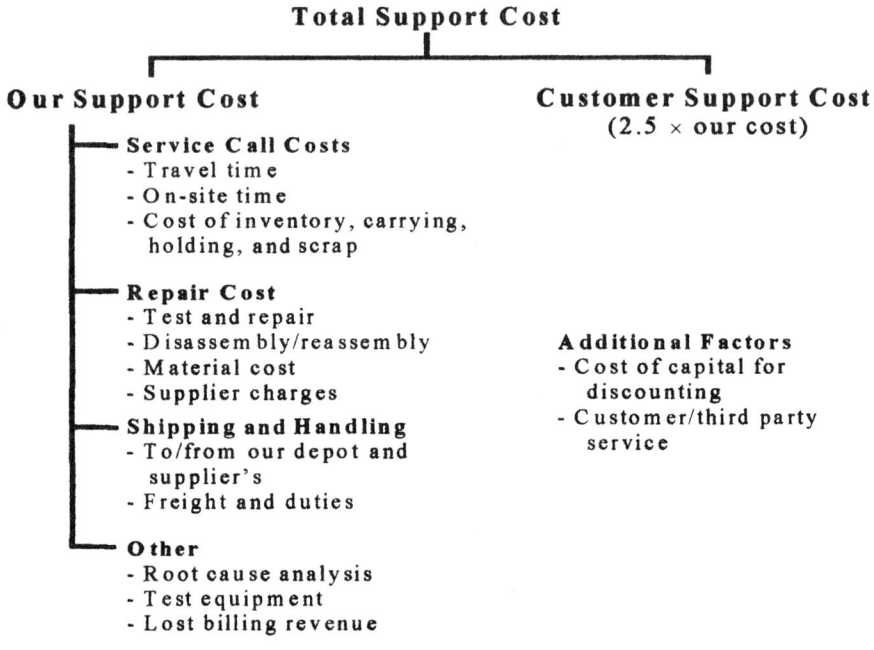

FIGURE 4 Support cost model.

TABLE 7 Power Supply Life Cycle Cost Comparison

Item	Supplier A	Supplier B
Input data		
Cost per unit	$1000	$1300
Expected service calls in 5 years	0.6	0.1
Cost per service call (OEM cost only)	$850	$1100
Cost per service call (OEM cost + customer cost)	$2550	$3600
Cost calculations		
Development cost	Minimal	Minimal
Production cost per unit	$1000	$1300
Support cost per unit (OEM cost only)	$510	$110
Support cost per unit (OEM cost + customer cost)	$1530	$360
Other costs	Minimal	Minimal
Life cycle cost		
Total cost (OEM cost only)	$1510	$1410
Total cost (OEM cost + customer cost)	$2530	$1660

Note: Actual dollar amounts altered.

only involved in a support role (acting as a consultant) in the specification and qualification of the power supplies.

The cost per service call is actually more for the higher-reliability power supply because the shipping and repair costs are higher. However, the number of service calls is significantly lower. The life cycle cost analysis convinced Purchasing to keep supplier B as a second source. Supplier B's power supplies have performed so reliably since product introduction that they quickly became the preferred power supply supplier and have been contracted for a majority of the end product.

3.12 STRESS ANALYSIS AND PART DERATING

Stress analysis consists of calculating or measuring the critical stresses applied to a component (such as voltage applied to a capacitor or power dissipation in a resistor) and comparing the applied stress to some defined criteria. Traditional stress/strength theory indicates that a margin should exist between the applied stresses on a component and the rated capabilities of that component. When sufficient margin exists between the applied stress and the component strength, the probability of failure is minimal. When the safety margin is missing, a finite probability exists that the stresses will exceed the component strength, resulting in failure. The real question that must be answered is how much margin is enough. Often the answer depends on the circuit function, the application, and the product operating environment.

The result of a stress analysis is a priority list requiring design action to eliminate the failure modes by using more robust components with greater safety margins, to minimize the effects of the failures, to provide for routine preventive maintenance or replacement, and/or to assure that repairs can be accomplished easily.

Part derating is essential to achieve or maintain the designed-in reliability of the equipment. Derating assures the margin of safety between the operating stress level and the actual rated level for the part and also provides added protection from system anomalies that are unforeseen during system design. These anomalies may occur as power surges, printed circuit board hot spots, unforeseen environmental stresses, part degradation with time, and the like.

Derating levels are not absolute values, and engineering judgment is required for resolving critical issues. Derating is simply a tradeoff between factors such as size, weight, cost, and failure rate. It is important to note that excessive derating can result in unnecessary increases in part count, part size, printed circuit board size, and cost. This can also increase the overall predicted failure rate; therefore, engineering judgment is necessary to choose the most effective level of derating. The derating guidelines should be exceeded only after evaluating all of the possible tradeoffs and using sound engineering principles and judgment.

Appendix A is one company's parts derating guidelines document at the end of this book.

The rules for derating a part are logical and applied in a specific order. The following is a typical example of such rules:

1. Use the part type number to determine the important electrical and environmental characteristics which are reliability sensitive, such as voltage, current, power, time, temperature, frequency of operation, duty cycle, and others.
2. Determine the worst case operating temperature.
3. Develop derating curves or plots for the part type.
4. Derate the part in accordance with the appropriate derating plots. This becomes the *operational parameter derating*.
5. Using a derating guideline such as that in Appendix A of Chapter 3 or military derating guideline documents [such as those provided by the U.S. Air Force (AFSCP Pamphlet 800-27) and the Army Missile Command] to obtain the derating percentage. Multiply the operational derating (the value obtained from Step 4) by this derating percentage. This becomes the *reliability derating*.
6. Divide the operational stress by the reliability derating. This provides the *parametric stress ratio* and establishes a theoretical value to determine if the part is overstressed. A stress ratio of 1.0 is considered to be critical, and for a value of >1.0 the part is considered to be overstressed.
7. If the part is theoretically overstressed, then an analysis is required and an engineering judgment and business decision must be made whether it is necessary to change the part, do a redesign, or continue using the current part.

3.11.1 Derating Examples

Several examples are presented to demonstrate the derating process.

Example 1: Power Resistor

This example involves derating an established reliability fixed wire-wound (Power Type) MIL-R-39007/5C resistor. This family of resistors has a fixed resistance value, rated for 2 W power dissipation at 25°C, derated to 0 W power dissipation at 275°C, and are designed for use in electrical, electronic communication, and associated equipment. In this example the designer assumed that resistor style PWR 71 meets the design criteria. An evaluation is required to assure that the resistor is not being overstressed. The known facts are that the resistor is being used at a worst case system temperature of 105°C and during operation will dissipate 1.2 W at this temperature. This resistor style derates to 0 W at 275°C. The problem solution follows the previously stated rules:

1. The power (rated wattage) versus temperature derating plot is shown in Figure 5. For a worst case temperature of 105°C, the power derates to approximately 1.36 W.
2. The resistor derating information is found in the MIL-R-39007 standard. The two important stress characteristics are power and temperature. The recommended power derating is 50%, while the maximum usage temperature is 275°C − 25°C, or 250°C.
3. The resistor is power derated to 1.36 W. It must now be derated an additional 50% (considered a reliability derating), or 1.36 W × .50 = 0.68 W.
4. The operating power dissipation or operating stress is 1.2 W. This divided by the reliability derating gives the safety factor stress ratio. In this case, 1.2/0.68 = 1.77. The stress ratio exceeds 1.0 by 0.77 and requires an engineering evaluation.
5. The engineering evaluation would consider the best case, which is an operational power dissipation of 1.2 W divided by the temperature derating of 1.6 W, or a ratio of 0.882. This indicates that the best available case with no reliability derating applied is a ratio of 0.882 versus a maximal acceptable ratio of 1.0. The largest reliability derating which

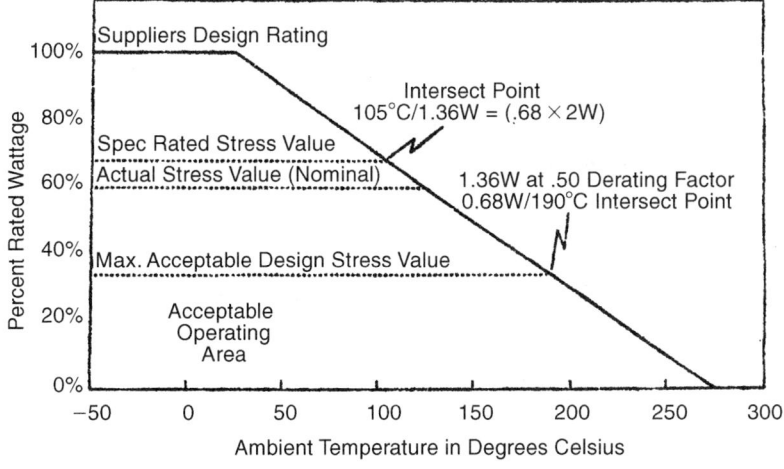

FIGURE 5 Typical power resistor derating curve.

then could be applied would be approximately 11.8% versus the recommended 50% derating guideline. This would derate the resistor power dissipation to 1.36 W ×0.118 = 160 mW. As a check: 1320 mW − 160 mW = 1200 mW, or 1.2 W. The safety factor stress ratio would be equal to 1.0. Reliability engineering should then perform a design reevaluation prior to allowing the use of this resistor in the circuit design. If it were used, it would be a reliability critical item and should appear on the reliability critical items parts list.
6. The typical engineering evaluation must ensure that there are no hot spot temperatures on the PWA which impact the resistor power dissipation, localized hot spots on the resistor, and that the surrounding air temperature is less than the temperature of the resistor body. These items must be confirmed by actually making the necessary temperature measurements and constructing a temperature profile.

Misconception of Part Versus System Operating Temperatures

A common misconception exists concerning the meaning of temperature versus derating. This misconception causes reliability problems. For example, the system operating temperature may be stated as being 75°C. To most people not familiar with derating concepts, this means that all parts in the system are operating at or below 75°C. This temperature of 75°C applies only to the external ambient temperature (worst case) which the system will encounter and does not imply that parts operating internally in the system reach 75°C worst case. It is typical for the stress ratio and derating to be calculated using a worse case temperature of 105°C. Again, this is a guideline temperature which provides an adequate operating reliability safety margin for the vast majority of parts operating in the system. For good reason, an analysis of each part in the system is still required to highlight parts that exceed a temperature of 105°C. Parts that exceed a stress ratio of 1 are considered to be high-risk reliability items and should be placed on a critical items list, resulting in an engineering analysis and judgement being made regarding appropriate corrective action for any potential problem area.

Example 2: Derating a Simple Digital Integrated Circuit

Integrated circuit description:

Triple-3 input NOR gate, commercial part number CD4025
Package: fourteen lead DIP
Supply voltage range: −0.5 to 18 V
Input current (for each input): +10 mA
Maximum power dissipation (PD): 200 mW
Maximum junction temperature: 175°C

Robust Design Practices

Recommended operating conditions: 4.5 to 15 V
Ambient operating temperature range: −55 to 125°C
Load capacitance: 50 pF maximum
Power derating begins at 25°C and derates to 0 at 175°C

Circuit design conditions:

Power supply voltage: 15 V
Operating worst case temperature: 105°C
Output current within derating guidelines
Output power dissipation: 55 mW
Junction temperature (T_j): 110°C

Often we want to find the junction to case thermal resistance. If it is not specified on the IC manufacturer's data sheet, we can calculate it as follows, given that the case temperature (T_c) is 104.4°C:

$$\theta_{jc} = \frac{T_j - T_c}{Pd} = \frac{110 - 104.4}{200} = 0.028 = 28°C/W$$

The junction temperature is calculated as follows:

$$T_j = T_c + (Pd)(\theta_{jc}) = 104.4 + (200)(28)$$
$$T_j = 104.4 + 5.6 = 110°C$$

A note of caution: the 28°C/W can vary depending upon the bonding coverage area between the die and the substrate.

Power Derating. There are several methods that can be used to determine the derated power. One is by using the formula

$$Pd = \frac{Pd(max)}{T_j(max) - 25°C}$$

In the present case, Pd = 200/150 = 1.33 mW/°C

An alternative method is to construct a power derating curve and solve the problem graphically. From the derating curve of Figure 6, the device dissipates 200 mW from −55 to 25°C and then rolls off linearly to 0 W at 175°C.

From the IC manufacturer maximum ratings and from the intended application, it is determined that the maximum junction temperature is 110°C. Going to Figure 6, we proceed as follows.

1. Locate 110°C on the power derating curve and proceed up the 110°C line until it intersects the derating curve.
2. From this point, proceed on a line parallel to the temperature axis to the point where the line intersects the power axis.

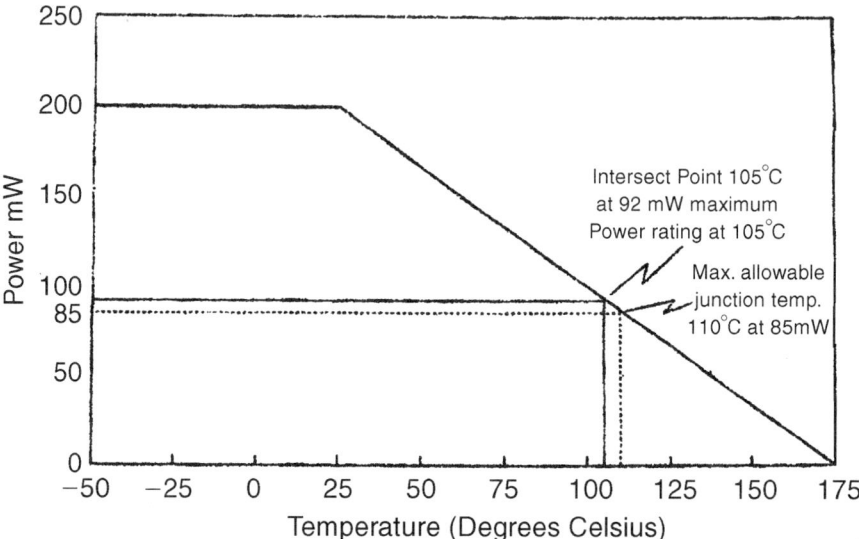

Figure 6 Power versus temperature derating curve for CD 4025 CMOS gate.

3. It is seen that at 110°C junction temperature, the maximum power dissipation is 85 mW (maximum).
4. The power has been derated to 42.5% of the maximum rated value. It was stated that the operating temperature is 105°C. At this temperature the power dissipation is 92 mW.

The next step requires calculating the power dissipation stress ratio. Since power dissipation is directly proportional to the voltage multiplied by the current, the power has to be derated to 90% (due to the intended application) of the derated value of 92 mW: 92 × 0.9 = 82.8 mW. The stress ratio is calculated by dividing the actual power dissipation by the derated power: 55/92 = 0.5978. This value is less than 1.0; thus the IC used in this design situation meets the built-in reliability design criteria.

There are several other conditions that must be satisfied. The case temperature must be measured and the junction temperature calculated to ensure that it is 105°C or less. The worst case voltage, current, and junction temperature must be determined to ensure that the worst case power dissipation never exceeds the stress ratio power dissipation of 92 mW.

It is important to understand the derating curve. Limiting the junction temperature to 110°C appears wasteful when the device is rated at 175°C. In this case the power dissipation has already been decreased to 50% of its 200-mW rating at approximately 100°C. Several of the concerns that must be evaluated

Robust Design Practices

deal with PWA thermal issues: hot spots and heat generating components and their effect on nearest neighbor components; voltage and current transients and their duration and period; the surrounding environment temperature; PWA workmanship; and soldering quality. There are other factors that should be considered, but those listed here provide some insight as to why it is necessary to derate a part and then also apply additional safety derating to protect against worst case and unforeseen conditions. In reliability terms this is called designing in a safety margin.

Example 3: Power Derating a Bipolar IC

Absolute maximum ratings:

> Power supply voltage range: -0.5 to 7.0 V
> Maximum power dissipation per gate (Pd): 50 mW
> Maximum junction temperature (T_j): 175°C
> Thermal resistance, junction to case (θ_{jc}): 28° C/W

Recommended operating conditions:

> Power supply voltage range: -0.5 to 7.0 V
> Case operating temperature range: -55 to 125°C

Circuit design conditions:

> Power supply voltage: 5.5 V
> Operating junction temperature: 105°C
> Output current within derating guidelines
> Output power dissipation: 20 mW/gate
> Maximum junction temperature for application: 110°C

Since the maximum junction temperature allowed for the application is 110°C and the estimated operating junction temperature is less than this (105°C), the operating junction temperature is satisfactory.

The next step is to draw the derating curve for junction temperature versus power dissipation. This will be a straight-line linear derating curve, similar to that of Figure 6. The maximum power dissipation is 50 mW/gate. Plotting this curve in Figure 7 and utilizing the standard derating method for ICs, we see that the derating curve is flat from -55 to 25°C and then rolls off linearly from 25 to 175°C. At 175°C the power dissipation is 0 W.

Using the derating curve of Figure 7 we proceed in the following manner:

1. Find the 105°C temperature point on the horizontal temperature axis. From this point, draw a vertical line to the point where the 105°C temperature line intersects the derating line.

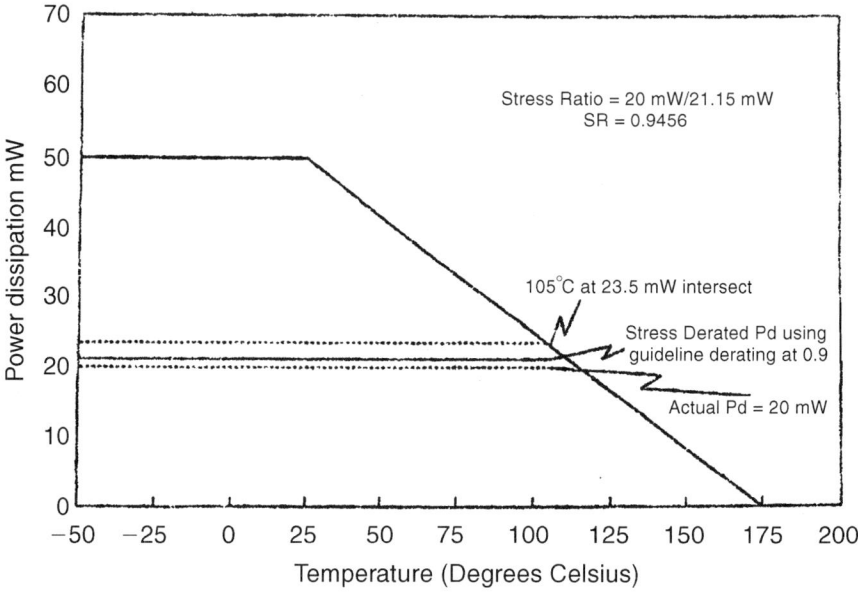

FIGURE 7 Bipolar IC power derating curve.

2. From the point of intersection, draw a line parallel to the temperature axis until it intersects the power dissipation axis.
3. The point of intersection defines the maximum power dissipation at 105°C. From Figure 7, the maximum power dissipation is 23.5 mW.
4. The output power dissipation was given as 20 mW/gate; therefore, the actual power dissipation is less than the derated value.

The stress ratio is calculated as follows:

1. For the intended application environment the power is derated to 90% (given condition) of the derated value. This becomes the stress derating.
2. The derated power at 105°C is 23.5 mW and is stress derated to 90% of this value: $0.90 \times 23.5 = 21.15$ mW.
3. The stress ratio is then calculated by dividing the actual power dissipated (20 mW) by the stress derated power (21.15 mW): $20/21.15 = 0.9456$.
4. Since the stress ratio of 0.9456 is less than 1.0, the design use of this part is marginally satisfactory. Since the junction temperature was estimated to be 105°C, the actual value should be determined by calcula-

tion or measurement since a temperature of 107°C (which is very close to the estimated 105°C) will yield a stress ratio of 1.0.

3.13 PCB DESIGN, PWA LAYOUT, AND DESIGN FOR MANUFACTURE

Integrating the many available advanced ICs into one reliable, manufacturable PCB has become more difficult. The PWA designer must work with the industrial designer and the mechanical designer to find a way to fit the necessary functionality into the desired ultraportable package for products such as cell phones and personal digital assistants (PDAs). Then, too, products such as cell phones and wireless LANs depend on radiofrequency (RF) technology, ranging from a few megahertz up to a few gigahertz. In the past the RF and microwave sections of a product could be put into a specially packaged and shielded area. However, with products shrinking and with the need to have a relatively powerful computer on the same tiny PCB as the RF transmit and receive circuitry, new challenges exist. Component placement, electromagnetic interference, crosstalk, signal integrity, and shielding become extremely critical.

Traditional board technologies that evolved in the mid-1980s and into the 1990s are reaching their limits. For example, conventional etching gets linewidths down to 100 μm. Drilled holes can get down to 200 μm. It's not just that the RF section has high frequency. Personal computers contain 1-GHz processors and bus standards are at 400 MHz and moving to 800 MHz. But it's not the raw clock speed that matters when it comes to designing the PCB. The real issue is the very high edge rates on the signal pulses which have harmonic components in the 8 to 9-GHz range. For these circuits to work, trace impedance, signal transmit times, and termination impedance all need to be very tightly controlled. Technologies requiring the use of differential pairs are becoming more common.

Board designers need to become more knowledgeable about managing these high speed issues during board design, as well as being able to use the new high speed design tools. Board designers are asked to put a radio station combined with a PC into a form factor that's becoming much smaller. Switching speeds, impedance matching, EMI, crosstalk, thermal problems, and very restricted three-dimensional enclosures all combine to exacerbate the problem. They need to do this while working very closely with their extended design teams, and the ICs that they need to put on the board are in multihundred- to multithousand-pin packages that are constantly changing pin-outs. At the same time manufacturing needs must be met.

The industry is responding to these changes with new PCB technologies. Controlling impedance requires new dielectrics that minimize signal loss at high frequencies. The very high pin and trace densities require a new class of etching technology to support the extremely fine linewidths and drilling technology to support the blind and buried microvias. The new high-density interconnect (HDI) technology generally uses laser drilling for the basic hole structures. Included in these technologies is the ability to put passive components such as printed resistors between layers. Buried capacitors and inductors will be available as well. The HDI boards require precision automated assembly machinery.

A product design with good manufacturability will move through the production (manufacturing) environment seamlessly and thus contain an efficient labor content and have high yields, resulting in a high-quality high-reliability product. Just as important as the circuit design itself, so is the PCB design and PWA layout. The PWA serves as the interconnection medium between the various individual components placed on the PCB and between these and the rest of the system/product as well as the external world. As such, the reliability of a product or equipment is dependent on the reliability of the solder joints. Through-hole components provided a robust mechanical connection to the PCB with less stress on the solder joints as compared with surface mount technology (SMT). It is surface mount packages, however, that have been the key enablers in the drive to reduce the overall size of electronic systems and components.

It is these surface mount packages that have placed an increasing demand on solder interconnect reliability. For example, solder joints of leadless ball grid array (BGA) and chip scale package (CSP) assemblies are more susceptible to early wearout than solder joints in compliant, CTE-matched leaded assemblies. Establishing the reliability of area array (BGA and CSP) SMT assemblies requires robust design and assembly practices, including high-yield and high-quality solder interconnects, characterization of the package and board materials, evaluation of the structural response of the entire assembly, and statistical characterization of solder joint failure distributions. Component-dependent attachment reliability trends have been established over the years based on the results of both modeling and accelerated testing programs. Figure 8 shows the impact of

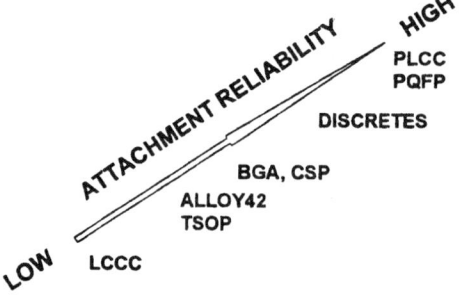

FIGURE 8 Assembly-related trends: relative ranking for SMT components on organic PCBs.

component types on attachment reliability and gives a relative ranking of attachment reliability for several types of surface mount packages on FR-4 PCBs. This ranking is not absolute since the reliability is application dependent and design parameters vary within a family of packages. Thus, the package styles, types, physical dimensions, construction, and materials that will be used in the physical implementation of the design must be considered during PCB design and PWA layout.

Printed circuit board design begins with the materials that constitute its construction. Problems encountered with bare PCBs by equipment and contract manufacturers alike include thin copper traces, exposed inner layer copper, insufficient copper, solder on gold tabs, outgassing and voids, flaking resist, bow and warp, poor solderability, missing solder mask, inner/outer opens, and poor hole quality. A "known good" PCB is thus required and forms the foundation for a robust and manufacturable PWA. This mandates that reliable PCB manufacturers are selected after extensive audits of their design, material supplier selection, materials analysis, and verification and manufacturing processes.

Design for manufacture (DFM) is gaining more recognition as it becomes clear that manufacturing engineers alone cannot develop manufacturable and testable PWAs and PCAs. Design for manufacture is the practice of designing board products that can be produced in a cost-effective manner using existing manufacturing processes and equipment. It is a yield issue and thus a cost issue. It plays a critical role in printed wiring or card assemblies. However, it must be kept in mind that DFM alone cannot eliminate all PWA defects. Defects in PWAs generally fall into three categories: design related problems; incoming material related problems (PCB, adhesive, solder paste, etc.); and problems related to manufacturing processes and equipment. Each defect should be analyzed to its root cause to permit appropriate corrective action to be taken as part of the design process.

The benefits of a manufacturable design are better quality, lower labor and material costs, increased profitability, faster time to market, shorter throughput time, fewer design iterations, more successful product acceptance, and increased customer satisfaction. This happens by thoroughly understanding the manufacturing process.

Design for manufacture and design for assembly (DFA) are integral to PCB design and are the critical links between design and volume manufacturing. Utilizing DFM/A helps bridge the performance gap between the myriad functional improvements being made to packaged silicon solutions (ICs). For example, real-time automated design systems gather feedback from design, test, and manufacturing and assimilate these data with the latest revisions to performance specifications and availability from component suppliers. Design then analyzes this information to enhance both the testability and manufacturability of the new product.

Design for manufacture and assembly is essential to the design of electronic products for the following reasons:

1. Products have become increasingly complex. In the last few years the sophistication of printed circuit packaging has increased dramatically. Not only is surface mount now very fine pitch, but ball grid array and chip scale packages and flip chip technologies have become commercially viable and readily available. This plus the many high-density interconnect structures (such as microvia, microwiring, buried bump interconnection, buildup PCB, and the like) available has made the design task extremely complex.
2. Minimizing cost is imperative. The use of DFM/A has been shown in benchmarking and case studies to reduce assembly costs by 35% and PWA costs by 25%.
3. High manufacturing yield are needed. Using DFM/A has resulted in first-pass manufacturing yields increasing from 89 to 99%.
4. In the electronic product design process, 60–80% of the manufacturing costs are determined in the first stages of design when only 35% or so of the design cost has been expended.
5. A common (standard) language needs to be established that links manufacturing to design and R&D. This common language defines producibility as an intrinsic characteristic of a design. It is not an inspection milestone conducted by manufacturing. The quantitative measure of producibility directly leads to a team approach to providing a high-quality cost-competitive product.

The traditional serial design approach, where the design proceeds from the logic or circuit designer to physical designer to manufacturing and finally to the test engineer for review is not appropriate because each engineer independently evaluates and selects alternatives. Worse is a situation where the manufacturing engineer sees the design only in a physical form on a PCB. This normally is the case when contract manufacturers only perform the component assembly (attachment to the PCB) process.

How should a product be designed? As mentioned previously, the design team should consist of representatives from the following functional organizations: logic design; analog design; computer-aided design (CAD) layout; manufacturing and process engineering; mechanical, thermal, component, reliability, and test engineering; purchasing; and product marketing. Alternatives are discussed to meet thermal, electrical, real estate, cost, and time-to-market requirements. This should be done in the early design phases to evaluate various design alternatives within the boundaries of the company's in-house self-created DFM document. This team should be headed by a project manager with good technical and people skills who has full team member buy-in and management support.

Manufacturing engineering plays an important role during the design phase and is tasked with accomplishing the following:

Robust Design Practices

1. PC board layout maximizing the use of automation
2. Controlling the cost of raw PCB fabrication
3. Implementing design-for-test techniques
4. Creating procurement bills of materials
5. Designing and ordering SMT stencils
6. Programming the manufacturing automation equipment from design files

For the design team to design a manufacturable product, it is important to establish guidelines. Guidelines for DFM/A are essential to establishing a design baseline throughout the company. Engineering designs to a certain set of specifications or requirements, and manufacturing has a certain set of capabilities. Synchronizing requirements and capabilities sets expectations for both functions. The DFM/A guidelines form a bridge between engineering and manufacturing and become a communication vehicle. They can start out as a simple one-page list of sound practices and then evolve into a more complex and comprehensive manual, defining every component and process available. As an example, typical DFM/A guidelines would include:

Component selection criteria
Component orientation requirements
Preferred components and packages
Component spacing requirements (keep-out zones)
Designator and naming conventions
PCB size and shape requirements
Land, pad, and barrel size requirements
PCB edge clearance requirements
Paneling and depaneling information
Trace width, spacing, and shaping requirements
Solder mask and silkscreen requirements
Printing and dispensing considerations
Placement and reflow considerations
Wave soldering and cleaning considerations
Inspection and rework considerations
Fiducial and tooling hole requirements
Test pad size and spacing requirements
Production machine edge clearance
Environmental considerations

Once a DFM guideline document has been created, it must be available immediately to everyone who needs the information or else it is useless. As with any document, the guidelines must be maintained and updated so that they accurately reflect manufacturing's current capabilities. This is especially important as production automation is replaced or upgraded, new technologies and component (IC) package styles are introduced, and the manufacturing activity is outsourced.

The DFM guidelines must be verified on prototype assemblies before an item is released for high-volume production. Validation of DFM should not pose a major problem because most designs go through one or two revisions, in the beginning stages, to fine tune the electrical performance. During those revisions, manufacturing defects should be detected. Some typical examples of PCB design guidelines and the documentation required are listed in Tables 8 and 9, respectively.

TABLE 8 PCB Design Guidelines

Vias not covered with solder mask can allow hidden shorts to via pads under components.

Vias not covered with solder mask can allow clinch shorts on DIP and axial and radical components.

Specify plated mounting holes with pads, unplated holes without pads.

For TO-220 package mounting, avoid using heat sink grease. Instead use sil-pads and stainless hardware.

Ensure that polarized components face in the same direction and have one axis for PTH automation to ensure proper component placement and PWA testing.

Align similar components in the same direction/orientation for ease of component placement, inspection, and soldering.

PTH hole sizes need adequate clearance for automation, typically 0.0015 in. larger than lead diameter.

Fiducial marks are required for registration and correct PCB positioning.

For multilayer PCBs a layer/rev. stack-up bar is recommended to facilitate inspection and proper automated manufacture.

Obtain land pattern guidelines from computer-aided design libraries with CAD programs, component manufacturers, and IPC-SM-782A. SMT pad geometry controls the component centering during reflow.

Provide for panelization by allowing consideration for conveyor clearances (0.125 in. minimum on primary side; 0.200 in. minimum on secondary side), board edge clearance, and drill/route breakouts.

Maximum size of panel or PCB should be selected with the capabilities of the production machine in mind as well as the potential warp and twist problems in the PCB.

PCBs should fit into a standard form factor: board shape and size, tooling hole location and size, etc.

To prevent PCB warpage and machine jams the panel width should not exceed $1.5\times$ the panel length.

Panels should be designed for routing with little manual intervention.

Robust Design Practices

TABLE 9 PCB Design Required Documentation

CAD and gerber data
Soft copy of bill of materials
Gold (functional and identical) PWA
Raw PCB for troubleshooting
Soft and hard copies of detailed schematic diagrams
Program data for programmable logic: PLDs, FPGAs, etc.
PDF data sheet for all ICs
Data for custom ICs
Functional test requirements

A DFM feedback process is necessary in order to effectively relay lessons learned in manufacturing to engineering. One effective technique is to have all engineering prototypes built by production personnel using the intended manufacturing processes. This is a proven method to transfer feedback on the success of building a product. Also, there are no surprises when a product is released for production because those same production processes were used throughout the design cycle. Feedback must be delivered quickly and accurately so the design team can immediately correct any problems observed by the production personnel on the prototypes.

A production readiness review that answers the questions when is an engineering product design done and when is manufacturing ready to accept a design is needed as well. This is an important step because the marketplace is competitive and customers demand high-quality, competitively priced, quick-to-market products designed for manufacturability at the onset of a new product introduction. The production readiness review measures the completeness of each deliverable from each functional group throughout the design cycle of a new product. This is an important crossfunctional design team responsibility and enables all issues to be resolved essentially in real time rather than tossing them over the wall from one functional organization to another. For those companies that don't use crossfunctional design teams, product readiness reviews take much time and can be frustrating events.

An example of an engineering deliverable might be all drawings released to manufacturing in order to buy materials with sufficient lead time. Therefore, engineering may have x deliverables, and manufacturing may have y deliverables. Each of these deliverables is measured at certain gates (i.e., checkpoints) throughout the design cycle. The crossfunctional new product team determines when these deliverables must be completed, and the performance of the readiness is then measured. Now it becomes a very objective measure of when engineering is done and manufacturing is ready. If there are 10 different gates in a 10-month development cycle, for example, and engineering reports 50 of 75 deliverables complete while manufacturing reports 100 of 100 deliverables at Gate 10, then engineering is clearly not done and manufacturing is ready and waiting.

Design for manufacture and assembly is predicated on the use of accurate and comprehensive computer-integrated manufacturing (CIM) tools and sophisticated software. These software programs integrate all relevant data required to design, manufacture, and support a product. Data such as simulation and models; CAD and computer-aided engineering (CAE) data files; materials, processes, and characteristics; specifications and documents; standards and regulations; and engineering change orders (ECOs), revisions, parts, etc. The software programs efficiently

> Communicate information both ways between design engineering and manufacturing.
> Automate CAD data exchange and revision archiving.
> Provide product data tracking and packaging completeness checking and support standard industry networking protocols.
> Allow design for assembly by analyzing part placement, supporting multiple machine configurations, analyzing machine capacity, and providing production engineering documentation.

By having these design files available in an integrated form, PWA design and manufacturing engineers have the necessary information available in one place to develop a cost-effective design implementation, including analyzing various tradeoff scenarios such as

> Product definition and system partitioning (technology tradeoff)
> Layout and CAD system setup
> PWA fabrication design rules, yield optimization, and cost tradeoffs
> SMT assembly process, packaging, component, and test tradeoffs

An example of such a tool is GenCAM (which stands for Generic Computer-Aided Manufacturing). GenCAM is an industry standard written in open ASCII format for electronic data transfer from CAD to computer-aided manufacturing (CAM) to assembly to test in a single file. This file may contain a single board to be panelized for fabrication or subpanelized for assembly. The fixture descriptions in the single GenCAM file allow for testing the assemblies in an array or singular format, as shown in Figure 9. Some of the features and benefits of GenCAM are listed in Table 10. A detailed description is documented in IPC-2511.

GenCAM contains 20 sections (Table 11) that convey design requirements and manufacturing details. Each section has a specific function or task, is independent of the other sections, and can be contained within a single file. The relationship between sections is very important to the user. For example, classic information to develop land patterns is important to both the assembly and in-circuit test

Robust Design Practices

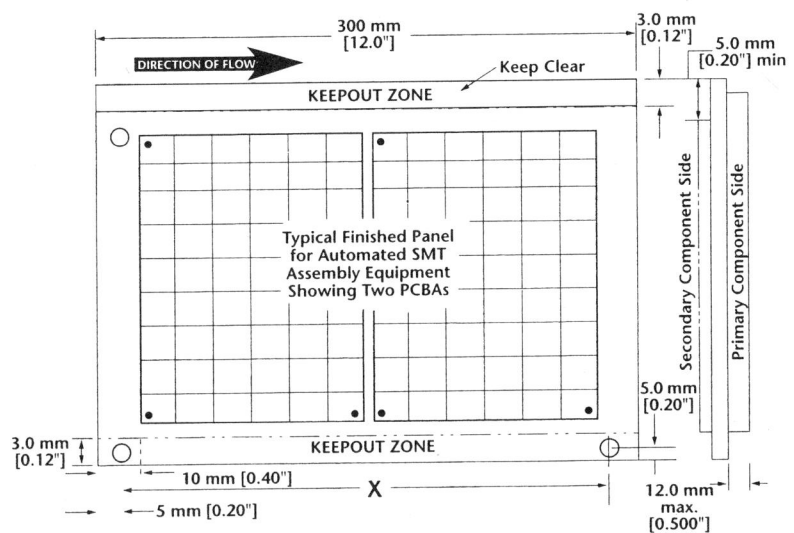

FIGURE 9 Fixture description of assembly subpanel with two identical assemblies. Within the single file, several unique assemblies can be described. (From Ref. 1.)

(ICT) functions. GenCAM files can be used to request quotations, to order details that are specifically process-related, or to describe the entire product (PWA) to be manufactured, inspected, tested, and delivered to the customer.

The use of primitives and patterns provides the information necessary to convey desired final characteristics and shows how, through naming conventions, one builds upon the next, starting with the simplest form of an idea, as shown in Figure 10. Figure 11 shows an example of various primitives. Primitives have no texture or substance. That information is added when the primitive is referenced or instanced.

When textured primitives are reused and named they can become part of an artwork, pattern, or padstack description. When primitives are enhanced, there are many ways in which their combinations can be reused. Primitives can also become symbols, which are a specific use of the pattern section. Figure 12 shows the use of primitives in a pattern to surface mount small-outline ICs (SOICs). In this instance, the symbol is given intelligence through pin number assignment. Thus, the logic or schematic diagram can be compared to the net list identified in the routes section.

GenCAM can handle both through-hole and surface mount components. GenCAM accommodates through-hole components (dual in line packages, pin grid array packages, and DC/DC converters, for example) by including holes

TABLE 10 Benefits of the GenCAM Data Transfer Format

Recipient	Advantages
User	Improves cycle time by reducing the need to spoon-feed the supply chain; supply-chain management, necessary as more services are outsourced; equipment reprocurement capability. Also establishes a valuable archiving capability for fabrication and assembly tooling enhancement; and segmentation of the GenCAM file avoids the need to distribute proprietary product performance data.
Designer	Features ability to provide complete descriptions of one or more assemblies; a direct correlation with CAD library methodology. GenCAM establishes the communication link between design and manufacturing; facilitates reuse of graphical data; permits descriptions of tolerances for accept/reject criteria; brings design into close contact with DFM issues.
Manufacturer	Provides complete description of PCB topology; opportunity to define fabrication panel, assembly subpanel, coupons, and other features; layering description for built-up and standard multilayer construction; ease of reference to industry material specifications; design rule check (DRC) or DFM review and feedback facilitation. Also, data can be extracted to supply input to various manufacturing equipment, e.g., drill, AOI, router.
Assembler	Provides complete integrated bill of materials. Identifies component substitution allowances. Accommodates several BOM configurations in a single file. Establishes flexible reuse of component package data. Supports subpanel or assembly array descriptions. Considers all electrical and mechanical component instances, including orientation, board-mount side.
Electrical bare board and ICT	Identifies one or more fixtures needed for electrical test requirements and specific location of pads or test points; describes test system power requirements, complete net list to establish component connectivity and net association, component values and tolerances. Provides reference to component behavior, timing, and test vectors.

Source: Ref. 1.

used in the CAD system in the padstack section to make connections to all layers of the PCB. For surface mount components, the relationship of vias, as shown in Figure 13, becomes an important element for design for assembly.

GenCAM handles components intermixing by combining components, padstacks, patterns, and routes imformation to position parts on the individual

TABLE 11 Descriptions of GenCAM Sections

Section keyword	Purpose and content
Header	Beginning of each file, includes name, company file type, number, revision, etc.
Administration	Describes ordering information necessary for identifying responsibility, quantity of ordered parts, and delivery schedule.
Fixtures	Describes fixturing for bare- and assembled-board testing.
Panels (panelization)	Includes description of manufacturing panels of PCBs and description of assembly arrays.
Boards	Description of boards and coupons. Includes outline of board/coupon, cutouts, etc.
Drawings	Describes engineering and formatting requirements for complete PCB and assembly descriptions.
Primitives	Describes simple and complex primitive physical shapes. Includes lands, holes, and standard patterns.
Artworks	Functional and nonfunctional geometries developed by the user, e.g., user macros. Includes shapes, logos, features not part of the circuit, and other user-defined figures.
Patterns	Descriptions to build libraries of reuseable packs and padstack.
Mechanicals	Provides information for handles, nuts, bolts, heat sinks, fixtures, holes, etc.
Layers	Board manufacturing descriptions. Includes conductive/nonconductive layer definition of silkscreens, details of dielectric tolerances, separations, and thickness.
Padstacks	CAD system data. Includes pads and drilling information through and within the board.
Packages	Describes a library of component packages. Includes true package dimensions.
Families	Describes logic families of components.
Devices	Component descriptions. Includes device part number.
Components	Identifies parts. Includes reference designators where appropriate.
Power (and ground)	Includes power injection types permitted.
Routes	Conductor location information. Includes location of conductors on all layers.
Test connects	Test-point locations. Includes probe points, single-name test-point types, etc.
Changes	Shows data changed from the design and sent to the manufacturing site.

Source: Ref. 1.

Primitives Section
Provides the shape and dimensions of basic primitives such as circles, squares, or user-defined primitives such as polygons, logos. The primitive section contains an identifier that correlates the primitive to its use.

Artworks Section
Derives data from the primitives section. This data gets texture and color added to it. The original identification is modified via line descriptions or paint descriptions. Thus, a single primitive could have several types of artwork descriptions as to how they might be used. An example of that would be a circle used as a land, a circle on a drawing, or a hole.

Patterns Section
Combinations of artwork primitives that have had the texture added. These patterns would be very domain usage-dependent. Patterns would be referred to in board, drawings, or anywhere patterns contain an obvious choice.

FIGURE 10   Definitions and relationships among the three primary sections of GenCAM.

PCB and on the subpanel assembly array. Since many assembly operations use wave soldering, the general description of the component identified in the packages section can be transformed through (X, Y) positioning, rotation, and mirror imaging. This permits a single description of a package to be positioned in many forms to meet the requirements, shown in Figure 14.

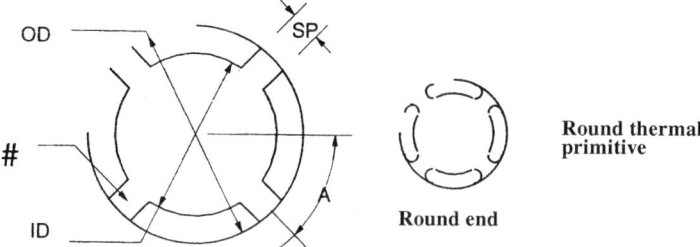

FIGURE 11   Examples of GenCAM primitive shapes.

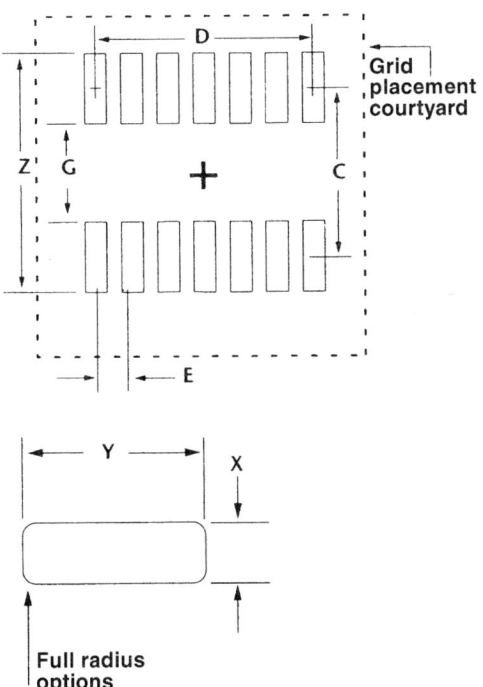

FIGURE 12 Use of primitives in patterns. Primitives can be used as symbols, as shown here, and are given intelligence by means of pin number. (From Ref. 1.)

IPC is an industry association that has taken the responsibility for generating, publishing, and maintaining extensive guidelines and standards for PCB design, artwork requirements, assembly and layout, qualification, and test—facilitating DFM. Table 12 provides a list of some of these documents.

Another change in PWA design and manufacturing that is driven by fast time to market is that PWAs are designed from a global input/output (I/O) perspective. This means that a given PWA is designed and the first article manufactured using embedded field programmable gate arrays (FPGAs) and programmable logic devices (PLDs) without the core logic being completed. After the PWA is manufactured, then the core logic design is begun. However, choosing to use FPGAs in the final design gives the circuit designers flexibility and upgradeability through the manufacturing process and to the field (customer), throughout the product's life. This provides a very flexible design approach that allows changes

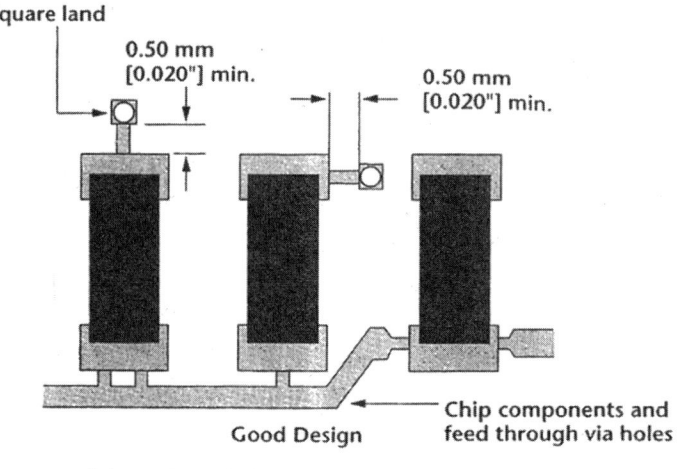
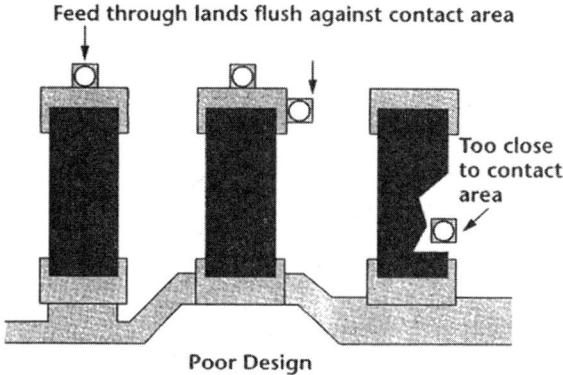

FIGURE 13 Land pattern to via relationship.

to be easily made (even extending programmability through the Internet)—all in serving the marketplace faster.

3.13.1 Some Practical Considerations

In the PWA manufacturing process there are numerous opportunities for potential issues to develop that could impact solderability and functionality. For illustrative purposes, three situations are presented for consideration.

Example 1

Suppose that a large-physical-mass, high-current component is placed next to a smaller component that is critical for circuit timing and one of the components

Robust Design Practices

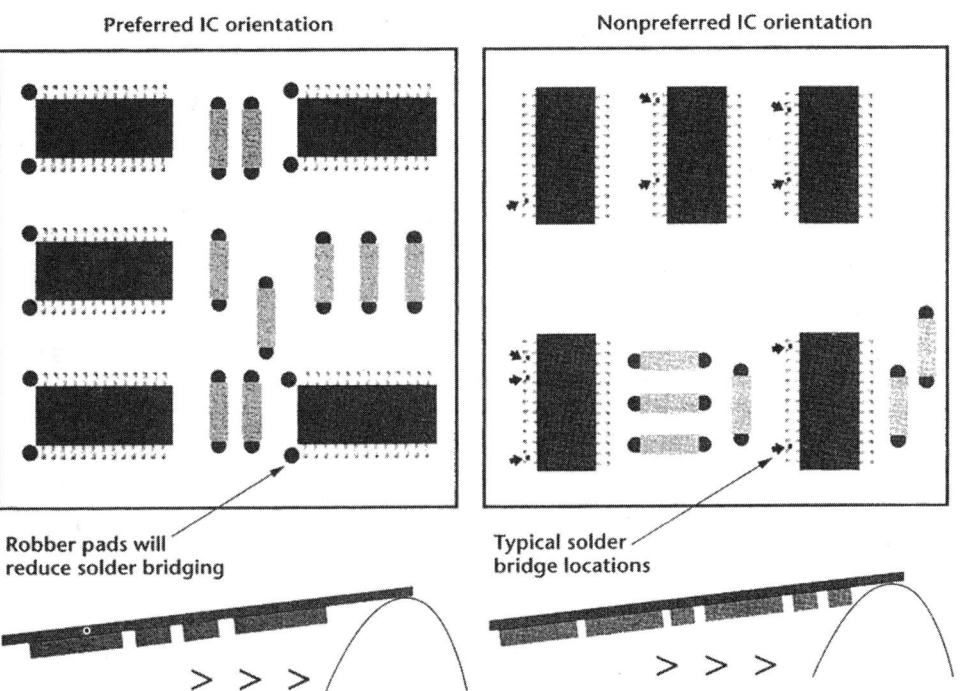

FIGURE 14 Component orientation for wave solder applications.

is not soldered properly. Which component gets soldered properly is dependent on which component the solder profile was set up for. The large part could be taking all the heat during the soldering process (due to its large mass), preventing the smaller component from receiving sufficient solder.

Example 2

From a soldering perspective we don't want components placed too close together. However, from a signal integrity perspective we want the components as close together as possible to minimize the signal parasitic effects.

Example 3

If we have a component with palladium leads close to a component with Alloy 42 leads, the part with palladium leads doesn't get soldered properly.

What these examples show is that careful attention must be paid by experienced engineering and manufacturing personnel to the components that are placed

TABLE 12 Listing of IPC Interconnect Design Documents

IPC document	Description
SMC-WP-004 Design for Excellence	
IPC-T-50 Terms and Definitions for Interconnecting and Packaging Electronic Circuits	
IPC-CC-110 Guidelines for Selecting Core Constructions for Multilayer Printed Wiring Board Applications	Defines guidelines for selecting core constructions in terms of fiberglass fabric style and configuration for use in multilayer printed wiring board applications.
IPC-D-279 Design Guidelines for Reliable Surface Mount Technology Printed Board Assemblies	Establishes design concepts, guidelines, and procedures for reliable printed wiring assemblies. Focuses on SMT or mixed technology PWAs, specifically addressing the interconnect structure and the solder joint itself. Discusses substrates, components, attachment materials and coatings, assembly processes, and testing considerations. In addition, this document contains appendices covering Solder attachments Plated through via structures Insulation resistance Thermal considerations Environmental stresses Coefficient of thermal expansion Electrostatic discharge Solvents Testability Corrosion aerospace and high-altitude concerns

Robust Design Practices

IPC-D-316 High-Frequency Design Guide
Addresses microwave circuitry, microwaves which apply to radiowaves in the frequency range of 100 MHz to 30 GHz. It also applies to operations in the region where distributed constant circuits enclosed by conducting boundaries are used instead of conventional lumped-constant circuit elements.

IPC-D-317A Design Guidelines for Electronic Packaging Utilizing High-Speed Techniques
Provides guidelines for design of high-speed circuitry. Topics include mechanical and electrical considerations and performance testing.

IPC-D-322 Guidelines for Selecting Printed Wiring Board Sizes Using Standard Panel Sizes

IPC-D-325 Documentation Requirements for Printed Boards, Assemblies, and Support Drawings

IPC-D-330 Design Guide Manual
Contains industry approved guidelines for layout, design, and packaging of electronic interconnections. Provides references to pertinent specifications: commercial, military, and federal.
The design guide contains the latest information on materials, design, and fabrication of rigid single- and double-sided boards; multilayers; flexible printed wiring; printed wiring assemblies; and others.

IPC-D-350 Printed Board Description in Digital Form
Specifies record formats used to describe printed board products with detail sufficient for tooling, manufacturing, and testing requirements. These formats may be used for transmitting information between a printed board designer and a manufacturing facility. The records are also useful when the manufacturing cycle includes computer-aided processes and numerically controlled machines.

IPC-D-351 Printed Board Drawings in Digital Form
Describes an intelligent, digital format for transfer of drawings between printed wiring board designers, manufacturers, and customers. Also conveys additional requirements, guidelines, and examples necessary to provide the data structures and concepts for drawing description in digital form. Supplements ANSI/IPC-D-350.
Pertains to four basic types of drawings: schematics, master drawings, assembly drawings, and miscellaneous part drawings.

TABLE 12 Continued

IPC document	Description
IPC-D-354 Library Format Description for Printed Boards in Digital Form	Describes the use of libraries within the processing and generation of information files. The data contained within cover both the definition and use of internal (existing within the information file) and external libraries. The libraries can be used to make generated data more compact and facilitate data exchange and archiving. The subroutines within a library can be used one or more times within any data information module and also in one or more data information modules.
IPC-D-355 Printed Board Automated Assembly Description in Digital Form	Describes an intelligent digital data transform format for describing component mounting information. Supplements IPC-D-350 and is for designers and assemblers. Data included are pin location, component orientation, etc.
IPC-D-356A Bare Substrate Electrical Test Information in Digital Form	Describes a standard format for digitally transmitting bare board electrical test data, including computer-aided repair. It also establishes fields, features, and physical layers and includes file comment recommendations and graphical examples.
IPC-D-390A Automated Design Guidelines	This document is a general overview of computer-aided design and its processes, techniques, considerations, and problem areas with respect to printed circuit design. It describes the CAD process from the initial input package requirements through engineering change.
IPC-C-406 Design and Application Guidelines for Surface Mount Connectors	Provides guidelines for the design, selection, and application of soldered surface mount connectors for all types of printed boards (rigid, flexible-rigid) and backplanes.
IPC-CI-408 Design and Application Guidelines for the Use of Solderless Surface Mount Connectors	Provides information on design characteristics and the application of solderless surface mount connectors, including conductive adhesives, in order to aid IC package-to-board interconnection.
IPC-D422 Design Guide for Press Fit Rigid Printed Board Backplanes	Contains back-plane design information from the fabrication and assembly perspective. Includes sections on design and documentation, fabrication, assembly, repair, and inspection.

Robust Design Practices

IPC-SM-782A	Surface Mount Design and Land Pattern Standard	Covers land patterns for all types of passives and actives: resistors, capacitors, MELFs, SOTs, SOPs, SOICs, TSOPs, SOJs, QFPs, SQFPs, LCCs, PLCCs, and DIPs. Also included are Land patterns for EIA/JEDEC registered components Land patterns for wave or reflow soldering Sophisticated dimensioning system Via location V-groove scoring
IPC-2141	Controlled Impedance Circuit Boards and High-Speed Logic Design	The goal in packaging is to transfer a signal from one device to one or more other devices through a conductor. High-speed designs are defined as designs in which the interconnecting properties affect circuit performance and require unique consideration. This guide is for printed circuit board designers, packaging engineers, printed board fabricators, and procurement personnel so that they may have a common understanding of each area.
IPC-221	Generic Standard on Printed Board Design	Establishes the generic requirements for the design of organic printed boards and other forms of component mounting or interconnecting structures.
IPC-2222	Sectional Standard on Rigid PWB Design	Establishes the specific requirements for the design of rigid organic printed boards and other forms of component mounting and interconnecting structures.
IPC-2224	Sectional Standard for Design of PWBs for PC Cards	This standard establishes the requirements for the design of printed boards for PC card form factors. The organic materials may be homogeneous, reinforced, or used in combination with inorganic materials; the interconnections may be single, double, or multilayered.
IPC-2511	Generic Requirements for Implementation of Product Manufacturing Description Data and Transfer Methodology (GenCAM)	IPC-2511 establishes the rules and protocol of describing data for electronic transfer in a neutral format. GenCAM helps users transfer design requirements and manufacturing expectations from computer-aided design systems to computer-aided manufacturing systems for printed board fabrication, assembly, and test.

next to or in close proximity to each other during the PWA design (to the size of the components and the materials with which they are made). Unfortunately, these experienced manufacturing personnel are getting rarer and lessons learned have not been documented and passed on, and it takes far too long to gain that experience. The difficulties don't end there. Often, the manufacturer is separated by long geographical distances which serve to create a local on-site technical competence shortage. Suffice it to say that PWA manufacturing is in a turbulent state of flux.

3.14 THERMAL MANAGEMENT

The most reliable and well-designed electronic equipment will malfunction or fail if it overheats. Considering thermal issues early in the design process results in a thermally conscious system layout that minimizes costs through the use of passive cooling and off-the-shelf components. When thermal issues are left until completion of the design, the only remaining solution may be costly and drastic measures, such as the design of a custom heat sink that requires all the space available. Incorporating a heat sink or fan into a product after it has been developed can be expensive, and still may not provide sufficient cooling of the product. I address thermal issues from two perspectives: from that of the individual ICs and other heat-generating components placed on the PWA and from that of a system or complete equipment/enclosure.

Today's high-speed CMOS integrated circuits operate at or above 1 GHz clock speeds and generate between 60 and 100 W! There is nothing low power about these circuits. Also, the junction temperatures have been steadily declining from 150°C to about 85–90°C for leading edge ICs. What all this means is that these ICs are operating in an accelerated manner (similar to that previously encountered during IC burn-in) in their intended ambient application. Integrated circuit suppliers estimate that for every 10°C rise of the junction temperature, the device failure rate doubles. If the heat generated inside the IC is not removed, its reliability is compromised. So there is a real challenge here in using leading edge ICs.

According to Moore's law, the amount of information stored in an IC (expressed as density in terms of the number of on-chip transistors) doubles every 18 months. This has been a valid measure of IC improvement since the 1970s and continues today. Moore's law also applies to thermal management. As chip technology becomes increasingly smaller and more powerful, the amount of heat generated per square inch increases accordingly. Various system level power management techniques like low-power quiescent modes, clock gating techniques, use of low-power circuits and low-power supply voltage, and power-versus-performance tradeoffs are widely used to reduce the generated heat. However, it is not all good news. Activation of an IC that is in its quiescent or quiet mode to its normal operating mode causes a large current spike, resulting in rapid

Robust Design Practices

heating. This produces a large thermal gradient across the surface of the IC die (or across several areas of the die), potentially cracking the die or delaminating some of the material layers. New assembly and packaging technology developments make the situation even more complex, requiring new approaches to cooling.

The ability of an electronic system to dissipate heat efficiently depends on the effectiveness of the IC package in conducting heat away from the chip (IC) and other on-board heat-generating components (such as DC/DC converters) to their external surfaces, and the effectiveness of the surrounding system to dissipate this heat to the environment.

The thermal solution consists of two parts. The first part of the solution is accomplished by the IC and other on-board component suppliers constructing their packages with high thermal conductivity materials. Many innovative and cost effective solutions exist, from the tiny small outline integrated circuit and chip scale packages to the complex pin grid array and ball grid array packages housing high-performance microprocessors, FPGAs, and ASICs.

Surface mount technology, CSP and BGA packages and the tight enclosures demanded by shrinking notebook computers, cell phones, and personal digital assistant applications require creative approaches to thermal management. Increased surface mount densities and complexities can create assemblies that are damaged by heat in manufacturing. Broken components, melted components, warped PWAs, or even PWAs catching on fire may result if designers fail to provide for heat buildup and create paths for heat flow and removal. Stress buildup caused by different coefficients of thermal expansion (CTE) between the PWA and components in close contact is another factor affecting equipment/system assembly reliability. Not only can excessive heat affect the reliability of surface mount devices, both active and passive, but it can also affect the operating performance of sensitive components, such as clock oscillators and mechanical components such as disk drives. The amount of heat generated by the IC, the package type used, and the expected lifetime in the product combine with many other factors to determine the optimal heat removal scheme.

In many semiconductor package styles, the only thing between the silicon chip and the outside world is high thermal conductivity copper (heat slug or spreader) or a thermally equivalent ceramic or metal. Having reached this point the package is about as good as it can get without resorting to the use of exotic materials or constructions and their associated higher costs. Further refinements will happen, but with diminishing returns. In many applications today, the package resistance is a small part (less than 10%) of the total thermal resistance.

The second part of the solution is the responsibility of the system designer. High-conductivity features of an IC package (i.e., low thermal resistance) are wasted unless heat can be effectively removed from the package surfaces to the external environment. The system thermal resistance issue can be dealt with by breaking it down into several parts: the conduction resistance between the IC package and the PWA; the conduction resistance between the PWAs and the external surface of the product/equipment; the convection resistance between the PWA, other PWAs, and the equipment enclosure; and the convection resistance between these surfaces and the ambient. The total system thermal resistance is the sum of each of these components. There are many ways to remove the heat from an IC: placing the device in a cool spot on the PWA and in the enclosure; distributing power-generating components across the PWA; and using a liquid-cooled plate connected to a refrigerated water chiller are among them.

Since convection is largely a function of surface area (larger means cooler), the opportunities for improvement are somewhat limited. Oftentimes it is not practical to increase the size of an electronic product, such as a notebook computer, to make the ICs run cooler. So various means of conduction (using external means of cooling such as heat sinks, fans, or heat pipes) must be used.

The trend toward distributed power (DC/DC converters or power regulators on each PWA) is presenting new challenges to the design team in terms of power distribution, thermal management, PWA mechanical stress (due to weight of heat sinks), and electromagnetic compatibility. Exacerbating these issues still further is the trend toward placing the power regulator as close as possible to the microprocessor (for functionality and performance reasons), even to the point of putting them together in the same package. This extreme case causes severe conflicts in managing all issues. From a thermal perspective, the voltage regulator module and the microprocessor should be separated from each other as far as possible. Conversely, to maximize electrical performance requires that they be placed as close together as possible. The microprocessor is the largest source of electromagnetic interference, and the voltage regulator module adds significant levels of both conducted and radiated interference. Thus, from an EMI perspective the voltage regulator and microprocessor should be integrated and encapsulated in a Faraday cage. However, this causes some serious thermal management issues relating to the methods of providing efficient heat removal and heat sinking. The high clock frequencies of microprocessors requires the use of small apertures to meet EMI standards which conflict with the thermal requirement of large openings in the chassis to create air flow and cool devices within, challenging the design team and requiring that system design tradeoffs and compromises be made.

A detailed discussion of thermal management issues is presented in Chapter 5.

3.15 SIGNAL INTEGRITY AND DESIGN FOR ELECTROMAGNETIC COMPATIBILITY

Intended signals need to reach their destination at the same time all the time. This becomes difficult as microprocessor and clock speeds continue to increase, creating a serious signal integrity issue. Signal integrity addresses the impact of ringing, overshoot, undershoot, settling time, ground bounce, crosstalk, and power supply noise on high-speed digital signals during the design of these systems. Some symptoms that indicate that signal integrity (SI) is an issue include skew between clock and data, skew between receivers, fast clocks (less setup time, more hold time) and fast data (more setup time, less hold time), signal delay, and temperature sensitivities, Figure 15 shows a signal integrity example as it might appear on a high-bandwidth oscilloscope. The clock driver has a nice square wave output wave form, but the load IC sees a wave form that is distorted by both overshoot and ringing. Some possible reasons for this condition include the PCB trace may not have been designed as a transmission line; the PCB trace transmission line design may be correct, but the termination may be incorrect; or a gap in either ground or power plane may be disturbing the return current path of the trace.

As stated previously, signal integrity is critical in fast bus interfaces, fast microprocessors, and high throughput applications (computers, networks, telecommunications, etc.). Figure 16 shows that as circuits get faster, timing margins decrease, leading to signal integrity issues. In a given design some of the ways in which faster parts are used, thus causing SI problems, include

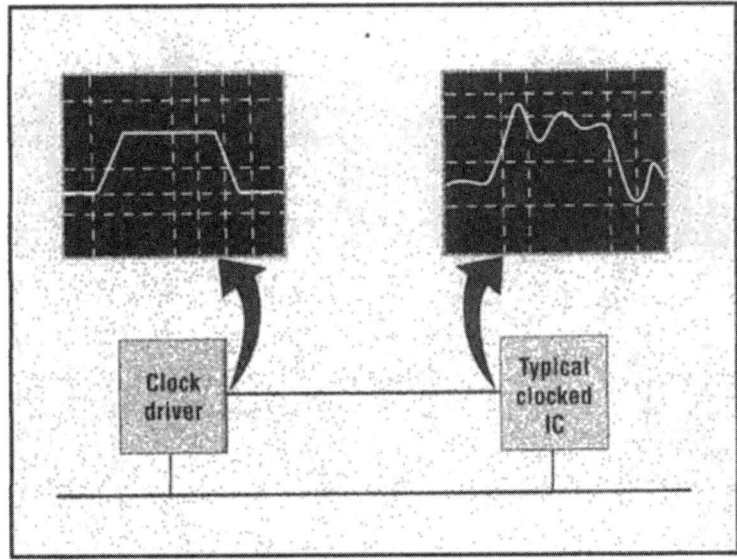

FIGURE 15 The signal integrity issue as displayed on an oscilloscope. (From Ref. 2, used with permission from *Evaluation Engineering,* November 1999.)

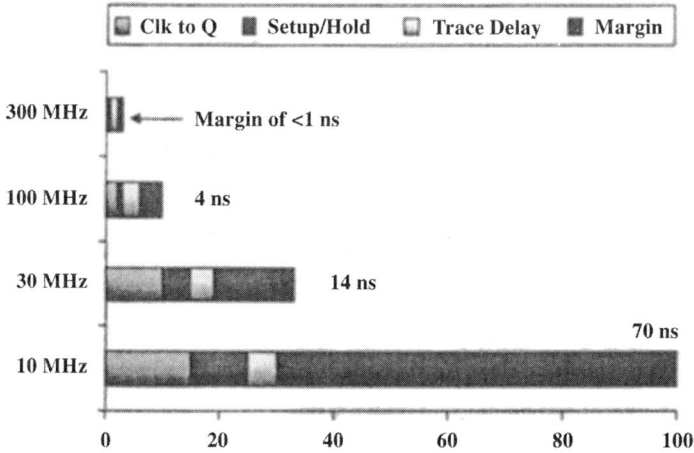

FIGURE 16 Impact of faster ICs on timing margins. (From Ref. 2, used with permission from *Evaluation Engineering,* November 1999.)

A faster driver is chosen for a faster circuit.

A slow part is discontinued, being replaced by a new and faster version.

An original part is replaced by a faster "die-shrunk" part to reduce manufacturing costs.

An IC manufacturer develops one part for many applications or has faster parts with the same part number as the older (previous) generation parts.

Without due consideration of the basic signal integrity issues, high-speed products will fail to operate as intended.

Signal integrity wasn't always important. In the 1970–1990 time frame, digital logic circuitry (gates) switched so slowly that digital signals actually looked like ones and zeroes. Analog modeling of signal propagation was not necessary. Those days are long gone. At today's circuit speeds even the simple passive elements of high-speed design—the wires, PC boards, connectors, and chip packages—can make up a significant part of the overall signal delay. Even worse, these elements can cause glitches, resets, logic errors, and other problems.

Today's PC board traces are transmission lines and need to be properly managed. Signals traveling on a PCB trace experience delay. This delay can be much longer than edge time, is significant in high-speed systems, and is in addition to logic delays. Signal delay is affected by the length of the PCB trace and any physical factors that affect either the inductance (L) or capacitance (C), such as the width, thickness, or spacing of the trace; the layer in the PCB stack-up; material used in the PCB stack-up; and the distance to ground and VCC planes.

Robust Design Practices 119

Reflections occur at the ends of a transmission line unless the end is terminated in Zo (its characteristic impedance) by a resistor or another line. Zo = $\sqrt{L/C}$ and determines the ratio of current and voltage in a PCB trace. Increasing PCB trace capacitance by moving the traces closer to the power plane, making the traces wider, or increasing the dielectric constant decreases the trace impedance. Capacitance is more effective in influencing Zo because it changes faster than inductance with cross-sectional changes.

Increasing PCB trace inductance increases trace impedance; this happens if the trace is narrow. Trace inductance doesn't change as quickly as capacitance does when changing the cross-sectional area and is thus less effective for influencing Zo. On a practical level, both lower trace impedances and strip lines (having high C and low Zo) are harder to drive; they require more current to achieve a given voltage.

How can reflection be eliminated?

Slow down the switching speed of driver ICs. This may be difficult since this could upset overall timing.
Shorten traces to their critical length or shorter.
Match the end of the line to Zo using passive components.

Signal integrity and electromagnetic compatibility (EMC) are related and have an impact on each other. If an unintended signal, such as internally or externally coupled noise, reaches the destination first, changes the signal rise time, or causes it to become nonmonotonic, it's a timing problem. If added EMC suppression components distort the waveform, change the signal rise time, or increase delay, it's still a timing problem. Some of the very techniques that are most effective at promoting EMC at the PWA level are also good means of improving SI. When implemented early in a project, this can produce more robust designs, often eliminating one prototype iteration. At other times techniques to improve EMC are in direct conflict with techniques for improving SI.

How a line is terminated determines circuit performance, SI, and EMC. Matched impedance reduces SI problems and sometimes helps reduce EMC issues. But some SI and EMC effects conflict with each other. Tables 13 and 14 compare termination methods for their impact from signal integrity and EMC perspectives, respectively. Notice the conflicting points between the various termination methods as applicable to SI and EMC.

If fast-switching ICs were not used in electronic designs and we didn't have signal transitions, then there would be no SI problems, or products manufactured for that matter. The faster the transitions, the bigger the problem. Thus, it is important to obtain accurate models of each IC to perform proper signal integrity and EMC analysis. The models of importance are buffer rather than logic models because fast buffer slew times relative to the line lengths cause most of the trouble.

TABLE 13 Comparison of Line Termination Methods from a Signal Integrity Perspective

Type	Advantage	Disadvantage
Series	Single component Low power Damps entire circuit	Value selection difficult Best for concentrated receiver loads
Pull-up/down	Single component Value choice easy Okay for multiple receivers	Large DC loading Increased power
AC parallel	Low power Easy resistor choice Okay for multiple receivers	Two components Difficult to choose capacitor
Diode	Works for a variety of impedances	Two components Diode choice difficult Some over/undershoot

There are two widely used industry models available, SPICE and IBIS. SPICE is a de facto model used for modeling both digital and mixed-signal (ICs with both digital and analog content) ICs. IBIS is used for modeling digital systems under the auspices of EIA 656. It is the responsibility of the IC suppliers (manufacturers) to provide these models to original equipment manufacturers (OEMs) for use in their system SI analysis.

In summary, as operating speeds increase the primary issues that need to be addressed to ensure signal integrity include

1. A greater percentage of PCB traces in new designs will likely require terminators. Terminators help control ringing and overshoot in transmission lines. As speeds increase, more and more PCB traces will begin to take on aspects of transmission line behavior and thus will re-

TABLE 14 Comparison of Line Termination Methods from an EMC Perspective

Type	Summary	EMC effect
Series	Best	Reduced driver currents give good performance. Works best when resistor is very close to driver.
DC pull-up/down	So-so	Less ringing generally reduces EMI. Some frequencies may increase.
AC parallel	So-so	Similar to DC parallel, but better if capacitor is small.
Diode	Worst	Can generate additional high-frequency emissions.

quire terminators. As with everything there is a tradeoff that needs to be made. Since terminators occupy precious space on every PC board and dissipate quite a bit of power, the use of terminators will need to be optimized, placing them precisely where needed and only where needed.
2. The exact delay of individual PCB traces will become more and more important. Computer-aided tool design manufacturers are beginning to incorporate features useful for matching trace lengths and guaranteeing low clock slew. At very high speeds these features are crucial to system operation.
3. Crosstalk issues will begin to overwhelm many systems. Every time the system clock rate is doubled, crosstalk intensifies by a factor of two, potentially bringing some systems to their knees. Some of the symptoms include data-dependent logic errors, sudden system crashes, software branches to nowhere, impossible state transitions, and unexplained interrupts. The dual manufacturing/engineering goal is to compress PCB layout to the maximum extent possible (for cost reasons), but without compromising crosstalk on critical signals.
4. Ground bounce and power supply noise are big issues. High powered drivers, switching at very fast rates, in massive parallel bus structures are a sure formula for power system disaster. Using more power and ground pins and bypass capacitors helps, but this takes up valuable space and adds cost.

Chapter 6 presents an in-depth discussion of EMC and design for EMC.

3.16 DESIGN FOR TEST

The purpose of electrical testing is to detect and remove any ICs or PWAs that fail operational and functional specifications. Integrated circuits and PWAs fail specifications because of defects that may be introduced during the manufacturing process or during subsequent handling operations. Testing an IC or PWA involves applying the voltage, current, timing conditions, and functional patterns it would see in a real system and sequencing it through a series of states, checking its actual against its expected responses.

Testability is concerned with controlling all inputs simultaneously and then trying to observe many outputs simultaneously. It can be defined as a measure of the ease with which comprehensive test programs can be written and executed as well as the ease with which defects can be isolated in defective ICs, PWAs, subassemblies, and systems. A digital circuit with high testability has the following features:

The circuit can be easily initialized.

The internal state of the circuit can be easily controlled by a small input vector sequence.

The internal state of the circuit can be uniquely and easily identified through the primary outputs of the circuit or special test points.

Complicating the problem of testability at the IC level is the use of mixed analog and digital circuitry on the same chip. Table 15 lists some of the testability issues of analog and digital ICs. As can be seen, these issues become very complex and severely impact the ability to test an IC when functional structures incorporating both digital and analog circuits are integrated on the same chip. These same mixed-signal circuit issues are relevant at PWA test as well.

Shrinking product development cycles require predictable design methodologies including those for test, at both the individual IC and PWA levels. The pace of IC and PWA design and level of complexity is increasing so rapidly that

The cost of developing an IC test program is approaching the cost of developing the IC itself.

TABLE 15   Testability Issues of Analog and Digital ICS

Analog circuitry
   Hard to test.
   Use analog waveforms.
   Apply a variety of test signals, wait for settling, and average several passes to reduce noise.
   Affected by all types of manufacturing process defects.
   Must check full functionality of the device within very precise limits.
   Defect models not well defined.
   Sensitive to external environment (60 Hz noise, etc.)
   ATE load board design, layout, and verification; calibration; and high-frequency calibration are critical issues.
   Synchronization issues between device and tester.
   Initialization results in going to an unknown state. This is a difference between analog and digital functions.

Digital circuitry
   More testable, less susceptible to manufacturing process defects, and easier to produce.
   Allow testing at real system clock(s) using industry standard test methodologies.
   Susceptible to spot defects but unaffected by global manufacturing defects.

Compatibility
   Must consider coupling effects of digital versus analog signals.
   Lack of well-defined interface between digital and analog circuitry and technology.
      Normally digital and analog circuitry are segmented.

Robust Design Practices

These higher levels of IC and PWA complexity and packing density integration result in reduced observability and controllability (decreased defect coverage).

The task of generating functional test vectors and designing prototypes is too complex to meet time-to-market requirements.

Tossing a net list over the wall to the test department to insert test structures is a thing of the past.

Traditional functional tests provide poor diagnostics and process feedback capability.

Design verification has become a serious issue with as much as 55% of the total design effort being focused on developing self-checking verification programs plus the test benches to execute them.

What's the solution? What is needed is a predictable and consistent design for test (DFT) methodology. Design for test is a structured design method that includes participation from circuit design (including modeling and simulation), test, manufacturing, and field service inputs. Design for test provides greater testability; improved manufacturing yield; higher-quality product; decreased test generation complexity and test time; and reduced cost of test, diagnosis, troubleshooting, and failure analysis (due to easier debugging and thus faster debug time). Design for test helps to ensure small test pattern sets—important in reducing automated test equipment (ATE) test time and costs—by enabling single patterns to test for multiple faults (defects). The higher the test coverage for a given pattern set, the better the quality of the produced ICs. The fewer failing chips that get into products and in the field, the lower the replacement and warranty costs.

Today's ICs and PWAs implement testability methods (which include integration of test structures and test pins into the circuit design as well as robust test patterns with high test coverage) before and concurrent with system logic design, not as an afterthought when the IC design is complete. Designers are intimately involved with test at both the IC and PWA levels. Normally, a multidisciplinary design team approaches the technical, manufacturing, and logistical aspects of the PWA design simultaneously. Reliability, manufacturability, diagnosability, and testability are considered throughout the design effect.

The reasons for implementing a DFT strategy are listed in Table 16. Of these, three are preeminent:

Higher quality. This means better fault coverage in the design so that fewer defective parts make it out of manufacturing (escapes). However, a balance is required. Better fault coverage means longer test patterns. From a manufacturing perspective, short test patterns and thus short test times are required since long test times cost money. Also, if it takes too long to generate the test program, then the product cycle is impacted

TABLE 16 Design for Test Issues

Benefits	Concerns
Improved product quality.	Initial impact on design cycle while DFT techniques are being learned.
Faster and easier debug and diagnostics of new designs and when problems occur.	Added circuit time and real estate area.
Faster time to market, time to volume, and time to profit.	Initial high cost during learning period.
Faster development cycle.	
Smaller test patterns and lower test costs.	
Lower test development costs.	
Ability to tradeoff performance versus testability.	
Improved field testability and maintenance.	

initially and every time there is a design change new test patterns are required. Designs implemented with DFT result in tests that are both faster and of higher quality, reducing the time spent in manufacturing and improving shipped product quality level.

Easier and faster debug diagnostics when there are problems. As designs become larger and more complex, diagnostics become more of a challenge. In fact, design for diagnosis (with the addition of diagnostic test access points placed in the circuit during design) need to be included in the design for test methodology. Just as automatic test pattern generation (ATPG) is used as a testability analysis tool (which is expensive this late in the design cycle), diagnostics now are often used the same way. Diagnosis of functional failures or field returns can be very difficult. An initial zero yield condition can cause weeks of delay without an automated diagnostic approach. However, diagnosing ATPG patterns from a design with good DFT can be relatively quick and accurate.

Faster time to market.

Design for test (boundary scan and built-in self-test) is an integrated approach to testing that is being applied at all levels of product design and integration, shown in Figure 17: during IC design, PWA (board) design and layout, and system design. All are interconnected and DFT eases the testing of a complete product or system. The figure shows built-in self-test (BIST) being inserted into large complex ICs to facilitate test generation and improve test coverage, primarily at the IC level but also at subsequent levels of product integration. Let's look at DFT from all three perspectives.

Robust Design Practices

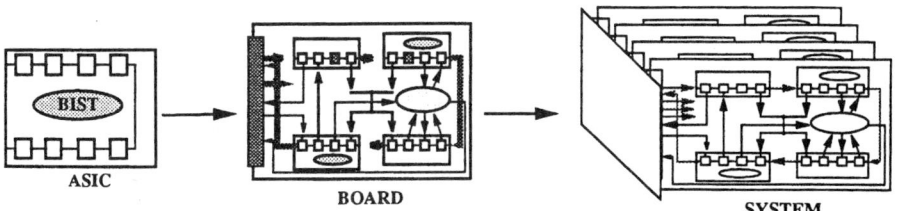

FIGURE 17 Applying BIST and boundary scan at various levels of product integration.

3.16.1 Design for Test at the IC Level

Integrated circuit designers must be responsible for the testability of their designs. At Xilinx Inc. (an FPGA and PLD supplier in San Jose, CA.), for example, IC designers are responsible for the testability and test coverage of their designs, even for developing the characterization and production electrical test programs.

The different approaches used to achieve a high degree of testability at the IC level can be categorized as ad hoc, scan based, and built-in self-test methods. In the *ad hoc method*, controllability and observability are maintained through a set of design-for-test disciplines or guidelines. These include

    Partitioning the logic to reduce ATPG time
    Breaking long counter chains into shorter sections
    Never allowing the inputs to float
    Electrically partitioning combinatorial and sequential circuits and testing them separately
    Adding BIST circuitry

A more comprehensive list of proven design guidelines/techniques that IC designers use to make design more testable is presented in Figure 18. From an equipment designer or systems perspective there isn't a thing we can do about IC testing and making an IC more testable. *However, what the IC designer does to facilitate testing and putting scan and BIST circuitry in the IC significantly impacts PWA testing.* Since implementing DFT for the PWA and system begins during IC design, I will spend some time discussing DFT during the IC design discussion in order to give PWA and system designers a feel for the issues.

Implementing DFT requires both strong tools and support. Test synthesis, test analysis, test generation, and diagnostic tools must handle a variety of structures within a single design, work with various fault/defect models, and quickly produce results on multimillion gate designs. Design for test helps to speed the ATPG process. By making the problem a combinatorial one, ensuring non-RAM memory elements are scannable and sectioning off areas of the logic that may

☞ **MAKE SURE CIRCUIT CAN BE RESET EASILY.**

○ ATPG Software may not be clever enough to discover special initialization sequences.

○ A straightforward RESET mechanism is also useful during chip simulation; no need to "kluge" the starting state during simulation.

○ Simple RESET mechanism helps prototype system debug.

☞ Disable all internal Oscillators and Clocks
  − Must be able to apply clocks only when required.

○ Corollary is "All clocks must be directly controllable by the test resource (e.g. Automatic Test Equipment)".

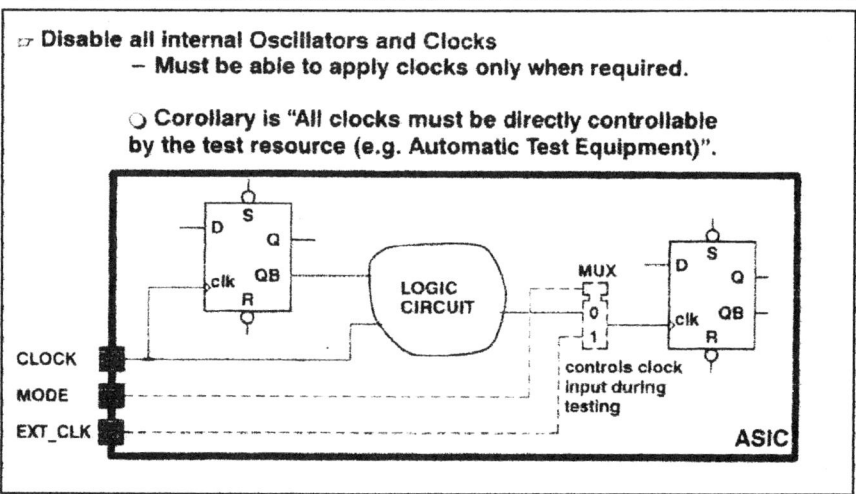

FIGURE 18 Design for test guidelines for the IC designer.

require special testing, the generation of test patterns for the chip logic can be rapid and of high quality.

The use of scan techniques facilitates PWA testing. It starts with the IC itself. Scan insertion analyzes a design, locates on chip flip flops and latches, and replaces some (partial scan) or all (full scan) of these flip flops and latches with scan-enabled versions. When a test system asserts those versions' scan-enable lines, scan chains carry test vectors into and out of the scan compatible flip flops, which in turn apply signals to inputs and read ouputs from the combinatorial logic connected to those flip flops. Thus, by adding structures to the IC itself, such as D flip flops and multiplexers, PWA testing is enhanced through better

Robust Design Practices

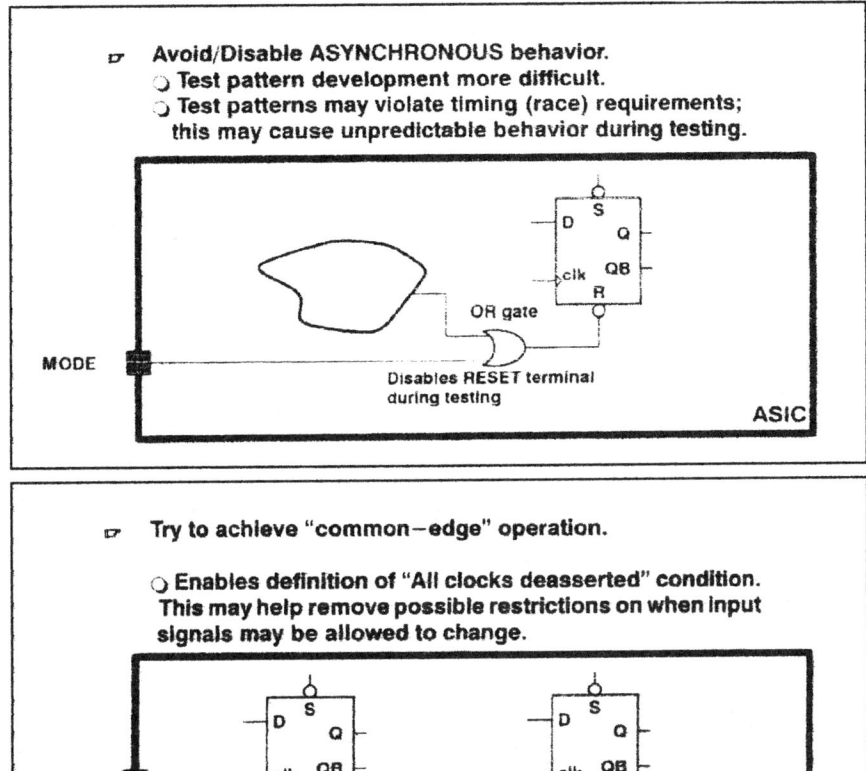

FIGURE 18   Continued

controllability and observability. The penalty for this circuitry is 5–15% increased silicon area and two external package pins.

Scan techniques include level sensitive scan design (LSSD), scan path, and boundary scan. In the scan path, or scan chain, technique, DQ flip flops are inserted internally to the IC to sensitize, stimulate, and observe the behavior of combinatorial logic in a design. Testing becomes a straightforward application of scanning the test vectors in and observing the test results because sequential

☞ **Provide BYPASS around embedded RAM blocks.**

○ Enables testing of logic downstream from the RAM; ATPG software may not be able to pass test signals through the RAM.

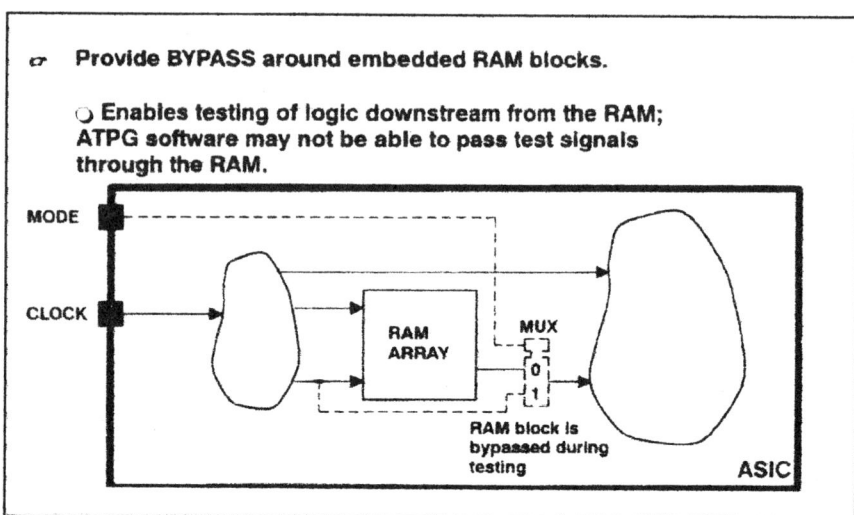

☞ **Provide access to inputs/outputs of embedded logic structures, such as RAM, PLA, mega-cell, hierarchical-block, etc.**

○ Enables testing of structures for which only stand-alone (probably manually-generated) test patterns may exist. This may be especially important when hierarchical blocks have been designed by different designers.

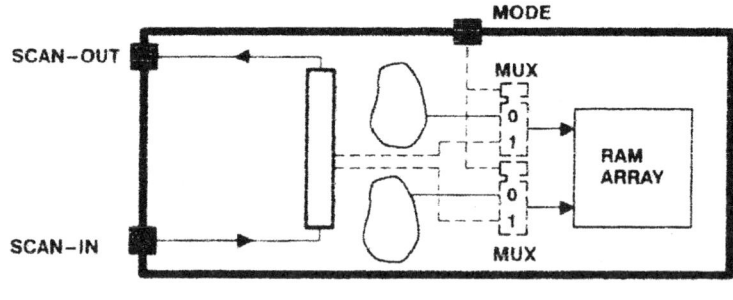

FIGURE 18   Continued

Robust Design Practices

☞ **Add Test Points (Control–inputs and/or Observe–outputs) to enhance controllability and observability of circuit nodes.**

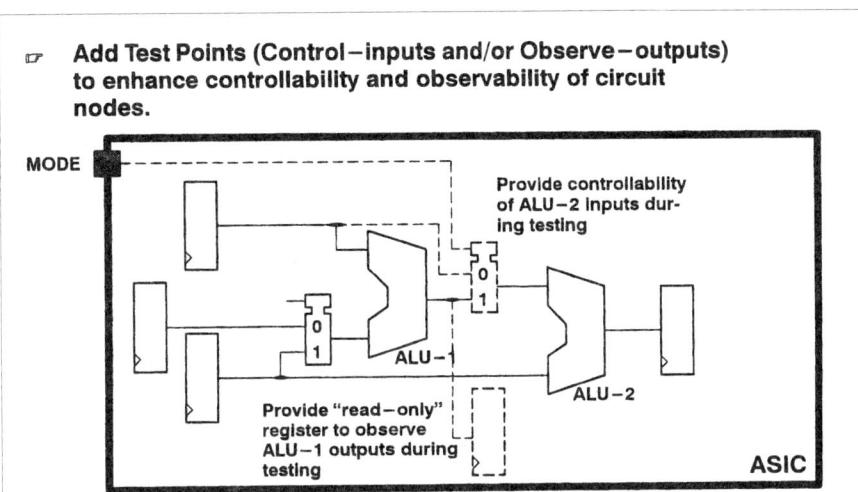

☞ **Eliminate ILLEGAL states from your design.**

○ ATPG software does not understand what is ILLEGAL.
○ Side effects of tests may prevent their use.

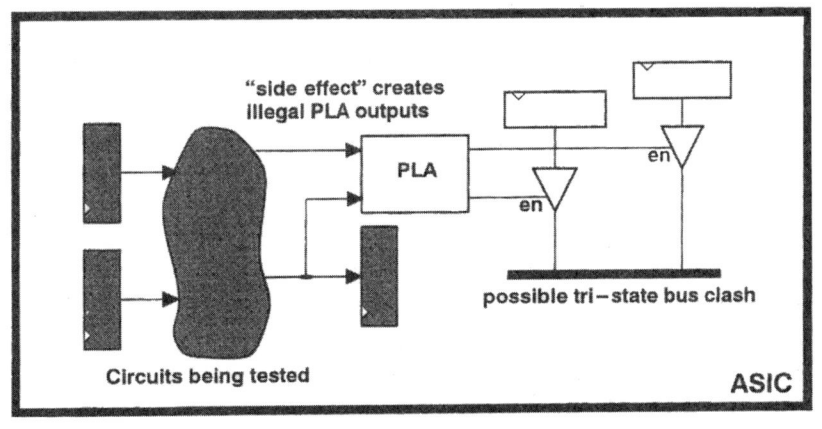

FIGURE 18   Continued

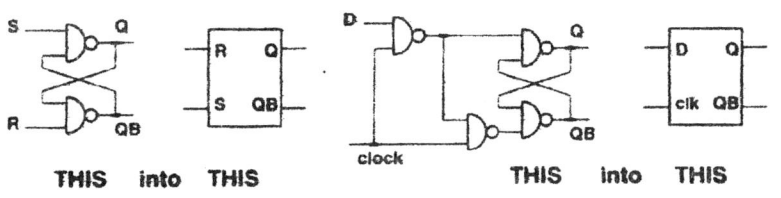

- AVOID feedback in combinational–only feedback paths.
  - Claims of CAE software suppliers that their products can handle such cases notwithstanding, this feature is one of the most problem–causing features in digital design.
  - Encapsulate "cross–coupled" NAND or NOR structures as primitive latches.

- AVOID redundant logic; Test generation through logic containing redundancy is very time consuming and low in fault–coverage; Redundant logic itself is untestable!

- BEWARE of redundant logic! Triple–redundancy for fault tolerance, Error Checking and Correcting (ECC) logic is by definition untestable as the circuit is expected to "cover–up" for a fault. Must add additional control inputs (e.g. disable ECC or, conversely, inject faults at inputs) or observe outputs (e.g. observe key signals internal to a block).

FIGURE 18   Continued

logic is transformed to combinational logic for which ATPG programs are more effective. Automatic place and route software has been adapted to make all clock connections in the scan path, making optimal use of clock trees.

The boundary scan method increases the testability over that of the scan path method, with the price of more on-chip circuitry and thus greater complexity. With the boundary scan technique, which has been standardized by IEEE 1149.1, a ring of boundary scan cells surrounds the periphery of the chip (IC). The boundary scan standard circuit is shown in Figures 19 and 20, and the specific characteristics and instructions applicable to IEEE 1149.1 are listed in Tables 17 and 18, respectively.

Each boundary scan IC has a test access port (TAP) which controls the shift–update–capture cycle, as shown in Figure 19. The TAP is connected to a

Robust Design Practices

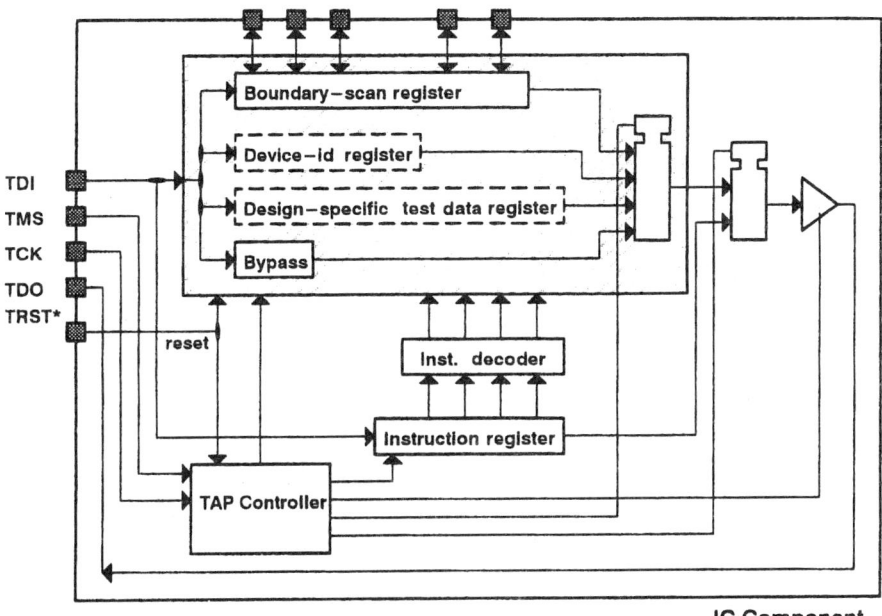

FIGURE 19   IEEE 1149.1 boundary scan standard circuit implementation.

test bus through two pins, a test data signal, and a test clock. The boundary scan architecture also includes an instruction register, which provides opportunities for using the test bus for more than an interconnection test, i.e. component identity. The boundary scan cells are transparent in the IC's normal operating mode. In the test mode they are capable of driving predefined values on the output pins and capturing response values on the input pins. The boundary scan cells are linked as a serial register and connected to one serial input pin and one serial output pin on the IC.

It is very easy to apply values at IC pins and observe results when this technique is used. The tests are executed in a shift–update–capture cycle. In the shift phase, drive values are loaded in serial into the scan chain for one test while the values from the previous test are unloaded. In the update phase, chain values are applied in parallel on output pins. In the capture phase, response values are loaded in parallel into the chain.

Boundary scan, implemented in accordance with IEEE 1149.1, which is mainly intended for static interconnection test, can be enhanced to support dynamic interconnection test (see Fig. 21). Minor additions to the boundary scan cells allow the update–capture sequence to be clocked from the system clock

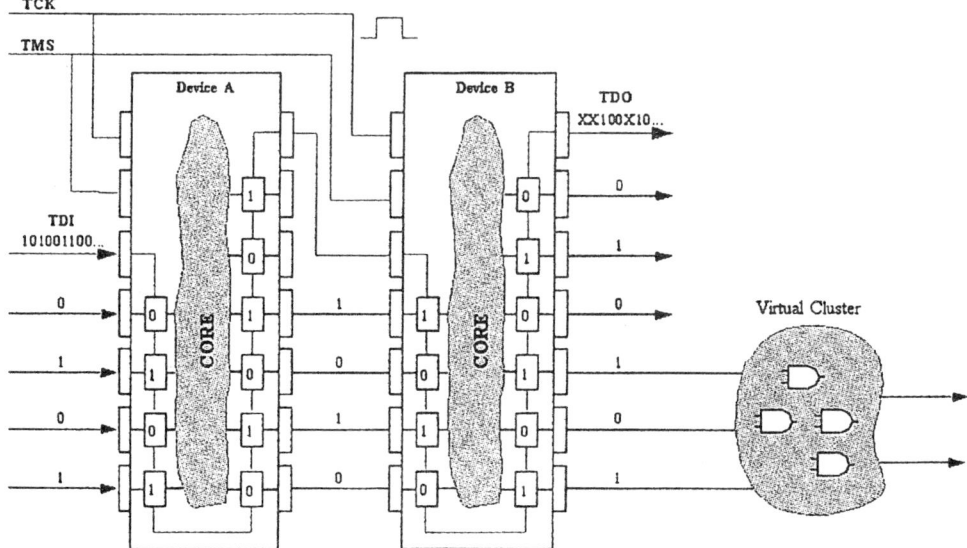

FIGURE 20 Boundary scan principles.

rather than from the test clock. Additional boundary scan instruction and some control logic must be added to the ICs involved in the dynamic test. The drive and response data are loaded and unloaded through the serial register in the same way as in static interconnection test. There are commercially available tools that support both static and dynamic test.

For analog circuits, boundary scan implemented via the IEEE 1149.4 test standard simplifies analog measurements at the board level (Fig. 22). Two (alternatively four) wires for measurements are added to the boundary scan bus. The

TABLE 17 IEEE 1149.1 Circuit Characteristics

Dedicated TAP pins (TDI, TDO, TMS, TCK, and TRST).
Dedicated Boundary scan cells. Includes separate serial–shift and parallel–update stages.
Finite-state machine controller with extensible instructions. Serially scanned instruction register.
Main target is testing printed circuit board interconnect. Philosophy is to restrict boundary cell behavior as necessary to safeguard against side effects during testing.
Second target is sampling system state during operation. Dedicated boundary scan cells and test clock (TCK).
Difficulties applying in hierarchical implementations.

Robust Design Practices

TABLE 18 IEEE 1149.1 Instructions for Testability

Bypass
  Inserts a 1-bit bypass register between TDI and TDO.
Extest
  Uses boundary register first to *capture*, then *shift*, and finally to *update* I/O pad values.
Sample/preload
  Uses boundary register first to *capture* and then *shift* I/O pad values without affecting system operation.
Other optional and/or private instructions
  Defined by the standard or left up to the designer to specify behavior.

original four wires are used as in boundary scan for test control and digital data. Special analog boundary scan cells have been developed which can be linked to the analog board level test wires through fairly simple analog CMOS switches. This allows easy setup of measurements of discrete components located between IC pins. Analog and digital boundary scan cells can be mixed within the same device (IC). Even though the main purpose of analog boundary scan is the test of interconnections and discrete components, it can be used to test more complex board level analog functions as well as on-chip analog functions.

After adding scan circuitry to an IC, its area and speed of operation change. The design increases in size (5–15% larger area) because scan cells are larger than the nonscan cells they replace and some extra circuitry is required, and

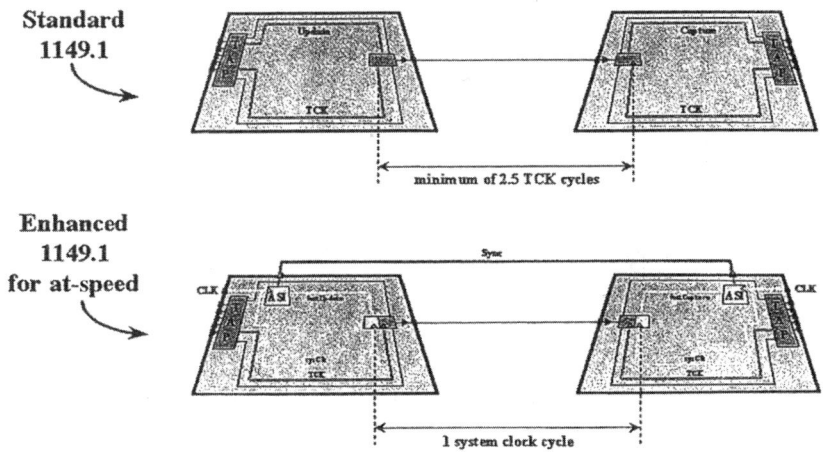

FIGURE 21 At-speed interconnection test. (From Ref. 4.)

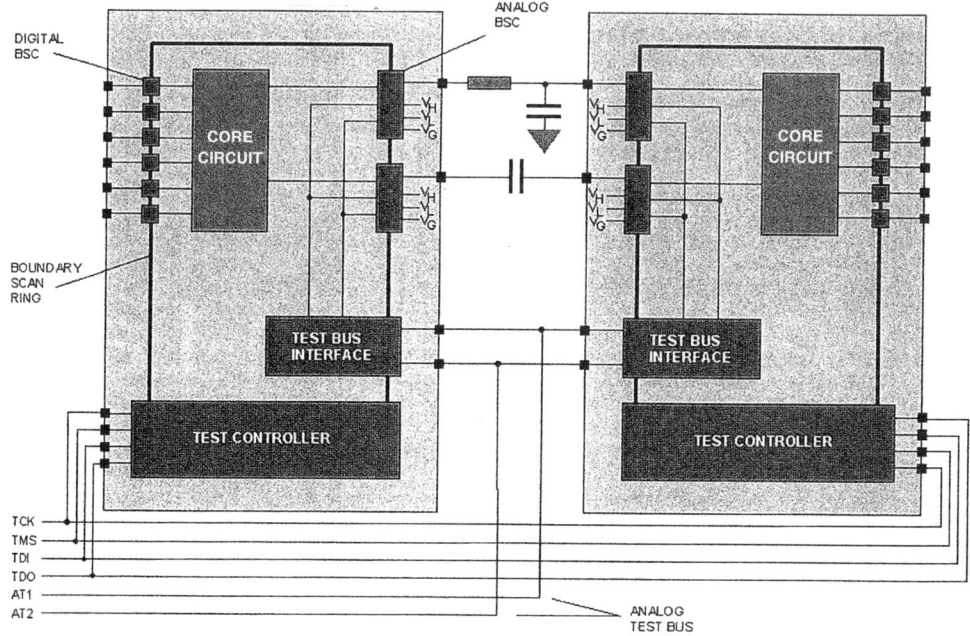

FIGURE 22 Analog boundary scan. (From Ref. 4.)

the nets used for the scan signals occupy additional area. The performance of the design will be reduced as well (5–10% speed degradation) due to changes in the electrical characteristics of the scan cells that replaced the nonscan cells and the delay caused by the extra circuitry.

Built-in self-test is a design technique in which test vectors are generated on-chip in response to an externally applied test command. The test responses are compacted into external pass/fail signals. Built-in self-test is usually implemented through ROM (embedded memory) code instructions or through built-in (on-chip) random word generators (linear feedback shift registers, or LSFRs). This allows the IC to test itself by controlling internal circuit nodes that are otherwise unreachable, reducing tester and ATPG time and date storage needs.

In a typical BIST implementation (Fig. 23) stimulus and response circuits are added to the device under test (DUT). The stimulus circuit generates test patterns on the fly, and the response of the DUT is analyzed by the response circuit. The final result of the BIST operation is compared with the expected result externally. Large test patterns need not be stored externally in a test system since they are generated internally by the BIST circuit. At-speed testing is possi-

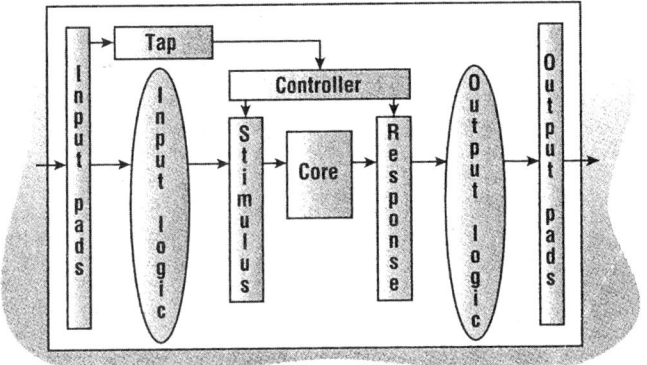

FIGURE 23 Built-in self-test can be used with scan ATPG to enable effective system-on-chip testing. (From Ref. 4.)

ble since the BIST circuit uses the same technology as the DUT and can be run off the system clock.

Built-in self-test has been primarily implemented for testing embedded memories since highly effective memory test algorithms can be implemented in a compact BIST circuit but at a cost of increased circuit delay. The tools for implementing digital embedded memory BIST are mature. Because of the unstructured nature of logic blocks, logic BIST is difficult to implement but is being developed. The implementation of analog BIST can have an impact on the noise performance and accuracy of the analog circuitry. The tools to implement analog BIST are being developed as well.

Both BIST and boundary scan have an impact on product and test cost during all phases of the product life cycle: development, manufacturing, and field deployment. For example, boundary scan is often used as a means to rapidly identify structural defects (e.g., solder bridges or opens) during early life debugging. Built-in self-test and boundary scan may be leveraged during manufacturing testing to improve test coverage, reduce test diagnosis time, reduce test capital, or all of the above. In the field, embedded boundary scan and BIST facilitate accurate system diagnostics to the field replacement unit (FRU, also called the customer replaceable unit, or CRU). The implementation of BIST tends to lengthen IC design time by increasing synthesis and simulation times (heavy computational requirements), but reduces test development times.

Design for test techniques have evolved to the place where critical tester (ATE) functions (such as pin electronics) are embedded on the chip being tested. The basic idea is to create microtesters for every major functional or architectural block in a chip during design. A network of microtesters can be integrated at the chip level and accessed through the IEEE 1149.1 port to provide a complete test solution. Embedded test offers a divide and conquer approach to a very complex problem. By removing the need to generate, apply, and collect a large number of test vectors from outside the chip, embedded test promises to both facilitate test and reduce the cost of external testers (ATE). The embedded test total silicon penalty is on the order of 1–2% as demonstrated by several IC suppliers.

First silicon is where everything comes together (a complete IC) and where the fruits of DFT start to pay dividends. At this point DFT facilitates defect detection diagnostics and characterization. Diagnostics can resolve chip failures both quickly and more accurately. Whether it is model errors, test pattern errors, process (wafer fab) problems, or any number of other explanations, diagnostics are aided by DFT. Patterns can be quickly applied, and additional patterns can be generated if needed. This is critical for timely yield improvement before product ramp-up.

During chip production, DFT helps ensure overall shipped IC quality. High test coverage and small pattern counts act as the filter to judge working and nonworking wafers. For the working (yielding) wafers, the diced (separated) and packaged chips are tested again to ensure working product.

From an IC function perspective, circuits such as FPGAs due to their programmability and reprogrammability can be configured to test themselves and thus ease the testing problem.

3.16.2 Design for Test PWA Level

The previous section shows that DFT requires the cooperation of all personnel involved in IC design. However, from a PWA or system perspective all we care about is that the IC designers have included the appropriate hooks (test structures) to facilitate boundary scan testing of the PWA.

At the PWA level, all boundary scan components (ICs) are linked to form a scan chain. This allows daisy chained data-in and data-out lines of the TAP to carry test signals to and from nodes that might be buried under surface mount devices or be otherwise inaccessible to tester probes. The boundary scan chain is then connected to two edge connectors.

The best manner to present DFT at the PWA level is by means of design hints and guidelines. These are listed in Table 19.

Many board level DFT methods are already supported by commercially available components, building blocks, and test development tools, specifically boundary scan. Since many testers support boundary scan test, it is natural to use the boundary scan test bus (including the protocol) as a general purpose test bus at the PWA level. Several commercially supported DFT methods use the boundary scan bus for test access and control. Additionally, several new DFT

Robust Design Practices 137

TABLE 19 Examples of Design Hints and Guidelines at the PWA Level to Facilitate Testing

Electrical design hints	PWA test point placement rules	Typical PWA test points
Disable the clocks to ease testing.	All test points should be located on single side of PWA.	Through leads. Uncovered and soldered via pads (bigger).
Provide access to enables.	Distribute test points evenly.	Connectors.
Separate the resets and enables.	Minimum of one test point per net.	Card-edge connectors. Designated test points.
Unused pins should have test point access.	Multiple VCC and ground test pads distributed across PWA.	
Unused inputs may require pull-up or pull-down resistors.	One test point on each unused IC pin.	
Batteries must have enabled jumpers or be installed after test.	No test points under components on probe side of PWA.	
Bed-of-nails test fixture requires a test point for every net, all on the bottom side of the board.		

methods are emerging that make use of the standardized boundary scan bus. This activity will only serve to facilitate the widespread adoption of DFT techniques.

The myriad topics involved with IC and board (PWA) tests have been discussed via tutorials and formal papers, debated via panel sessions at the annual International Test Conference, and published in its proceedings. It is suggested that the reader who is interested in detailed information on these test topics consult these proceedings.

3.16.3 Design for Test at the System Level

At the system level, DFT ensures that the replaceable units are working properly. Often, using a BIST interface, frequently assessed via boundary scan, components can test themselves. If failures are discovered, then the failing components can be isolated and replaced, saving much system debug and diagnostics time. This can also result in tremendous savings in system replacement costs and customer downtime.

In conclusion, DFT is a powerful means to simplify test development, to decrease manufacturing test costs, and to enhance diagnostics and process feedback. Its most significant impact is during the product development process, where designers and test engineers work interactively and concurrently to solve the testability issue. Design for test is also a value-added investment in improving testability in later product phases, i.e., manufacturing and field troubleshooting.

3.17 SNEAK CIRCUIT ANALYSIS

Sneak circuit analysis is used to identify and isolate potential incorrect operating characteristics of a circuit or system. A simplified example of a sneak circuit, which consists of two switches in parallel controlling a light, illustrates one type of unwanted operation. With both switches open, either switch will control the light. With one switch closed, the other switch will have no effect. Such problems occur quite often, usually with devastating results. Often the sneak circuit analysis is included in the various CAD libraries that are used for the design.

3.18 BILL OF MATERIAL REVIEWS

Many large equipment manufacturers conduct periodic bill of material reviews from conceptual design throughout the physical design process. The BOM review is similar to a design review, but here the focus is on the parts and the suppliers. These reviews facilitate the communication and transfer of knowledge regarding part, function, supplier, and usage history between the component engineers and the design team. The purpose of periodic BOM reviews is to

   Identify risks with the parts and suppliers selected
   Communicate multifunctional issues and experiences regarding parts and suppliers (DFT, DFM, quality, reliability, and application sensitivities).
   Identify risk elimination and containment action plans
   Track the status of qualification progress

Typical BOM review participants include design engineering, component engineering, test engineering, manufacturing engineering, reliability engineering, and purchasing. The specific issues that are discussed and evaluated include

   Component (part) life cycle risk, i.e., end of life and obsolescence
   Criticality of component to product specification
   Availability of SPICE, timing, schematic, simulation, fault simulation, and testability models
   Test vector coverage and ease of test
   Construction analysis of critical components (optional)
   Part availability (sourcing) and production price projections
   Failure history with part, supplier, and technology
   Supplier reliability data
   Known failure mechanisms/history of problems

Robust Design Practices

TABLE 20 BOM Review Process Flow

Component engineers work up front with the design team to understand their needs and agree on the recommended technology and part choices (see Table 4).
Determine who should be invited to participate in the review and develop an agenda stating purpose and responsibilities of the review team (see previously listed functional representatives).
Send out preliminary BOM, targeted suppliers, and technology choices.
Develop and use standard evaluation method.
Discuss issues that arise from the evaluation, and develop solutions and alternatives.
Develop an action plan based on the evaluation.
Meet periodically (monthly) to review actions and status as well as any BOM changes as the design progresses.

Responsiveness, problem resolution, previous experience with proposed suppliers
Financial viability of supplier
Part already qualified versus new qualification and technology risks
Compatibility with manufacturing process
Supplier qualification status
Application suitability and electrical interfacing with other critical components

A typical BOM process flow is presented in Table 20. A note of caution needs to be sounded. A potential problem with this process is that part and supplier needs do change as the design evolves and the timeline shortens, causing people to go back to their nonconcurrent over-the-wall habits (comfort zone). An organization needs to have some group champion and drive this process. At Tandem/Compaq Computer Corp., component engineering was the champion organization and owned the BOM review process.

3.19 DESIGN REVIEWS

Design reviews, like BOM reviews, are an integral part of the iterative design process and should be conducted at progressive stages throughout the design cycle and prior to the release of the design to manufacturing. Design reviews are important because design changes made after the release of a design to manufacturing are extremely expensive, particularly, where retrofit of previously manufactured equipment is required. The purpose of the design review is to provide an independent assessment (a peer review) to make sure nothing has been overlooked and to inform all concerned parties of the status of the project and the risks involved.

A design review should be a formally scheduled event where the specific design or design methodology to be used is submitted to the designer's/design team's peers and supervisors. The members of the design review team should come from multiple disciplines: circuit design, mechanical design, thermal design, PWA design and layout, regulatory engineering (EMC and safety), test engineering, product enclosure/cabinet design, component engineering, reliability engineering, purchasing, and manufacturing. This ensures that all viewpoints receive adequate consideration. In small companies without this breadth of knowledge, outside consultants may be hired to provide the required expertise.

Each participant should receive, in advance, copies of the product specification, design drawings, schematic diagrams and data, the failure modes and effects analysis (FMEA) report, the component derating list and report, current reliability calculations and predictions, and the BOM review status report. The product manager reviews the product specification, the overall design approach being used, the project schedule, the design verification testing (DVT) plan, and the regulatory test plan, along with the schedules for implementing these plans. Each peer designer (electrical, thermal, mechanical, EMC, and packaging) evaluates the design being reviewed, and the other team members (test engineering, EMC and safety engineering, manufacturing, service, materials, purchasing, etc.) summarize how their concerns have been factored into the design. The component engineer summarizes the BOM review status and open action items as well as the supplier and component qualification plan. The reliability engineer reviews the component risk report, the FMEA report, and the reliability prediction. Approval of the design by management is made with a complete understanding of the work still to be accomplished, the risks involved, and a commitment to providing the necessary resources and support for the required testing.

At each design review an honest, candid, and detailed appraisal of the design methodology, implementation, safety margins/tolerances, and effectiveness in meeting stated requirements is conducted. Each of the specified requirements is compared with the present design to identify potential problem areas for increased attention or for possible reevaluation of the need for that requirement. For example, one of the concerns identified at a design review may be the need to reapportion reliability to allow a more equitable distribution of the available failure rate among certain functional elements or components. It is important that the results of the design review are formally documented with appropriate action items assigned.

A final design review is conducted after all testing, analysis, and qualification tasks have been completed. The outcome of the final design review is concurrence that the design satisfies the requirements and can be released to manufacturing/production.

Small informal design reviews are also held periodically to assess specific aspects or elements of the design. These types of design reviews are much more prevalent in smaller-sized entrepreneurial companies.

3.20 THE SUPPLY CHAIN AND THE DESIGN ENGINEER

Many of the techniques for optimizing designs that were useful in the past are becoming obsolete as a result of the impact of the Internet. Bringing a product to market has traditionally been thought of as a serial process consisting of three phases—design, new product introduction (NPI), and product manufacturing. But serial methodologies are giving way to concurrent processes as the number and complexity of interactions across distributed supply chains increase. Extended enterprises mean more companies are involved, and the resulting communications issues can be daunting to say the least. Original equipment manufacturers and their supply chain partners must look for more efficient ways to link their operations. Because 80% of a product's overall costs are determined in the first 20% of the product development process, the ability to address supply chain requirements up front can significantly improve overall product costs and schedules.

Today's OEMs are looking for the "full solution" to make the move to supply chain–aware concurrent design. The necessary ingredients required to make this move include

1. Technology and expertise for integrating into multiple EDA environments.
2. Advanced Internet technologies to minimize supply chain latency.
3. Technologies that automate interactions in the design-to-manufacturing process.
4. Access to supply chain intelligence and other informational assets.
5. An intimate knowledge of customer processes.

The new services available in bringing a product to market collaboratively link the design and supply chain. OEMs and their supply chain partners will create new competitive advantages by integrating these technologies with their deep understanding of design-to-manufacturing processes.

Traditionally, interdependent constraints between design and supply chain processes have been addressed by CAD and material management functions with in-house solutions. As OEMs increasingly outsource portions of their supply chain functions, many in-house solutions that link design and supply chain functions need to be reintegrated. OEMs are working with supply chain partners to facilitate and streamline dialogue that revolves around product design, supply management, and manufacturing interdependencies. Questions such as the following need to be addressed: Which design decisions have the most impact on supply constraints? How will my design decisions have the most impact on supply constraints? How will my design decisions affect NPI schedules? What design decisions will result in optimizing my production costs and schedules?

3.20.1 Design Optimization and Supply Chain Constraints

As mentioned previously, the three phases of bringing a product to market (also called the product realization process) are design, new product introduction, and production manufacturing. Let's focus on design. The design phase consists of a series of iterative refinements (as discussed earlier in this chapter). These refinements are a result of successive attempts to resolve conflicts, while meeting product requirements such as speed, power, performance, cost, and schedules. Once these requirements are satisfied, the design is typically handed off to supply chain partners to address material management or production requirements.

Iterative refinements are an integral part of the design process. These iterations explore local requirements that are resolved within the design phase. Constraints that are explored late in the process contribute to a majority of product realization failures. Design iterations that occur when materials management or manufacturing constraints cannot be resolved downstream must be avoided as much as possible. These iterative feedback or learning loops are a primary cause of friction and delay in the design-to-manufacturing process.

A change in the product realization process introduces the notion of concurrent refinement of design, NPI, and production manufacturing requirements. This process shift recognizes the value in decisions made early in the design process that consider interdependent supply chain requirements. In the concurrent process, optimization of time to volume and time to profit occurs significantly sooner.

A big cause of friction in the product realization process is the sharing of incomplete or inconsistent design data. Seemingly simple tasks, such as part number cross-referencing, notification of part changes, and access to component life cycle information, become prohibitively expensive and time consuming to manage. This is especially true as the product realization process involves a greater number of supply chain partners.

This new process requires new technologies to collaboratively link design and supply chain activities across the distributed supply chain. These new technologies fall into three main categories:

1. Supply chain integration technology that provides direct links to design tools. This allows preferred materials management and manufacturing information to be made available at the point of component selection— the designer's desktop.
2. Bill of materials collaboration and notification tools that support the

Robust Design Practices

iterative dialogue endemic to concurrent methodologies. These tools must provide a solution that supports exploratory design decisions and allows partners to deliver supply chain information services early in the design-to-manufacturing process.

3. Data integration and data integrity tools to allow for automated sharing and reconciliation of design and supply chain information. These tools ensure that component selections represented in the bill of materials can be shared with materials management and manufacturing suppliers in an efficient manner.

3.20.2 Supply Chain Partner Design Collaboration

Electronics distributors and EMS providers have spent years accumulating supply chain information, building business processes, and creating customer relationships. For many of these suppliers, the questions they now face include how to use this wealth of information to enhance the dynamics of the integrated design process and ensure that the content remains as up-to-date as possible.

With this in mind, distributors, suppliers, and EMS providers are combining existing core assets with new collaborative technologies to transform their businesses. The transformation from part suppliers and manufacturers to high-value product realization partners focuses on providing services which allow their customers to get products designed and built more efficiently. One such collaborative effort is that between Cadence Design Systems, Flextronics International, Hewlett-Packard, and Avnet, who have partnered with SpinCircuit to develop new technologies that focus on supply chain integration with the design desktop. This collaboration was formed because of the need for a full solution that integrates new technology, supply chain services, information assets, and a deep understanding of design-to-manufacturing processes.

Electronic design automation (EDA) companies have provided concurrent design methodologies that link schematic capture, simulation, and PC board layout processes. In addition, some of these companies also provide component information systems (CIS) to help designers and CAD organizations to manage their private component information.

However, what has been missing is concurrent access to supply chain information available in the public domain, as well as within the corporate walls of OEMs and their supply chain partners. Because the design process involves repeated refinements and redesigns, supply chain information must be embedded into design tools in an unencumbering manner or it will not be considered.

Flextronics International, a leading EMS provider, has installed SpinCircuit's desktop solution. This solution can be launched from within EDA design tools and allows design service groups to access supply chain information from within their existing design environment. SpinCircuit currently provides seamless integration with Cadence Design Systems and Mentor Graphics schematic capture environments and is developing interfaces to other leading EDA tools.

The desktop solution provides designers and component engineers with access to both private and public component information. The results are views, side-by-side, in a single component selection window. Designers and component engineers can access component information such as schematic symbols, footprints, product change notifications (PCNs), pricing, availability, and online support. Users can also "punch out" to access additional information such as data sheets and other component-specific information available on supplier sites.

SpinCircuit's Desktop solution provides material management and manufacturing groups with the ability to present approved vendor lists (AVL's) and approved materials lists (AML's). Component selection preference filters can be enabled to display preferred parts status. These preference filters provide optimization of NPI and manufacturing processes at the point of design and prevent downstream "loopbacks."

Another critical challenge faced by supply chain partners is access to BOM collaboration tools and automated notification technology. Early in the design phase, this solution links partners involved in new product introduction and allows OEMs to share the content of their BOMs with key suppliers. The transmission of a bill of materials from designers to EMS providers and their distribution partners is an issue that needs to be addressed. Typically, component engineers must manually cross-reference parts lists to make sense of a BOM. In other words, the supply chain may be connected electronically, but the information coming over the connection may be incomprehensible.

These gaps break the flow of information between OEMs, distributors, parts manufacturers, and EMS providers. They are a significant source of friction in the design-to-manufacturing process. Seemingly simple but time-consuming tasks are cross-referencing part numbers, keeping track of PCN and end-of-life (EOL) notifications, and monitoring BOM changes. These tasks are especially time consuming when they are distributed throughout an extended enterprise.

Avnet, a leading electronics distributor, uses solutions from SpinCircuit to accelerate their new product introduction services and improve their customer's ability to build prototypes. SpinCircuit provides tools and technology to streamline bill-of-materials processing. Each time a BOM is processed, SpinCircuit's BOM analysis tools check for PCNs, EOL notifications, and design changes to identify and reconcile inconsistent or incomplete data that may impact NPI processes.

3.21 FAILURE MODES AND EFFECTS ANALYSIS

The purpose of the FMEA is to identify potential hardware deficiencies including undetectable failure modes and single point failures. This is done by a thorough,

Robust Design Practices 145

systematic, and documented analysis of the ways in which a system can fail, the causes for each failure mode, and the effects of each failure. Its primary objective is the identification of catastrophic and critical failure possibilities so that they can be eliminated or minimized through design change. The FMEA results may be either qualitative or quantitative, although most practitioners attempt to quantify the results.

In FMEA, each component in the system is assumed to fail catastrophically in one of several failure modes and the impact on system performance is assessed. That is, each potential failure studied is considered to be the only failure in the system, i.e., a single point failure. Some components are considered critical because their failure leads to system failure or an unsafe condition. Other components will not cause system failure because the system is designed to be tolerant of the failures. If the failure rates are known for the specific component failure modes, then the probability of system malfunction or failure can be estimated. The design may then be modified to make it more tolerant of the most critical component failure modes and thus make it more reliable. The FMEA is also useful in providing information for diagnostic testing of the system because it produces a list of the component failures that can cause a system malfunction.

The FMEA, as mentioned, can be a useful tool for assessing designs, developing robust products, and guiding reliability improvements. However, it is time consuming, particularly when the system includes a large number of components. Frequently it does not consider component degradation and its impact on system performance. This leads to the use of a modified FMEA approach in which only failures of the high risk or critical components are considered, resulting in a simpler analysis involving a small number of components. It is recommended that the FMEA include component degradation as well as catastrophic failures.

Although the FMEA is an essential reliability task for many types of system design and development, it provides limited insight into the probability of system failure. Another limitation is that the FMEA is performed for only one failure at a time. This may not be adequate for systems in which multiple failure modes can occur, with reasonable likelihood, at the same time. However, the FMEA provides valuable information about the system design and operation.

The FMEA is usually iterative in nature. It should be conducted concurrently with the design effort so that the design will reflect the analysis conclusions and recommendations. The FMEA results should be utilized as inputs to system interfaces, design tradeoffs, reliability engineering, safety engineering, maintenance engineering, maintainability, logistic support analysis, test equipment design, test planning activities, and so on. Each failure mode should be explicitly defined and should be addressed at each interface level.

The FMEA utilizes an inductive logic or bottom-up approach. It begins at the lowest level of the system hierarchy (normally at the component level) and using knowledge of the failure modes of each part it traces up through the system hierarchy to determine the effect that each potential failure mode will have on system performance. The FMEA focus is on the parts which make up the system. The FMEA provides

1. A method for selecting a design with a high probability of operational success and adequate safety.
2. A documented uniform method of assessing potential failure modes and their effects on operational success of the system.
3. Early visibility of system interface problems.
4. A list of potential failures which can be ranked according to their seriousness and the probability of their occurrence.
5. Identification of single point failures critical to proper equipment function or personnel safety.
6. Criteria for early planning of necessary tests.
7. Quantitative, uniformly formatted input data for the reliability prediction, assessment, and safety models.
8. The basis for troubleshooting procedures and for the design and location of performance monitoring and false sensing devices.
9. An effective tool for the evaluation of a proposed design, together with any subsequent operational or procedural changes and their impacts on proper equipment functioning and personnel safety.

The FMEA effort is typically led by reliability engineering, but the actual analysis is done by the design and component engineers and others who are intimately familiar with the product and the components used in its design. If the design is composed of several subassemblies, the FMEA may be done for each subassembly or for the product as a whole. If the subassemblies were designed by different designers, each designer needs to be involved, as well as the product engineer or systems engineer who is familiar with the overall product and the subassembly interface requirements. For purchased assemblies, like power supplies and disk drives, the assembly design team needs to provide an FMEA that meets the OEM's needs. We have found, as an OEM, that a team of responsible engineers working together is the best way of conducting a FMEA.

The essential steps in conducting an FMEA are listed here, a typical FMEA worksheet is shown in Table 21, and a procedure for critical components is given in Table 22.

1. *Reliability block diagram construction.* A reliability block diagram is generated that indicates the functional dependencies among the various elements of the system. It defines and identifies each required subsystem and assembly.
2. *Failure definition.* Rigorous failure definitions (including failure modes, failure mechanisms, and root causes) must be established for the entire system, the subsystems, and all lower equipment levels. A

Robust Design Practices

TABLE 21 Potential Failure Mode and Effects Analysis Summary

A.
Design/manufacturing responsibility: _____
Geometry/package no: _____
Other areas involved: _____

Process description and purpose	Potential failure mode (with regard to released engineering requirements or specific process requirements)	Potential effects of failure on "customer" (effects of failure on the customer)	Severity	
Briefly describe process being analyzed. Concisely describe purpose of process. *Note*: If a process involves multiple operations that have different modes of failure, it may be useful to list as separate processes.	Assume incoming parts/materials are correct. Outlines reason for rejection at specific operation. Cause can be associated with potential failure either upstream or downstream. List each potential failure mode in terms of a part or process characteristic, from engineer and customer perspectives. *Note*: Typical failure mode could be bent, corroded, leaking, deformed, misaligned.	Customer could be next operation, subsequent operation or location, purchaser, or end user. Describe in terms of what customer might notice or experience in terms of system performance for end user or in terms of process performance for subsequent operation. If it involves potential noncompliance with government registration, it must be indicated as such. Examples include noise, unstable, rough, inoperative, erratic/intermittent operation, excessive effort required, operation impaired.	Severity is an assessment of seriousness of effect (in Potential effects of failure column):	
			Severity of effect	Rank
			Minor No real effect caused. Customer probably will not notice failure.	1
			Low Slight customer annoyance. Slight inconvenience with subsequent process or assembly. Minor rework action.	2,3
			Moderate Some customer dissatisfaction. May cause unscheduled rework/repair/damage to equipment.	4,5,6
			High High degree of customer dissatisfaction due to the nature of the failure. Does not involve safety or noncompliance with government regulations. May cause serious disruption to subsequent operations, require major rework, and/or endanger machine or operator.	7,8
			Very High Potential failure mode affects safe operation. Noncompliance with government regulations. *Note*: Severity can only be affected by design.	9,10

Potential causes of failure (How could failure mode occur in terms of something that can be corrected or controlled?)	Occurrence		Current controls
List every conceivable failure cause assignable to each potential mode. If correcting cause has direct impact on mode, then this portion of FMEA process is complete. If causes are not mutually exclusive, a DOE may be considered to determine root cause or control the cause. Causes should be described such that remedial efforts can be aimed at pertinent causes. Only specific errors or malfunctions should be listed; ambiguous causes (e.g., operator error, machine malfunction) should not be included. Examples are handling damage, incorrect temperature, inaccurate gauging, incorrect gas flow.	Occurrence is how frequently the failure mode will occur as a result of a specific cause (from Potential causes of failure column). Estimate the likelihood of the occurrence of potential failure modes on a 1 to 10 scale. Only methods intended to prevent the cause of failure should be considered for the ranking: failure detecting measures are *not* considered here. The following occurrence ranking system should be used to ensure consistency. The possible failure rates are based on the number of failures that are anticipated during the process execution.		Describe the controls that either prevent failure modes from occurring or detect them should they occur. Examples could be process control (i.e., SPC) or postprocess inspection/testing.
	Probability	Rank	Possible failure rate
	Remote: failure unlikely; $Cpk \geq 1.67$	1	≤ 1 in 10^6, $\approx \pm 5$ sec
	Very low: in statistical control; $Cpk > 1.33$	2	>1 in 10^6, ≤ 1 in 20k, $\approx \pm 4$ sec
	Low: relatively few failures, in statistical control; $Cpk > 1.00$	3	>1 in 20k, ≤ 1 in 4k, $\approx \pm 3.5$ sec
	Moderate: occasional failures, in statistical control; $Cpk \leq 1.00$	4,5,6	>1 in 4k, ≤ 1 in 80, $\approx \pm 3$ sec
	High: repeated failures, not in statistical control; $Cpk < 1$	7,8	>1 in 80, ≤ 1 in 40, $\approx \pm 1$ sec
	Very high: failure almost inevitable	9,10	1 in 8, ≤ 1 in 40

Robust Design Practices

TABLE 21 Continued

Process description and purpose	Detection	Risk priority number (RPN)	Recommended actions
	Detection is an assessment of probability that the proposed controls (Current controls column) will detect the failure mode before the part leaves the manufacturing/assembly locations. Assume failure has occurred. Assess capability of current controls to prevent shipment. Do not assume detection ranking is low because occurrence is low. Do assess ability of process controls to detect low-frequency failures. Evaluation criteria based on likelihood defect existence will be detected prior to next process, subsequent process or before leaving manufacturing/assembly.	RPN = (O)(S)(D) = occurrence × severity × detection Use RPN to rank items in Pareto analysis fashion.	Once ranked by RPN, direct corrective action at highest-ranking or critical items first. Intent is to reduce the occurrence of severity and/or detection rankings. Also indicate if no action is recommended. Consider the following actions: 1. Process and/or design revisions are required to reduce probability of occurrence. 2. Only a design revision will reduce severity ranking. 3. Process and/or design revisions are needed to increase probability of detection. *Note*: Increased quality assurance inspection is not a positive corrective action—use only as a last resort or a temporary measure. Emphasis must be on preventing defects rather than detecting them, i.e., use of SPC and process improvement rather than random sampling or 100% inspection.

Likelihood of detection	Rank
Very high: process automatically detects failure.	1,2
High: controls have good chance of detecting failure.	3,4
Moderate: controls may detect failure.	5,6
Low: poor chance controls will detect failure.	7,8
Very low: controls probably won't detect failure.	9
Absolute certainty of nondetection: controls can't or won't detect failure.	10

B.		C.	
Prepared by: _____		FMEA date (orig.): _____	
(Rev. date): _____		Key production date: _____	
Eng. release date: _____		Area: _____	
Plant(s): _____			
		Action Results	
Area/individual responsible and completion date	Action taken and actual completion date	Severity Occurrence Detection	RPN
Enter area and person responsible. Enter target completion date.	After action has been taken, briefly describe actual action and effective or completion date.		Resulting RPN after corrective action taken. Estimate and record the new ranking for severity, occurrence, and detection resulting from corrective action. Calculate and record the resulting RPN. If no action taken, leave blank. Once action has been completed, the new RPN is moved over to the first RPN column. Old FMEA revisions are evidence of system improvement. Generally the previous FMEA version(s) are kept in document control. *Note*: Severity can only be affected by design.
	Follow-up: Process engineer is responsible for assuring all recommended actions have been implemented or addressed. "Living documentation" must reflect latest process level, critical (key) characteristics, and manufacturing test requirements. PCN may specify such items as process condition, mask revision level, packaging requirements, and manufacturing concerns. Review FMEAs on a periodic basis (minimum annually).		

properly executed FMEA provides documentation of all critical components in a system.

3. *Failure effect analysis.* A failure effect analysis is performed on each item in the reliability block diagram. This takes into account each different failure mode of the item and indicates the effect (consequences) of that item's failure upon the performance of the item at both the local and next higher levels in the block diagram.

4. *Failure detection and compensation.* Failure detection features for each failure mode should be described. For example, previously known symptoms can be used based on the item behavior pattern(s) indicating that a failure has occurred. The described symptom can cover the operation of the component under consideration or it can cover both the component and the overall system or evidence of equipment failure.

Robust Design Practices 151

TABLE 22 FMEA Procedure for Critical Components

1. The reliability engineer prepares a worksheet listing the high-risk (critical) components and the information required.
2. The product engineer defines the failure thresholds for each of the outputs of the subassemblies/modules, based on subassembly and product specifications.
3. Each design engineer, working with the appropriate component engineer, analyzes each of the components for which he or she is responsible and fills in the worksheet for those components listing the effects of component failure on the performance at the next level of assembly.
4. Each design engineer analyzes each of the components for which he or she is responsible and estimates the amount of component degradation required to cause the subassembly to fail, per the definitions of Step 2 above, and then fills in the appropriate sections of the worksheet. If the design is tolerant of failure of a specific component because of redundancy, the level of redundancy should be noted.
5. The design engineers consider each critical component and determine whether the design should be changed to make it more tolerant of component failure or degradation. They add their comments and action items to the report.
6. Then reliability engineering analyzes the completed worksheets, prepares a report listing the critical components (those whose failure causes system failure), and summarizes the potential failure modes for each high-risk component and the definitions of failure for each failure mode.

A detected failure should be corrected so as to eliminate its propagation to the whole system and thus to maximize reliability. Therefore, for each element provisions that will alleviate the effect or malfunction or failure should be identified.

5. *Recordkeeping.* The configurations for both the system and each item must be properly identifed, indexed, and maintained.
6. *Critical items list.* The critical items list is generated based on the results of Steps 1 through 3.

It is important to note that both the FMEA and reliability prediction have definite shelf lives. Being bottom-up approaches, every time that a component (physical implementation) changes, the FMEA and reliability prediction become less effective. It must be determined at what point and how often in the design phase these will be performed: iterative throughout, at the end, etc. Nonetheless, an FMEA should be conducted before the final reliability prediction is completed to provide initial modeling and prediction information. When performed as an integral part of the early design process, it should be updated to reflect design changes as they are incorporated. An example of an FMEA that was conducted for a memory module is presented in Appendix B at the back of this book. Also provided is a list of action items resulting from this analysis, the implementation of which provides a more robust and thus more reliable design.

3.22 DESIGN FOR ENVIRONMENT

The push for environmentally conscious electronics is increasing. It is being fueled by legal and regulatory requirements on a global level. Most of the directives being issued deal with (1) the design and end-of-life management of electronic products, requiring manufacturers to design their products for ease of disassembly and recycling, and (2) banning the use of specific hazardous materials such as lead, mercury, cadmium, hexavalent chromium, and flame retardants that contain bromine and antimony oxide.

The adoption of ISO 14000 by most Japanese OEMs and component suppliers is putting pressure on all global product/equipment manufacturers to establish an environmental management system. The ISO 14000 series of standards for environmental management and certification enables a company to establish an effective environmental management system and manage its obligations and responsibilities better. Following the adoption of the ISO 14000 standards, European and Asian companies are beginning to require a questionnaire or checklist on the environmental management system status of supplier companies. OEMs must obtain as much environmentally related information as possible from each of their suppliers, and even from their suppliers' suppliers. To make this job easier, OEMs are developing DFE metrics and tools that can be used by the supply base.

While most of the materials used in electrical and electronic products are safe for users of the products, some materials may be hazardous in the manufacturing process or contribute to environmental problems at the end of the product life. In most cases, these materials are used in electronic products because functional requirements cannot be met with alternative materials. For example, the high electrical conductivity, low melting point, and ductility of lead-based solder make it ideal for connecting devices on PWAs. Similarly, the flame-retardant properties of some halogenated materials make them excellent additives to flammable polymer materials when used in electrical equipment where a spark might ignite a fire.

Continued improvements in the environmental characteristics of electrical and electronic products will require the development and adoption of alternative materials and technologies to improve energy efficiency, eliminate hazardous or potentially harmful materials (where feasible), and increase both the reuseability and the recyclability of products at their end of life. The challenge facing the electronics industry with regard to environmentally friendly IC packaging is to make a switch to materials that have comparable reliability, manufacturability, price, and availability.

The electronics industry at large has established a list of banned or restricted materials, often called *materials of concern*. These materials include those prohibited from use by regulatory, legislative, or health concerns, along with materials that have been either banned or restricted by regulation or industrial customers or for which special interest groups have expressed concern. Several of the identified materials of concern are commonly found in electronic products, and eliminating them will require significant efforts to identify, develop, and qualify alternatives. These materials include

> *Flame retardants.* Flame retardants are found in PWAs, plastic IC packages, plastic housings, and cable insulation. The most common approach to flame retardancy in organic materials is to use halogenated, usually brominated, materials. Some inorganic materials, such as antimony trioxide are also used either alone or in conjunction with a brominated material.
> *Lead.* Lead is found in solder and interconnects, batteries, piezoelectric devices, discrete components, and cathode ray tubes.
> *Cadmium.* Cadmium is found in batteries, paints, and pigments and is classified as a known or suspected human carcinogen. Most major electronics companies are working to eliminate its use, except in batteries where there are well-defined recycling procedures to prevent inappropriate disposal.
> *Hexavalent chromium.* This material is found in some pigments and paints (although these applications are decreasing) and on fasteners and metal parts, where it is used for corrosion resistance. Automotive OEMs are either banning its use or strongly encouraging alternatives. But these alternatives cannot consistently pass corrosion-resistance specifications.
> *Mercury.* Mercury is found in the flat panel displays of laptop computers, digital cameras, fax machines, and flat panel televisions. Mercury is highly toxic and there are few alternatives to its use in flat panel displays.

The growing demand for electrical and electronic appliances will at the same time create more products requiring disposal. Efforts to increase the reuse and recycling of end-of-life electronic products have been growing within the electronics industry as a result of the previously mentioned regulatory and legal pressures. There are also efforts to reduce packaging or, in some cases, provide reuseable packaging. Products containing restricted or banned materials are more costly and difficult to recycle because of regional restrictive legislation. All of this is adding complexity to the product designer's task.

In order to avoid landfill and incineration of huge amounts of discarded products, it will be necessary to develop a cost-effective infrastructure for reuse and recycling of electronic equipment. This trend will accelerate as more geographic regions pass "take-back" legislation to reduce the burden of landfills. While many of the materials commonly found in electronic products can be easily recycled (e.g., metals and PWAs, for example), several materials commonly found in electronic products present special challenges. These include plastics and leaded glass from televisions and computer monitors.

3.23 ENVIRONMENTAL ANALYSIS

Various environmental analyses, such as mechanical shock and vibration analyses, are employed when new, unusual, or severe environments are anticipated. Printed circuit boards and other structures can be modeled and resonant frequencies and amplitudes can be calculated, allowing any overstress conditions to be identified and alleviated. Other environments include temperature excursions, water and humidity, air pressure, sand and dust, etc. Long-term durability requires that cyclic stresses, particularly thermal cycling and vibration, be considered. These must be studied to assure that the proposed design will not be degraded by the anticipated environmental exposures.

3.24 DEVELOPMENT AND DESIGN TESTING

The purpose of conducting design evaluation tests is to identify design weaknesses and thus areas for improvement, resulting in a more robust product. A summary of the key test and evaluation methods is as follows:

1. Prototyping, design modeling, and simulation are used while the design is still fluid to validate the design tools; validate and verify the design; identify design weaknesses, marginalities, and other problems; and drive improvement. The use of both BOM and design reviews aids this process.
2. Design for test is used to design the product for easy and effective testing as well as for rapid product debug, leading to early problem resolution.
3. The techniques of test, analyze, and fix and plan–do–check–act–repeat (the Deming cycle) are used to assess the current product's robustness, identify how much margin exists with regard to performance parameters, and identify areas for improvement, maximizing the reliability growth process.
4. Software test and evaluation is used to assess the status of the current version of the system software, identify software bugs and areas for improvement, and drive the improvement.

5. Special studies and application tests are used to investigate the idiosyncracies and impact of unspecified parameters and timing condition interactions of critical ICs with respect to each other and their impact on the operation of the product as intended.
6. Accelerated environmental stress testing (such as HALT and STRIFE testing) of the PWAs, power supplies, and other critical components is used to identify weaknesses and marginalities of the completed design (with the actual production components being used) prior to release to production.

Some of these have been discussed previously; the remainder will now be addressed in greater detail.

3.24.1 Development Testing

Development is the best time to identify and correct problems in a product. Making changes is easier and cheaper during development than at any other stage of the product life. Anything that can be done to improve the product here will pay back the maximum benefit since 80% of product cost is usually locked in during this phase.

Development testing is conducted to detect any design errors or omissions overlooked by any previous analyses. This is also known as test, analyze, and fix (TAAF) testing. To be effective all three items must be addressed:

1. Tests must be severe and simulate the worst case expected environment.
2. All problems uncovered must be analyzed to identify the root cause.
3. Positive corrective action must be developed and incorporated to eliminate the root cause.

Testing must then be readministered to verify that the corrections are effective.

This results in building more robust products and improving next-generation design of products.

3.24.2 Design Verification Testing

Once final prototype units have been manufactured, design verification testing (DVT) is conducted to ensure that the product meets its performance specifications (including exposure to anticipated application environments such as temperature, humidity, mechanical shock, and vibration), to assess a product's design margins, and to determine its robustness.

In a typical DVT process the new product goes through a battery of tests created to force real-time design flaws and manufacturing incompatibilities. Electrical circuit design engineers verify the electrical performance of the design.

They verify the CAD model simulation results and rationalize them with actual hardware build and with the variability of components and manufacturing processes. Mechanical engineers model and remodel the enclosure design. Printed circuit board designers check the layout of the traces on the PCB, adjust pad and package sizes, and review component layout and spacing. Manufacturing process engineers check the PCB's chemistry to ensure it is compatible with production cells currently being built. If a PCB has too many ball grid array components or too many low profile ceramic components, it may force the use of the more expensive and time-consuming "no-clean" chemistry. Test engineers look for testability features such as net count and test point accessibility. A board that has no test points or exposed vias will make in-circuit testing impossible and thus require a costlier alternative such as functional test. Cable assembly engineers must look at interconnects for better termination and shielding opportunities. Today's products are challenged by higher transmission rates, where greater speeds can cause crosstalk, limiting or preventing specified performance. Finally, plastic/polymer engineers review for flow and thermal characteristics that will facilitate an efficient production cycle, and sheet metal engineers look for tooling and die compatibility.

A well-designed DVT provides a good correlation of measured reliability results to modeled or predicted reliability. Design verification testing delivers best on its objectives if the product has reached a production-ready stage of design maturity before submission to DVT. Major sources of variation (in components, suppliers of critical components, model mix, and the like) are intentionally built into the test population. Test data, including breakdown of critical variable measurements correlated to the known sources of variation, give the product design team a practical look at robustness of the design and thus the ability to produce it efficiently in volume.

In these ways test is an important aspect of defining and improving product quality and reliability, even though the act of performing testing itself does not increase the level of quality.

3.24.3 Thermography

Design verification testing is also a good time to check the product design's actual thermal characteristics and compare them with the modeled results, and to validate the effectiveness of the heat sinking and distribution system. A thermal profile of the PWA or module is generated looking for hot spots due to high power–dissipating components generating heat and the impact of this on nearby components.

Electronic equipment manufacturers have turned to the use of computational fluid dynamics (CFD) (discussed in Chapter 5) during the front end of the design process and thermography after the design is complete to help solve com-

Robust Design Practices

plex thermal problems. Thermography uses a thermal imaging camera to take a picture of a PWA, module, or product. Thermographic cameras view infrared (IR) energy, as opposed to visible light energy, and display the resultant temperatures as shades of gray or different colors. Figure 24 is an example of a typical thermal scan of a PWA showing the heat-generating components (see color insert).

Any thermal anomaly can indicate a fault or defect. Components running too hot or cold can indicate a short or an open circuit, a diode placed backward or bent IC pins, to name several defects. Elevated temperature operation shortens the life of ICs, while large temperature gradients between components and the PCB increase the stress that can cause early failure due to material delamination.

Thermal imaging and measurement systems provide an effective means for identifying and resolving thermal-related problems by giving a direct measurement of the actual component, PWA, or module as opposed to the modeled thermal profile provided by CFD before the design is committed to hardware build.

3.24.4 Accelerated Stress Testing

Accelerated stress testing is an effective method for improving product reliability since products often have hidden defects or weaknesses which cause failures during normal operation in the field. Product failures may occur when the statistical distribution for a product's strength, or its capability of withstanding a stress, overlaps with the distributions of the operating environmental stresses (Fig. 25). To prevent product failures, reliability may be achieved through a combination of robust design and tight control of variations in component quality and manu-

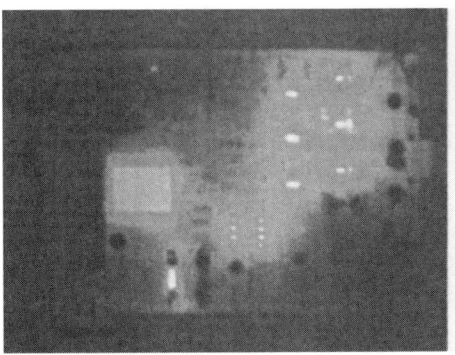 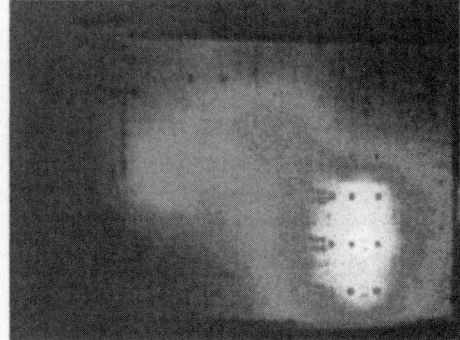

FIGURE 24 Examples of thermal imaging/scan applied to a DC control PWA. Left: temperature scan of component side of PWA showing operating temperature with 250W load. Right: temperature scan of back (solder) side of same PWA showing effect of thermal conduction from "warm" components on the top side. To understand the coordinates of the right scan, imagine the left scan is rolled 180° around its horizontal axis. (See color insert.)

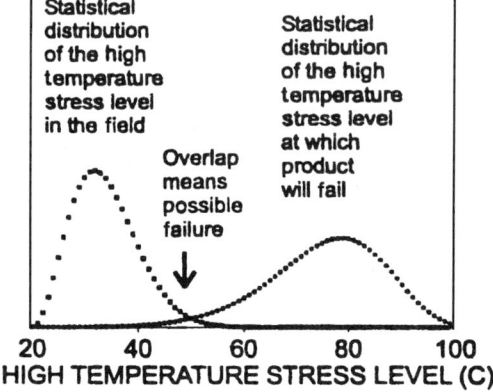

FIGURE 25 Typical environmental stress distribution versus product strength distribution.

facturing processes. When the product undergoes sufficient improvements, there will no longer be an overlap between the stresses encountered and product strength distributions (Fig. 26).

Accelerated stress testing (i.e., HALT and STRIFE), which is normally conducted at the end of the design phase, determines a product's robustness and detects inherent design and manufacturing flaws or defects. Accelerated stress

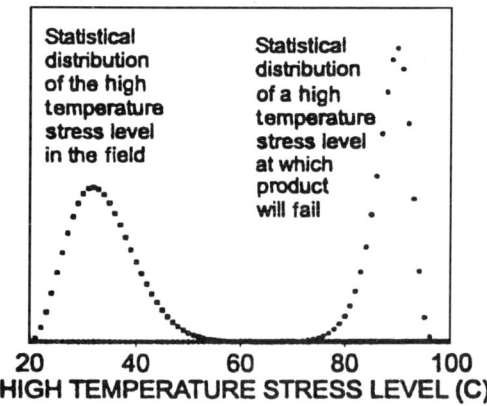

FIGURE 26 Ideal environmental stress and product strength distributions after product improvements.

Robust Design Practices

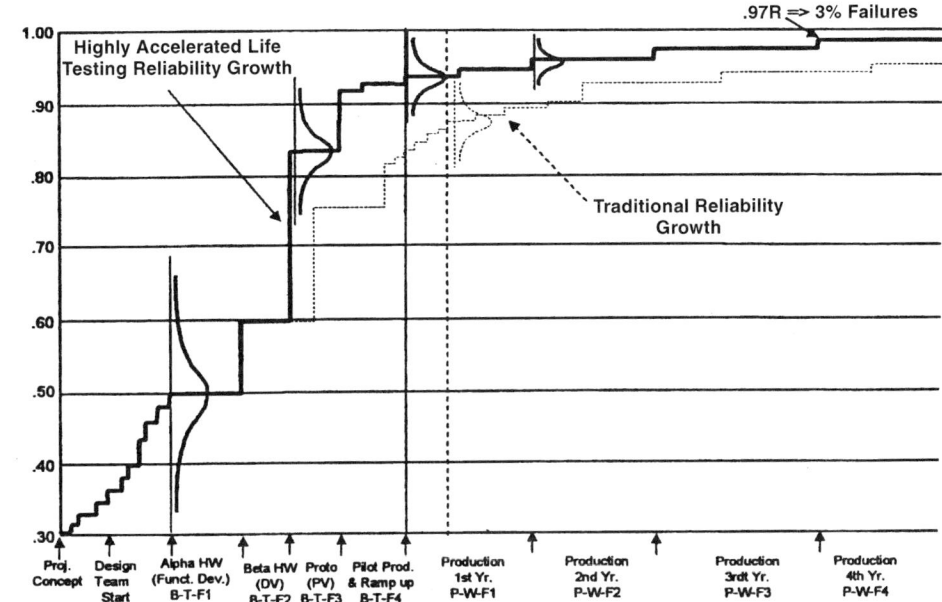

FIGURE 27 Faster reliability growth as a result of conducting accelerated stress testing. (From Ref. 3.)

testing during development is intended to identify weak points in a product so they can be made stronger. The increased strength of the product translates to better manufacturing yields, higher quality and reliability, and faster reliability growth (Fig. 27). The major assumption behind accelerated stress testing is that any failure mechanism that occurs during testing will also occur during application (in the field) if the cause is not corrected. Hewlett-Packard has claimed an 11× return on the investment in accelerated testing through warranty costs alone. Their belief is that it is more expensive to argue about the validity of a potential field problem than to institute corrective actions to fix it.

Typically, a series of individual and combined stresses, such as multiaxis vibration, temperature cycling, and product power cycling, is applied in steps of increasing intensity well beyond the expected field environment until the fundamental limit of technology is reached and the product fails.

HALT

Several points regarding HALT (an acronym for highly accelerated life test, which is a misnomer because it is an overstress test) need to be made.

1. Appropriate stresses must be determined for each assembly since each has unique electrical, mechanical, thermal mass, and vibration characteristics. Typical system stresses used during HALT include the following:

Parameter	Fault found
VCC voltage margining	Design faults, faulty components
Clock speed/frequency	Design faults, faulty components
Clock symmetry	Design faults, faulty components
Power holdup/cycling	Overloads, marginal components
Temperature	
Cold	Margins
Hot	Overloads, low-quality components
Cycling	Processing defects, soldering
Vibration	Processing defects, soldering

Selection of the stresses to be used is the basis of HALT. Some stresses are universal in their application, such as temperature, thermal cycling, and vibration. Others are suitable to more specific types of products, such as clock margining for logic boards and current loading for power components. Vibration and thermal stresses are generally found to be the most effective environmental stresses in precipitating failure. Temperature cycling detects weak solder joints, IC package integrity, CTE mismatch, PWA mounting problems, and PWA processing issues—failures that will happen over time in the field. Vibration testing is normally used to check a product for shipping and operational values. Printed wire assembly testing can show weak or brittle solder or inadequate wicking. Bad connections may be stressed to failure at levels that do not harm good connections.

2. HALT is an iterative process, so that stresses may be added or deleted in the sequence of fail–fix–retest.

3. In conducting HALT there is every intention of doing physical damage to the product in an attempt to maximize and quantify the margins of product strength (both operating and destruct) by stimulating harsher-than-expected end-use environments.

4. The HALT process continues with a test–analyze–verify–fix approach, with root cause analysis of all failures. Test time is compressed with accelerated stressing, leading to earlier product maturity. The results of accelerated stress testing are

> Fed back to design to select a different component/assembly and/or supplier, improve a supplier's process, or make a circuit design or layout change

Robust Design Practices

Fed back to manufacturing to make a process change, typically of a workmanship nature

Used to determine the environmental stress screening (ESS) profiles to be used during production testing, as appropriate

5. The importance of determining root causes for all failures is critical. Root cause failure analysis is often overlooked or neglected due to underestimation of resources and disciplines required to properly carry out this effort. If failure analysis is not carried through to determination of all root causes, the benefits of the HALT process are lost.

Figure 28 depicts the impact that accelerated stress testing can have in lowering the useful life region failure rate and that ESS can have in lowering the early life failure rate (infant mortality) of the bathtub curve.

3.25 TRANSPORTATION AND SHIPPING TESTING

Once a design is completed, testing is conducted to evaluate the capability of the product/equipment to withstand shock and vibration. Shock and vibration, which are present in all modes of transportation, handling, and end-user environments, can cause wire chafing, fastener loosening, shorting of electrical parts, component fatigue, misalignment, and cracking. Dynamic testing tools (which include both sine and random vibration, mechanical shock, and drop impact and simulation of other environmental hazards) are used to more effectively design and test products to ensure their resistance to these forces. The shipping package container design is verified by conducting mechanical shipping and package tests.

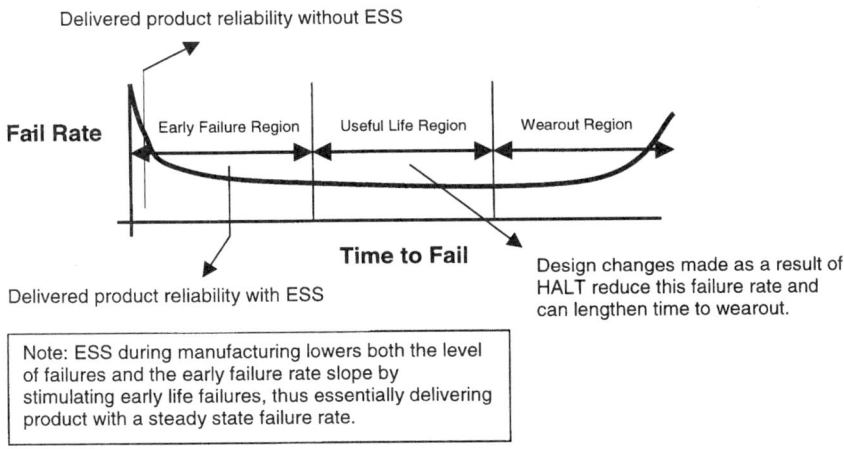

FIGURE 28 Impact of accelerated stress testing and ESS on failure rates.

3.26 REGULATORY TESTING

Appropriate reliability and regulatory tests are typically conducted at the conclusion of the design phase. Product regulations are the gate to market access. To sell a product in various geographical markets, the product must satisfy specific regulatory compliance requirements. To achieve this, the correct regulatory tests must be conducted to ensure that the product meets the required standards. For electronic-based products—such as computers, medical devices, and telecommunication products—safety, electromagnetic compatibility, and, as appropriate, telecommunications tests need to be performed. In all cases, certification of the product, usually by a regulating authority/agency in each country in which the product is sold, is a legal requirement.

3.26.1 Acoustic Measurements

The need for product acoustic noise emissions measurement and management is gaining increased importance. Information on acoustic noise emission of machinery and equipment is needed by users, planners, manufacturers, and authorities. This information is required for comparison of the noise emissions from different products, for assessment of noise emissions against noise limits for planning workplace noise levels, as well as for checking noise reduction achievements. Both sound pressure and sound power are measured according to ISO 7779, which is recognized as a standard for acoustic testing.

3.26.2 Product Safety Testing

Virtually all countries have laws and regulations which specify that products must be safe. On the surface, product safety testing appears to be a straightforward concept—a product should cause no harm. However, the issue gets complicated when trying to meet the myriad different safety requirements of individual countries when selling to the global market. Several examples are presented that make the point.

For electrical product safety standards in the United States and Canada, most people are familiar with Underwriters Laboratories (UL) and the Canadian Standards Association (CSA). The UL safety standard that applies to information technology equipment (ITE), for example, is UL1950. A similar CSA standard is CSA950. In this instance, a binational standard, UL1950/CSA950, also exists.

Standards governing electrical products sold in Europe are set up differently. The European Union (EU) has established European Economic Community (EEC) directives. The directive that applies to most electrical products for safety is the Low Voltage Directive (LVD), or 73/23/EEC. The LVD mandates CE marking, a requirement for selling your products in Europe. Furthermore, 73/23/EEC specifies harmonized European Norm (EN) standards for each product

Robust Design Practices

grouping, such as EN 60950 for ITE. Table 23 lists typical product safety test requirements.

3.26.3 Electromagnetic Compatibility Testing

Compliance to electromagnetic compatibility requirements is legally mandated in many countries, with new legislation covering emissions and immunity being introduced at an increasingly rapid rate. Electromagnetic compatibility requirements apply to all electrical products, and in most countries you cannot legally offer your product for sale without having the appropriate proof of compliance to EMC regulations for that country. This requires a staff of engineers who are familiar with myriad U.S. and international standards and regulations as well as established relationships with regulatory agencies. Table 24 lists the commonly used EMC test requirements for information technology equipment.

TABLE 23 Typical ITE Safety Tests

Rating test capabilities
   Purpose: determine the suitability of the product's electrical rating as specified in the applicable standard.
Temperature measurement capabilities
   Purpose: determine that the product's normal operating temperatures do not exceed the insulation ratings or the temperature limits of user-accessible surfaces.
Hi-pot testing capabilities
   Purpose: verify the integrity of the insulation system between primary and secondary as well as primary and grounded metal parts.
Humidity conditioning capabilities
   Purpose: introduce moisture into hydroscopic insulation prior to hi-pot testing.
Flammability tests to UL 1950/UL 94
   Purpose: determine the flame rating of insulation material or enclosures to determine compliance with the applicable end-use product standards.
Force measurements as required by IEC 950 and IEC 1010 standards
   Purpose: determine if enclosure mechanical strength and product stability complies with standard.
Ground continuity testing
   Purpose: determine if the ground impedance is low enough to comply with the applicable standard.
Leakage current instrumentation to IEC 950 and IEC 1010
   Purpose: determine if the chassis leakage current meets the standard limits.
X-radiation
   Purpose: verify that the X-radiation from a CRT monitor does not exceed standard limits.

TABLE 24 Typical ITE EMC Test Requirements and Standards

Test	Corresponding standard
Electrostatic discharge	EN 50082-1,2
	EN 61000-4-2
	IEC 61000-4-2
Radiated electric field immunity	EN 50082-1,2
	EN 61000-4-3
	IEC 6100-4-3
	ENV 50140
	ENV 50204
Electrical fast transient burst	EN 50082-1,2
	EN 61000-4-4
	IEC 61000-4-4
Surge immunity	EN 50082-1,2
	IEC 61000-4-5
	IEC 801-5
Conducted immunity	EN 50082-1,2
	EN 61000-4-6
	IEC 61000-4-6
	ENV 50141
Power frequency magnetic field immunity	EN 50082-1,2
	EN 61000-4-8
	IEC 61000-4-8
Voltage dips, sags, and interruptions	EN 50082-1,2
	EN 61000-4-11
	IEC 61000-4-11
Harmonic current emissions	EN 61000-3-2
Voltage fluctuations and flicker	EN 61000-3-3

3.27 DESIGN ERRORS

Design errors can occur in specifying the function, timing, and interface characteristics of an IC or in the logic and circuit design. They can also occur as a result of errors in the design models, design library, simulation and extraction tools, PWA layout software; using the wrong component (e.g., an SRAM with timing conditions that don't match the timing constraints required for interfacing it with the selected microprocessor); microprocessor and DSP code issues, etc.

In addition to being stored as a voltage, data and control signals are read as a voltage. If a signal voltage is above a certain threshold, then the data or control bit is read as a logic 1 below the threshold it is read as logic 0. When one bit is a logic 1 and the next bit is a logic 0, or vice versa, there is a transition period to allow the voltage to change. Because each individual device has slightly different signal delay (impedance) and timing characteristics, the length of that transition period varies. The final voltage value attained also varies slightly as a function of the device characteristics and the operating environment (temperature, humidity). Computer hardware engineers allow a certain period of time (called design margin) for the transition period to be completed and the voltage value to settle. If there are timing errors or insufficient design margins that cause the voltage to be read at the wrong time, the voltage value may be read incorrectly, and the bit may be misinterpreted, causing data corruption. It should be noted that this corruption can occur anywhere in the system and could cause incorrect data to be written to a computer disk, for example, even when there are no errors in computer memory or in the calculations.

The effect of a software design error is even less predictable than the effect of a hardware design error. An undiscovered software design error could cause both a processor halt and data corruption. For example, if the algorithm used to compute a value is incorrect, there is not much that can be done outside of good software engineering practices to avoid the mistake. A processor may also attempt to write to the wrong location in memory, which may overwrite and corrupt a value. In this case, it is possible to avoid data corruption by not allowing the processor to write to a location that has not been specifically allocated for the value it is attempting to write.

These considerations stress the importance of conducting both hardware and software design reviews.

ACKNOWLEDGMENTS

Section 3.10.1 courtesy of Reliability Engineering Department, Tandem Division, Compaq Computer Corporation. Portions of Section 3.12 used with permission from SMT magazine. Portions of Section 3.11 courtesy of Noel Donlin, U.S. Army (retired).

REFERENCES

1. Bergman D. CAD to CAM made easy, SMT, July 1999 and PennWell, 98 Spit Brook Rd., Nashua, NH 03062.
2. Brewer R. EMC design practices: preserving signal integrity. Evaluation Engineering, November 1999.
3. McLeish JG. Accelerated Reliability Testing Symposium (ARTS) USA, 1999.
4. Carlsson G. DFT Enhances PCB manufacturing. Future Circuits International. www.mriresearch.com.
5. Bergman D. GenCam addresses high density circuit boards. Future Circuits International, Issue No. 4. www.mriresearch.com.

Further Reading

1. Barrow P. Design for manufacture. SMT, January 2002.
2. Cravotta R. Dress your application for success. EDN, November 8, 2001.
3. Dipert B. Banish bad memories. EDN, November 22, 2001.
4. McDermott RE et al. The Basics of FMEA. Portland, OR: Productivity, Inc.
5. Nelson R. DFT lets ATE work MAGIC. Test & Measurement World, May 2001.
6. Parker KP and Zimmerle D. Boundary scan signals future age of test. EP&P, July 2002.
7. Sexton J. Accepting the PCB test and inspection challenge. SMT, April 2001.
8. Solberg V. High-density circuits for hand-held and portable products. SMT, April 2001.
9. Troescher M and Glaser F. Electromagnetic compatibility is not signal integrity. Item 2002.
10. Webb W. Designing dependable devices. EDN, April 18, 2002.
11. Williams P and Stemper M. Collaborative product commerce—the next frontier. Electronic Buyers News, May 6, 2002.

4

Component and Supplier Selection, Qualification, Testing, and Management

4.1 INTRODUCTION

This chapter focuses on an extremely important, dynamic, and controversial part of the design process: component and supplier selection and qualification. The selection of the right functional components and suppliers for critical components in a given design is the key to product manufacturability, quality, and reliability. Different market segments and system applications have different requirements. Table 1 lists some of these requirements.

4.2 THE PHYSICAL SUPPLY CHAIN

In the electronics industry the exponential growth of ICs coupled with competitive pressures to speed time to market and time to volume are placing increased demands on the supply chain for materials and components movement and intercompany information exchange. New tools such as Internet-based software and emerging supply chain standards such as RosettaNet and others developed by industry collaborations hold the promise of improving supply chain efficiencies and reducing cycle time.

Complex and high-performance products, just-in-time delivery, and the increasing sophistication of procurement systems necessitate the optimization of efficiency and effectiveness in analyzing the manufacturability of new designs,

TABLE 1 Requirements for Different Markets and Systems

Market system application	Volume	Component quality requirements	Components used — Std[a]	Components used — Well proven[b]	Components used — Leading edge	Technology leader or follower	Fault coverage requirement	Time to market	Component available at design	Sensitive to material cost	Component versus total system cost	System design style
Aerospace	Low	Extremely comprehensive	X	X		Follower	>99%	6–7 years	[d]	Low	Low	Custom
Mainframe computer	Low	Comprehensive	X		X[c]	Leader	>99%	2–3 years	[d,f]	Low	Medium/low	Revolutionary
High-end PC server	Medium	Comprehensive for core/critical components	X		X (core commodities)	Follower	>99%	6–18 months	[e]	Medium	Medium/high	Evolutionary
Automotive	High	Comprehensive	X		X[c]	In their own world	>99%	3–5 years	[f]	High	Low	Custom. Review every 3–5 years. Evolutionary in between major model changes.
PC	High	Comprehensive for core/critical components	X		X (core commodities)	Early adopter	95% okay	~6 months	[e]	High	High	Evolutionary
Consumer (toys, games, VCR, etc.)	High	Minimal	X		X[c]	Follower	Don't care	0.5–1 year	[d]	High	Medium/high	Evolutionary to breakthrough

[a] Off-the-shelf.
[b] "Old" or mature.
[c] Custom leading-edge component or chip set for heart of system (chip for autofocus camera; microcontroller for automotive, etc.).
[d] All components available at design stage.
[e] System designed before critical/core component is available, then dropped in at FCS.
[f] Work with supplier to have custom components available when needed.

Component and Supplier Selection

capacity planning, yield prediction/management, and inventory controls. The supply chain is indeed undergoing numerous and substantial changes simultaneously to improve operations, financial, and delivery performance. These fundamental changes in supply chain structure lead to a challenge: how to effectively manage the supply chain while increasing the collaboration between partner companies and delivering what the customer wants.

Fundamentally, a supply chain is responsible for delivering the right product at the right place at the right time for the right price. This requires close linkage between all parties across the supply chain. Figure 1 shows a high level supply chain for an original electronic equipment manufacturer (OEM) from raw materials to the consumer, including the necessary high-level wraparound communication requirements. This figure is easy to understand from a conceptual perspective. But practically the supply chain is a complex integration of the interlinkages of many companies across widespread geographical locations that is in a state of constant change—a work in progress. A more detailed supply chain flow including logistics, freight, and myriad other interlocking activities is shown in Figure 2.

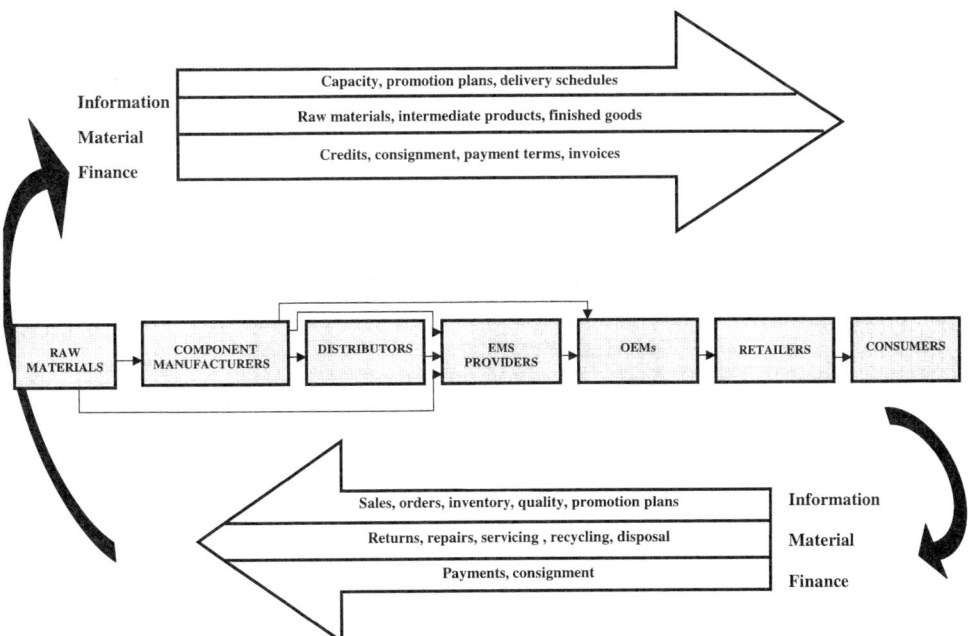

FIGURE 1 Supply chain for electronic equipment manufacturer with communication requirements included.

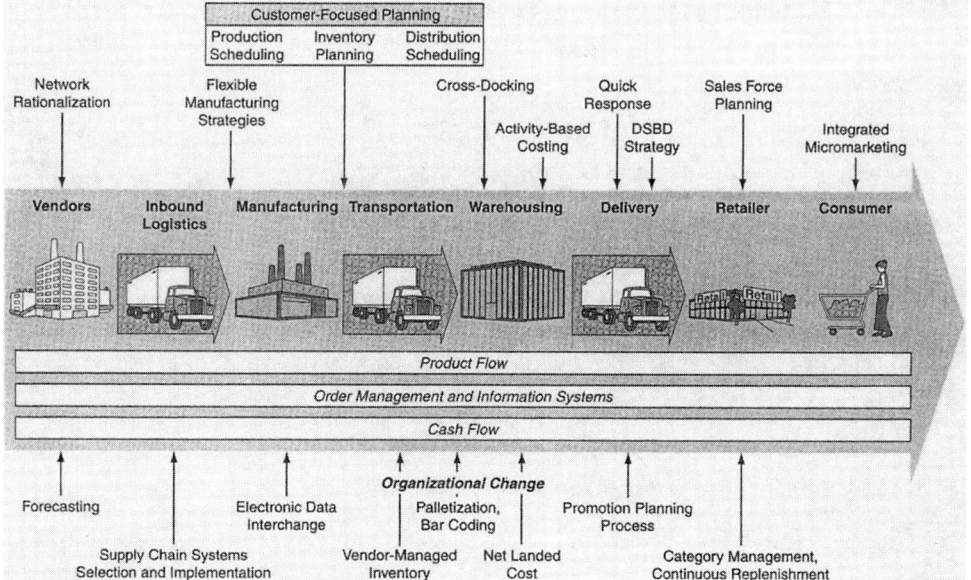

FIGURE 2 Detailed supply chain diagram showing logistics, freight, and systems interconnections.

The complexity of supplier/partner interrelationships, linkages, and dependencies—from wafer fab to finished IC and shipment to channel or product manufacturer and assembly into the product and shipment to the user—requires much care, relationship management, and attention to detail. It all starts with the selection of the right component part and the right supplier.

4.3 SUPPLIER MANAGEMENT IN THE ELECTRONICS INDUSTRY

4.3.1 Overview

The velocity of change is increasing world wide. The only constant is that we will have to continually reevaluate and change everything we do to adapt to these changes. Some of the drivers, or agents of change, include

> Increasing processing power. The processing power of the microprocessor is doubling every 18 months (Moore's law).
> The technology driver is changing from the PC to networking and wireless applications.
> Increase in electronic commerce.

Component and Supplier Selection

Instant global communications.
Booming growth in information technology (i.e., the Internet).
Smarter markets.
Customization of products (lot size of one).
Creation of virtual companies. The resources of many individual expert companies are being brought together to design, manufacture, and deliver a specific product for a specific marketplace and then disbanded when the project is complete. This involves "coopetition"—cooperating with competitors.
Microenterprises (small businesses created by individuals or groups of less than 20 individuals) will lead in the creation of jobs worldwide.

Change isn't easy. Michael Porter, writing in *The Competitive Advantage of Nations,* states that "Change is an unnatural act, particularly in successful companies; powerful forces are at work to avoid it at all costs." You don't bring about change by changing individuals. According to Michael Beer of Harvard University, "You must put individuals into a new organizational context, which imposes new roles, responsibilities, and relationships on them."

How do we anticipate, respond to, and manage the change, specifically as it relates to the supply base and supplier relationships? These agents of change bring about demands from the marketplace that have a profound impact on both the customers and the suppliers of components used in the design and manufacture of electronic equipment. In today's competitive electronic equipment market—be it industrial, consumer, telecommunications, or computer—the pressure is on to

Increase performance, resulting in more complex products with exponentially escalating design development and test development costs.
Reduce product development and test costs.
Focus on core competencies and outsource everything else.
Reduce cycle time, design time, time to market, time to volume, and product life cycle.
Reduce inventories. Increasingly build to order (BTO) rather than build to stock.
Increase quality.
Lower prices.
Exceed changing customer expectations.

Some unique organizational structures that add their own twist have been created to address these requirements in dealing with the supply base. All tend to complicate the relationships with the supply base. They include the following:

Industry consortia of competitors representing a given market segment (such as computer, automotive, telecommunication, and semiconductor, for example) have been formed to develop common areas of focus, requirements, and guidelines for suppliers of that industry segment with the benefit of improved supplier quality and business processes and reduced overhead.

Commodity (aka supply base) teams consisting of representatives of many functional disciplines own the responsibility for managing the suppliers of strategic or critical components.

Many traditional job functions/organizations will have to justify their existence by competing with open market (external) providers of those same services for the right to provide those services for their company (e.g., Purchasing, Human Resources, Shipping/Logistics, Design Engineering, Component Engineering, etc.). A company is no longer bound to use internal service providers.

Why has managing the supply base become so important? The largest cost to an OEM is external to the factory, i.e., the suppliers or supply base. It has been estimated that approximately 60% of the cost of a personal computer is due to the purchased components, housings, disk drives, and power supplies. Therein lies the importance of managing the supply base.

Companies are taking the supply base issue seriously. In today's competitive environment, original equipment manufacturers are continually evaluating their supply base. Nearly all companies have too many suppliers, not enough good suppliers, inadequate measurements for supplier performance, and supplier problems. They are evaluating their supply base to determine

1. If they have the right suppliers
2. Which poorly performing current suppliers are incapable of becoming good suppliers
3. Which poorly performing current suppliers have the potential of becoming good suppliers with OEM cooperation
4. How good current suppliers can be helped to become even better
5. Where can more good suppliers be found with the potential to become *best in class* for the components they provide

The OEM must develop a comprehensive and thorough process to select those suppliers whose technology road maps and business paths align themselves with those of the customer. World class customers require world class suppliers! Their destinies and successes are interrelated. Today's OEMs must adopt a supply strategy in conjunction with the company's business strategy—competition demands it! This requires a partnership/alliance with selected strategic suppliers encompassing all aspects of product development, manufacturing, delivery, and follow-up (i.e., across the supply chain). In effect, the suppliers and customers become mutual stakeholders in each other's success. This type of relationship cannot be managed effectively for a multitude of suppliers. The result is a reduced supply base to a vital few manageable suppliers. Concurrently, suppliers are becoming much more discriminating in selecting strategic customers with whom they wish to partner and support as part of their business strategies. They are reducing their customer base; it's too costly for them to deal with bad customers.

Supplier–OEM relationships are a critical part of successful business processes and are receiving increased attention because of the added value they bring to the design and manufacturing equation. As such, much energy is expended in nurturing these relationships, which are developed at the microorganizational level, i.e., person to person. What does this require? A new method of dealing with suppliers:

- Suppliers are selected with a strategic viewpoint in mind.
- Careful selection is made of a *few* critical suppliers. Relationships with a broad supplier base cannot be developed, managed, and given the nurturing required to be successful.
- Selected suppliers are treated as family members with their failures and successes tied to your failures and successes; this is the essence of mutual stakeholders. It should be a closely intertwined seamless relationship, one where it is hard to tell who is the customer and who is the supplier.
- Open kimono relations—there are no secrets.

As a result, the following issues are extremely vital in supplier management:

- Commitment to the relationship
- Trust
- Constant, open, and accurate communication
- Honesty and integrity
- Shared values, goals, and objectives
- Common ground
- Cultural fit/match
- Complementary and parallel technology road maps

Notice that most of these are soft relational issues dealing with basic values and human relations issues.

Managing suppliers presents some complex and unique issues and is different than managing a functional organization. There are no formal reporting and accountability structures, and people are typically working toward different business goals.

4.3.2 Historical Perspective

The traditional view of purchasing treats the supplier as an adversarial foe. There was no relationship with the supplier. The purchasing equation was strictly based on price, the lowest bidder got the business, in an arm's length and often adversarial relationship. There was no such thing as total cost of ownership. The entire sourcing decision was based on an antiquated cost accounting system. Delivery, quality, responsiveness, and technical expertise ran far behind price consideration. Companies felt that the more suppliers they had, the better. They would (and some still do) leverage (pit) suppliers against one another for price concessions. There are no metrics established for either the supplier or the purchasing function. Negotiated lead times were used to ensure delivery.

Purchasing was staffed as a tactical organization being reactive to Engineering's and Manufacturing's needs, rather than having a strategic forward-looking thrust. The sourcing activity was treated as an unimportant subactivity of the Purchasing Department, which was more internally focused than externally focused. The main focus was on activities such as manufacturing resource planning and inventory strategies. Purchasing was essentially a clerical function with no special skills or technical education required. It was viewed as providing no competitive advantage for the company.

Internally, Engineering generated the design essentially without utilizing the suppliers' technical expertise, Manufacturing's inputs, or Purchasing's involvement. Then Engineering would throw the specifications over the wall to Purchasing. Purchasing would obtain competitive quotations and place the order with the lowest bidder with guaranteed lead times. Every functional organization at the OEM's facility operated as an independent silo (Fig. 3) doing its own thing, as opposed to today's use of crossfunctional teams. The entire organizational structure was based on doing things (activities) rather than achieving measurable results, i.e., firefighting versus continuous improvement. All in all, it was a poor use of everyone's intellectual resources.

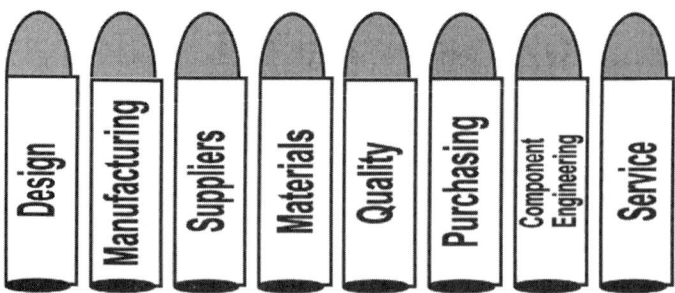

FIGURE 3 Traditional supplier communications. Everyone operates in separately functioning silos. (Courtesy of Ken Stork and Associates, Inc., Batavia, IL.)

4.3.3 Core Competencies

The need for change has resulted in a review of those activities which are core and those which are context to a company (see Ref. 1 for a detailed discussion on core and context activities). To be leaders, global manufacturers must identify their core competencies and use them to build a marketplace presence and image. A core competency is something a company does so well that it provides a competitive advantage in its targeted markets and perhaps serves as a barrier of entry against competitors. This is the ground upon which companies differentiate their products from those of their competitors. The goal of core work is to create and sustain that differentiation by assigning one's best resources to that work/challenge. In other words, core competencies are those tasks that add the defining or unique value to a product or service. The core competencies for a given company in reality are few in number, not easily emulated (or duplicated), and provide considerable value. By contrast every other activity in a company is referred to as context. Context tasks are to be executed as effectively and efficiently as possible in as standardized and undifferentiated manner as possible. They are thus prime candidates to be outsourced. In any given business category, one company's core may well be another company's context.

Wal-Mart	Low-cost distribution
	Employee relations
	Strategic management information service
	Supplier management of store inventory (automatic replenishment)
Compaq Computer Corp., Tandem Division	Nonstop data processing
	Design of scaleable network clustering solutions
Honda	Design and manufacturing of small engines and power trains for a host of equipment
Sony	Short design cycles
Southwest Airlines	Low-cost on-time flights
	Genuine care for employees and customers
Apple Computer	Ease of use (original core competence)

Sun Microsystems understands and concentrates on their core competencies and uses suppliers to gain competitive advantage by going to those that have the best technology and design support structure. Sun has transitioned to a model where they consider suppliers to be an extension of their own work force, even to the extent of having supplier representatives report to work (co-locate) at Sun. Sun believes that if suppliers are involved early and their expertise is utilized, then the final design and specifications are more in line with their (the supplier's) capabilities and core competencies, and Sun ends up getting better components with a lower total cost of ownership. Co-location also allows problems to be addressed quickly.

4.3.4 Types of Sourcing

Outsourcing

Outsourcing and strategic sourcing are key elements in today's supply management process. This section focuses on outsourcing in general. Outsourcing manufacturing functions is discussed in Chapter 7, and strategic sourcing is presented next.

Companies that undertake true reengineering efforts find that much management energy is spent in maintaining support and management processes that could better be maintained by companies for which they are the core business processes. A company must determine, enhance, and build upon its core competencies and outsource everything else in today's world. Peter Drucker predicted (in 1999) that in 10 or 15 years, corporations would be outsourcing all work that is not directly producing revenue or does not involve directly managing workers. Tom Peters suggests that if a company cannot sell its services in a particular process on the open market it should consider finding a contractor who does just that, i.e., runs "unsalable" operations.

Companies in all industry sectors have increasingly turned to outsourcing (transferring selected functions or services and delegating day-to-day management responsibility to a third party) in response to reengineering efforts of the late 1980s and early 1990s. They choose to outsource for either tactical reasons (to lower cost of goods sold) or strategic reasons (improve return on assets). Typically, OEMs that outsource from a tactical perspective focus on narrow short-term goals, while OEMs that outsource strategically have a long-term focus. Many companies originally turned to outsourcing primarily for short-term financial improvement (cost reduction). Some needed to stop a hemorrhaging bottom line because they couldn't control their own processes; they refused or were unable to address deficiencies; they wanted to get rid of a headache; or they just plain jumped on a fad bandwagon. What they often didn't do was to think of the ramifications of and support required to manage their outsourcing decision/strategy. These include removal of critical skills from one's company, the impact on the morale of both directly affected and remaining employees, learning curve loss, the ability to be innovative, and the ability to provide flexibility and fast response to customers.

Strategic outsourcing changes the OEM's way of doing business and focuses on long-term, vital strategic issues and direction. Strategically oriented OEMs seek to change their cost structure by moving fixed costs to variable costs in order to gain greatest return on assets. Those OEMs who manufacture complex electronic systems are often larger companies in volatile industries such as semiconductor process equipment. The industry volatility makes the management of

TABLE 2 Comparisons of Top Reasons for Tactical and Strategic Outsourcing

Reasons for tactical outsourcing	Reasons for strategic outsourcing
Reduce or control operating costs	Focus on core competencies
Reduce capital funds invested in noncore business functions	Provide access to world-class capabilities
Opportunity to receive an infusion of cash as the OEM's assets are transferred (sold) to the outsource service provider	Accelerate organizational change and free valuable resources to pursue new ideas
	Share risks with outsource service provider that provides for more flexible, dynamic, and adaptable responses to changing opportunities
Need for physical, technical or geographical resources not otherwise available	Free resources from noncore activities for other purposes such as research and development, customer satisfaction, and return on investment
Elimination of functions that are difficult to manage or that are out of control	

people and fixed assets difficult. This leads to the natural tendency to seek out qualified partners that can materially contribute to strategic goals. Table 2 lists some of the main tactical and strategic reasons for outsourcing.

I need to point out that outsourcing is not merely contracting out. What are the differences?

1. While contracting is often of limited duration, outsourcing is a long-term commitment to another company delivering a product and/or service to your company.
2. While providers of contracted services sell them as products, providers of outsourcing services tailor services to the customer's needs.
3. While a company uses external resources when it contracts out, when it employs outsourcing it usually transfers its internal operation (including staff) to the outsource supplier for a guaranteed volume support level over a specified period of time (5 years, for example).
4. While in a contracting relationship the risk is held by the customer and managed by the supplier, in an outsourcing relationship there is a more equal sharing of risk.
5. Greater trust is needed to engage successfully in an outsourcing arrangement than in a simple contracting situation. This is because in outsourcing the supplier also assumes the risk, but the customer is more permanently affected by any failure by the supplier.
6. While contracting is done within the model of a formal customer–supplier relationship, the model in an outsourcing relationship is one of true partnership and mutually shared goals.

What are the advantages of outsourcing noncore work? First, as already stated, are the simple cost issues. Fixed costs are voided (they become variable costs) in that the company does not have to maintain the infrastructure through peak and slack periods. This is a big benefit for the OEM since the market volatility on the cost of goods sold and overhead costs is now borne by the contract manufacturer (CM), making the OEM immune to this market variability.

Second, there are cost savings due to the fact that the outsource service provider can provide the services more efficiently and thus cheaper than the OEM can. If the service-providing company is at the cutting edge of technology, it is constantly reengineering its core processes to provide increased efficiency and reduced costs to its OEM customers.

Third is the issue of staffing. OEM in-house service processes are staffed to cope with all possible crises and peak demand. An outsourced support process can be staffed to meet day-to-day needs, secure in the knowledge that there is adequate staff in reserve for peak loads and adequate specialized staff to bring on board quickly and cost efficiently.

Fourth are service issues. Having an agreement with a provider of a particular service gives a company access to a wider skill base than it would have in-house. This access provides the OEM with more flexibility than it would have if it had to recruit or contract specific skills for specialized work. The OEM can change the scope of service any time with adequate notice, not having to face the issues of "ramping up" or downsizing in response to market conditions. Working with a company whose core business process is providing a particular service also improves the level of service to OEMs above what they are often able to provide for themselves.

Fifth, creating a good outsourcing relationship allows the OEM to maintain control over its needs and sets accountability firmly with the partnering service provider. The clear definition and separation of roles between OEM and CM ensures that service levels and associated costs can be properly identified and controlled to a degree rarely seen in-house. All of this can be done without the internal political issues that so often clutter relations between support and management process managers who are seen by their internal customers as merely service providers.

Finally, and most importantly, outsourcing allows the OEM to focus its energies on its core business processes. It focuses on improving its competitive position and on searching the marketplace for opportunities in which to compete.

Companies thinking about outsourcing some of their non-core processes have some fears or concerns. One set of concerns revolves around the loss of in-house expertise and the possible coinciding loss of competitiveness, and the loss of control over how the services will be provided. A second set of concerns revolves around becoming locked in with one supplier and his technologies. If that supplier does not keep pace with industry trends and requirements and develop new technologies to meet (and in fact drive) these changes, an OEM can be rendered noncompetitive.

Both sets of fears and concerns reflect the basic reticence of business leaders to engage in long-term relationships based on the kind of trust and "partnership" necessary to function in today's business environment. A third set of concerns revolves around the internal changes that will be necessary to effect the kind of business change that will occur when support and management processes are shed. The cost and logistics of planning and implementing changes to any process are considerable, but they must always be balanced against the opportunity of upgrading capability in another area.

Strategic Sourcing

The old-fashioned relationship to suppliers by companies who operated in a "get the product out the door" mindset doesn't work today. This method of operation (in which technologists find what the market needs and the operationally directed part of the company makes it) was characterized by customers who

> Exploited suppliers
> Found subordinate and easily swayed suppliers
> Limited the information provided to suppliers
> Avoided binding or long-term agreements
> Purchased in single orders, each time setting up competition among suppliers

The major factor driving change in the procurement process is the recognition among strategic thinking electronics manufacturers that the largest single expense category a company has is its purchases from suppliers (typically, corporations spend 20 to 80% of their total revenue on goods and services from suppliers). Thus, it is important to view suppliers strategically because they exert a great influence on the product manufacturing costs and quality through the materials and design methodology. An example of this is shown in Table 3 for Ford Motor Company.

TABLE 3 Impact of the Design and Material Issues on Manufacturing Cost at Ford

	Percent of product cost	Percent of influence on manufacturing costs	Percent of influence on quality
Material	50	20	55
Labor	15	5	5
Overhead	30	5	5
Design	5	70	35
Total	100%	100%	100%

Source: From the short course "Supplier Management" at the California Institute of Technology. Courtesy of Ken Stork and Associates, Inc., Batavia, IL.

Material, as might be expected, is the biggest contributor to product cost. As stated before, approximately 60% of the cost of a personal computer is due to the purchased components, housing (cabinet), disk drives, and power supplies.

Manufacturing companies are increasingly finding that they must engage their suppliers as partners to achieve their strategic intent. Supplier equity is a "new asset" and in some businesses may be a core competence providing competitive advantage. Competitive advantage means an advantage a company has due to one of the five customer values: cost, quality, service, time, or innovation.

Original equipment manufacturers can either focus internally on the company's operations and products or on identifying the business opportunities posed by the marketplace and translating them into products the company could make, given its processes and capabilities (external or market focus). But the latter is not enough if a company maintains an old-fashioned, rigid bureaucratic outlook.

A market-focused company needs to have a culture that is group oriented, rather than bureaucratic and hierarchical. It takes both medium and long-term views of the market and combines that with market analysis and a focus on innovation and product diversity. The basis of competition for market-focused companies is quality, lead time, and flexibility, and they engage their partners up and down the value chain in discussions regarding those competitive aspects.

The supply strategies that market-focused companies employ include

- Vertical integration of a logistics network to integrate suppliers and customers throughout the value chain
- Comakership or mutual stakeholder mindset and involvement not just in operations, but in product design and in manufacturing key components and technologies
- Reduction of the supplier base to a few suppliers, tightly integrated into the business, which provides significant operating efficiency, fast time to market and cost reduction
- Implementation of a common information system for operations, deliveries, planning, design, and change management
- Outsourcing appropriate manufacturing, support, and management processes to specialist companies

What Is Strategic Sourcing? Strategic sourcing can be defined as the skillful planning of those activities from which value originates. Stated another way, strategic sourcing is leveraging core competencies for competitive advantage. It is difficult for a company to be strategic when it is buried in tactical fire-fighting and administrative issues. Both strategic processes and tactical (daily) operations are necessary, but they must be separated. Strategic processes are proactive, externally focused, and forward looking, whereas tactical processes are reactive in nature, inwardly focused on today's events and problems. Most supplier sourcing is reactive (~80%), while only about 5–10% is strategic.

Component and Supplier Selection

Strategic sourcing isn't easy to implement. There are many barriers to overcome. Especially hard pressed to change are those companies who have an inspection/rework mindset based on mistrust of the adversarial arms-length transactions with their suppliers, such as government contractors. Strategic sourcing takes a lot of work to implement properly. It is neither downsizing (or rightsizing as some prefer to call it) nor is it a quick fix solution to a deeper problem. A strategic sourcing organization is not formed overnight. It takes a concerted effort and evolves through several stages (Table 4).

The goal of strategic sourcing is to develop and maintain a loyal base of critical (strategic) suppliers that have a shared destiny with the OEM. The supplier provides the customer with preferential support (business, technology, quality, responsiveness, and flexibility) that enables and sustains the customer's competitive advantages and ensures mutual prosperity.

This sourcing strategy can occur on several levels. Take the case of a candy bar manufacturer who purchases milk, cocoa, sweetener, and packaging. In this

TABLE 4 Stages of Strategic Sourcing

Stage	Goal	Actions
1	Minimize supplier's negative potential	Strategic decisions are made without considering supply issues. Only problems are communicated. Issues are price and delivery.
2	Achieve parity with competitors	Institute supplier reduction programs. Initiate cost reduction programs. Use internal organizations to fix problems.
3	Provide credible support to business	Active involvement of suppliers to support business strategy. Supplier relationship strategy is formulated and pursued. Long-term planning is established.
4	Add supply issues to a manufacturing-based strategy	Anticipate supplier's new technology. Supply issues considered up front by Engineering/Marketing/Manufacturing. Involve suppliers early on so they can anticipate needs and development capabilities.
5	Preferential service from best-in-class and world-class suppliers	Proactive, concentrated efforts to earn preferred service from a small supply base. Long-term commitment.

situation there is no one-size-fits-all sourcing strategy. Milk is sourced locally (to the manufacturing plant) for freshness and perishability reasons. Cocoa is sourced at the division level because of the price leverage gained. Sweetener is sourced at the corporate level because it leverages across divisions; it is used in more products than just candy bars. Packing is sourced locally due to its uniqueness and the flexibility provided.

Strategic sourcing is a process or series of processes, not a group of functional tasks. The functional silo diagram illustrated earlier (Fig. 3) is inefficient and ineffective in a strategic sourcing company. Chrysler, for example, created various platform teams (such as interior, exterior, body, chassis, and power train) that integrated the various silos' expertise horizontally in a cross-functional team implementation (see Fig. 4).

Strategic sourcing needs to be a well-thought-out process. It takes time to implement (to a large extent because it involves building personal relationships of mutual trust) but provides renewed vigor to a company. Let's take the example of Xerox in the 1980s as a brief case study. The situation: Xerox had lost its focus and market share. To improve they turned to strategic sourcing to turn the company around. Xerox was losing market share for many reasons:

1. They had an arrogant manner.
2. Xerox didn't understand its business
3. Xerox didn't pay attention to the needs of its served markets.
   They had a diminished concern for their customers.
   They lost their focus by buying a computer company to go head-to-head with IBM.
   Xerox failed to acknowledge changing conditions in the copier and

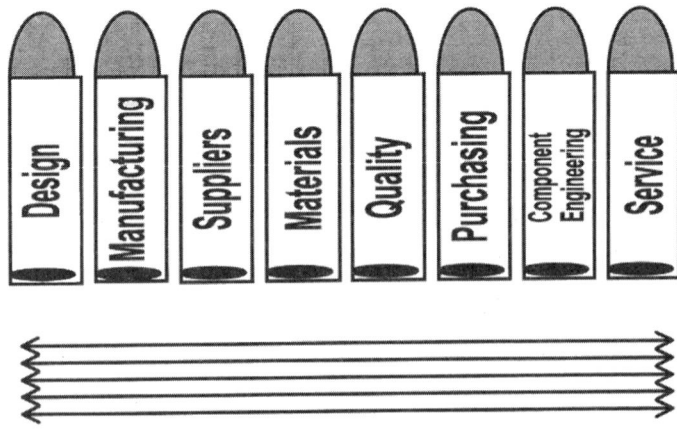

FIGURE 4 Chrysler platform teams utilizing functional silo expertise. (Courtesy of Ken Stork and Associates, Inc., Batavia, IL.)

computer businesses. (Canon shifted the competitive base for a product prone to break down but supported by a terrific service organization, i.e., Xerox, to copiers that didn't need to be serviced.)
4. Their manufacturing costs were bloated.
5. They had high equipment service rates and thus high costs due to poor quality.
6. They became inwardly focused and fought internal turf wars. They ignored advice from within and from specially formed task groups.
7. They centralized decisionmaking and didn't listen to employee input.
8. They were slow to respond to competition. Xerox was preoccupied with what Kodak was doing but ignored the upstart Japanese competition.

After losing significant market share, Xerox turned to benchmarking to understand what had changed both in the marketplace and with their competitors. The results of benchmarking led them to adopt strategic sourcing. They centralized materials management in order to centralize sourcing. They employed a commodity team structure and reduced their supplier base from approximately 5000 to 300. These remaining suppliers were welcomed and embraced as family members and were provided with the appropriate training. They focused on retraining their employees. The result of these actions allowed Xerox to improve its competitive position, increase market share, strengthen its bottom line, and build a more solid foundation for the future. But it is not over. They must continually seek to improve all they do, since customer expectations increase and new competitors provide new competitive advantages and capabilities.

Strategic sourcing, or supply base management (I use both terms interchangeably), represents a very advanced and complex methodology of dealing with the supply base. Some of the issues that add to this increased level of complexity include

Long-term visionary focus
Customer–supplier interconnectivity
Organizational connectivity—mutual stakeholders with shared destinies
Increased leverage
Higher quality
Reduced supply base
Lower total cost
Shorter lead times
Increased flexibility
Joint technology development
Shorter time to market
Open, direct, consistent, and transparent channels of communication at all levels within customer and supplier organizations (including IT resources)

> Empowered cross-functional commodity teams (see next section)
> Comingling of customer/supplier technological and manufacturing resources
> Focus on suppliers of critical or strategic components
> Required senior management sponsorship in both the customer's and supplier's organizations but implemented at the *microlevel* person-to-person
> Realistic expectations of clients and suppliers

Strategic sourcing means that a company

> Develops partnerships with a limited number of suppliers who are leaders in their respective industries
> Looks to these suppliers for quality products, leading-edge technology, competitive pricing, and flexible deliveries
> Bases these relationships on honesty, integrity, fairness, and a desire to do whatever it takes to be mutually successful.
> Promotes win/win cosharing

What are some of the key elements of strategic sourcing? To be successful, strategic sourcing

> Must be part of an overall business strategy and actively supported by top management. Being part of only the manufacturing strategy leads to suboptimal results.
> Is integrated with company strategy and technology planning.
> Focuses on critical procurements and develops appropriate strategies for each case.
> Understands cost and competitive drivers.
> Understands technology road maps.
> Uses cost-based pricing.
> Facilitates and uses cross-functional team processes, i.e., commodity teams.
> Constantly measures performance to its customers.
> Establishes annual breakthrough goals from benchmarking other companies for such items as material cost reductions, supplier quality improvements, on-time delivery, lead time reduction, inventory reduction, and quicker customer response.
> Implements continuous performance measurement and improvement practices.

All of these elements are in essence related. It's hard to separate and discuss them individually without referring to the others. As an example Figure 5 lists some of the strategic sourcing considerations that an electronics equipment manufacturer makes. The inner circle lists those few high-level critical corporate issues. Moving from the center outward the list of considerations expands, i.e., becomes more granular, or detailed.

Component and Supplier Selection

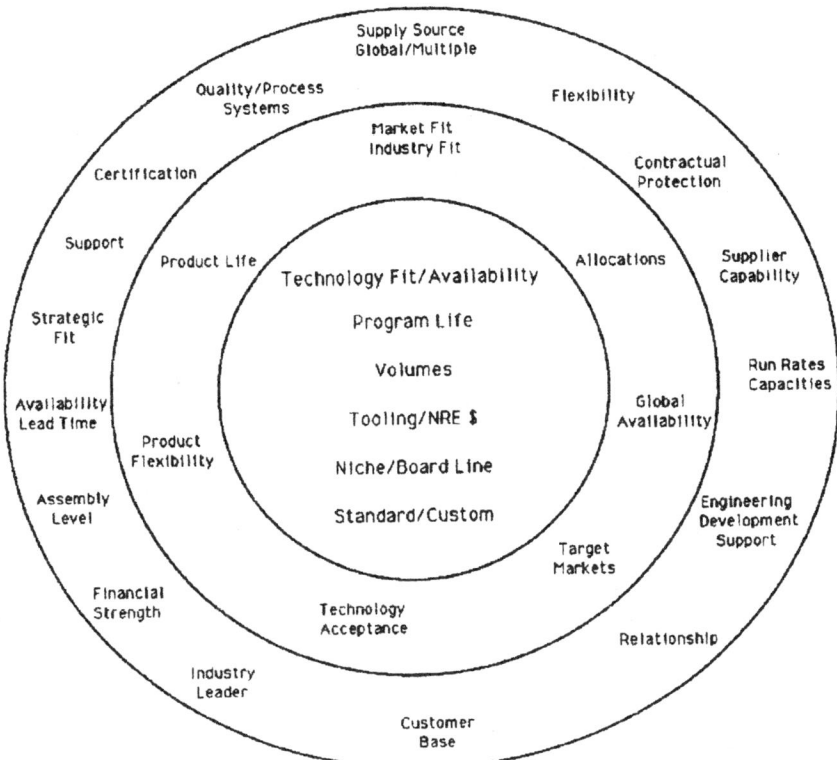

FIGURE 5 Example of sourcing strategy considerations of an electronics equipment manufacturer.

Where to Start. Before adopting a strategic sourcing methodology, a company's management must do some self-assessment and introspection by asking itself some extremely pointed questions, such as the following:

1. How do we compare to world class companies?
2. What do we need to do to be better than the competition?
3. Are internal best practices widely used across our company? If not, why not?
4. Do we spend our company's money as wisely as we spend our own?
5. Are we focused on the right priorities?
6. Are we organized correctly?
7. Do we have the right skills and expertise?
8. Have we consistently demonstrated the proper leadership and support for strategic sourcing and related issues, such as concurrent engineering?

9. Are our people empowered to make the correct decisions?
10. Have we identified our core competencies?
11. Have we identified the critical commodities that if faulty could spell disaster to our equipment in the field?

Once these questions are answered, it is wise to create a sourcing statement that clearly relates to both a company's corporate objectives and related functional strategies. It is important that this is effectively and consistently communicated along with the results obtained on a regular basis both internally and with a company's strategic suppliers.

Purchasing's New Role. Historically, most corporations had a purchasing department in name only, with limited or no input in overall corporate strategy and direction. This in spite of the fact that the purchasing function accounts for 40–60% of the total dollars a company spends with its outside suppliers.

The change in sourcing mindset represents a major culture change for an OEM. Top management support is essential if sourcing professionals are to add value as part of cross-functional sourcing teams. Strategic sourcing offers a company some real benefits: total cost reduction, supply chain process improvement, process integrity, a strong competitive philosophy, and more.

In a strategic sourcing model the role of Purchasing changes from a focus on price and pitting supplier against supplier to a more value added function: establishing and maintaining collaborative supplier relationships. In a strategic sourcing organization, the Purchasing function has a strategic rather than tactical mindset. Purchasing becomes a strategic driver for the company because today customers want the variety, velocity, quality, and cost competitiveness that can only be provided through collaboration. A study conducted in 1990 by Yankelovich, Clancy, and Shulinan (and updated more recently with the same results) showed the reduced importance of price as the main purchasing issue. When asked the question what constitutes quality, the respondent companies listed the following as being important (in ranked order): reliability, durability, ease of maintenance, ease of use, a known and trusted name, and lastly a low price.

One of the biggest challenges for the purchasing function is how to turn an individual accustomed to transactional purchasing into a skilled strategic sourcing professional. Purchasing departments have many individuals who are quite comfortable in the transactional world. But change is necessary. Purchasing is taking a more proactive and strategic role in component supplier selection and use, moving into the design space to provide added experience and value. To accomplish this, Engineering is being integrated into Purchasing to provide the reqired technical expertise. Job functions such as Commodity Engineer, Purchasing/Procurement Engineer, and Procurement Qualification Engineer are becoming commonplace. An increasing number of personnel populating the procurement ranks have BSEE degrees; some have MSEE degrees.

Component and Supplier Selection

In a strategic sourcing company, high-value opportunities are identified and resources are allocated to focus on them. Here the focus is on managing suppliers and supplier relationships, ensuring access to the top suppliers, and encouraging their best efforts by qualifying existing and potential suppliers, monitoring their performance, and upgrading or eliminating marginal performers. The Purchasing Department

- Is focused on developing and managing long-term strategic supplier/partner relationships. This process includes the proactive use of scorecard evaluations and links the assessment of suppliers' past performance to current and future partnering opportunities. Appendix A of Chapter 4 presents an example of a supplier scorecard measurement process.
- Reduces the effort associated with procuring items where the procurement function can add limited value.
- Eliminates low-value-added activities by streamlining the procurement process and outsourcing these activities.
- Has become a strategic resource by providing a competitive advantage to an electronics manufacturing company.
- Provides critical information about supply conditions and market trends, recommending suppliers before new product design begins.
- Is involved with marketing strategy and specific sales opportunities (winning business by enabling customers to meet equipment purchase budgets through creative component pricing and purchasing techniques).

So the basic elements of strategic sourcing that Purchasing now addresses include

1. Supplier selection
   Global strategies
   Qualified competitive sources
   Ensured supply
2. Supplier negotiation
   Industry-leading terms and conditions
   Flexible, continuous supply
3. Supplier development
   Relationship development at microlevel
   Continuous improvement and innovation goals
4. Supplier evaluation
   Performance measurement and reporting
   Total cost of ownership
   Technology road map match
5. Long-range supplier strategy
   Core commodities
   Maximizing leverage across commodities
   Matching technology road map with client product road map
6. Strategic resource to both Design and Marketing-Sales

I can't emphasize this enough: strategic sourcing is not a fixed process; it is flexible and adaptable to a specific company's needs. Even as this is being written, the sourcing responsibility is undergoing refinement and change to be more in tune with and effective in providing the design community with the suppliers and components required to produce a product that meets customer expectations. The focus must be external—on the customer.

Strategic Sourcing Summary. A 1995 study by A. T. Kearney of 26 multinational corporations in a variety of industries predicted that developing suppliers is going to be a part of a company's market strategy. The study found dramatic changes in how corporations view suppliers and the procurement function. Restructuring their relationships with suppliers was revealed in the study as one of the key ways that leading U.S. companies have found to compete, dramatically cut costs, and slash the development time for new products.

The OEMs focusing on strategic sourcing are able to offer products and services faster and at lower prices and thus invest more cash into the core competency of research and development. Other companies have used strategic sourcing methodology and discipline to find and lock up the best suppliers in long-term exclusive relationships, effectively shutting the competition out of participation in critical markets. A summary of the Kearney findings is presented in Table 5.

Thus, the company that revolutionizes its supplier relationships will have lower costs, faster response times, and happier customers. Other things (like making the right products) being equal, that must also mean higher profits. An open, honest relationship between equal partners in the supply chain is the only way of optimizing the performance of the supply chain. Similarly, within companies

TABLE 5 Strategic Sourcing Key Findings of 1995 A. T. Kearney Study

---

95% of corporations surveyed reported a significant change in the procurement organization in the past decade; 55% anticipated additional change in the future.

The use of outsourcing is growing rapidly. 86% of surveyed corporations reported outsourcing some function in 1995 versus 58% in 1992.

Supplier concentration is still being aggressively pursued, with corporations reducing their supply base by 28% between 1992 and 1995. The study predicted that another 35% reduction is expected by 1998.

Leading companies chopped their inventory levels by 4% using strategic sourcing; half the companies in the study expected inventory levels to fall at least another 40% by the year 2000.

Leading companies reduced product development time by 62% over a one-year period.

Source: From the short course "Supplier Management" at the California Institute of Technology. Courtesy of Ken Stork and Associates, Inc., Batavia, IL.

the only way of achieving velocity in design and manufacture is to break down the old barriers and rebuild new relationships by adopting simultaneous engineering. That means all functions working together in teams from the start; it is about relationships, not technology.

4.3.5 Categories of Suppliers

Where a given IC supplier stands in respect to specific business, technical, quality, and support issues (see Table 6) determines the level of effort required to support him. I have used IC suppliers here since specific IC functions normally provide an OEM with a competitive advantage.

TABLE 6 Categories of IC Suppliers

Tier 1 supplier	Tier 2 supplier	Tier 3 supplier
Large supplier	Medium-sized supplier	Small supplier
Well-defined and documented quality system	Well-defined and documented quality system	Minimal quality system
Full service supplier	Niche to full service supplier	Niche/boutique supplier
Comprehensive design practices with large R&D resources	Comprehensive design practices with medium R&D resources	Not as rigorous design practices and minimal R&D resources
Defined and controlled processes	Defined and controlled processes	Minimal process control
Well-developed support infrastructure with appropriate staff and equipment (e.g., CAD models, test resources, F/A and corrective action, reliability tests)	In-place support infrastructure with appropriate staff and equipment	Minimal support infrastructure with minimal staff and equipment
Provides breadth and depth of product offering	Focused on one or two key functional component categories	Provides performance advantage and/or unique benefit for one functional component type and/or family
Rigid rules for customer requests	Some rules with some flexibility	Very flexible
Select customers carefully for relationships	Open to meaningful partnerships	Interested in customers who need performance parts
80% of purchase spending 20% of problems	80% of purchase spending 20% of problems	20% of purchase spending 80% of problems

Not all suppliers are created equal. Nor should all suppliers be considered as strategic sources because

1. The components they provide are not critical/strategic nor do they provide a performance leverage to the end product.
2. The OEM has limited resources to manage a strategic sourcing relationship. It's simply a case of the vital few suppliers versus the trivial many.

The more tiers of suppliers involved, the more likely an OEM will get into the arena of dealing with very small companies that do not operate to the administrative levels that large companies do. With fewer staff available to oversee second- and third-tier suppliers, it has become increasingly difficult to maintain a consistent level of quality throughout the supply chain. Companies have to accept the commitment to work with second- and third-tier suppliers to develop the same quality standards as they have with their prime suppliers. It's part of managing the supply chain. Most companies in the electronics industry prefer to deal with tier 1 or tier 2 suppliers. However, often unique design solutions come from tier 3 suppliers, and they are then used after due consideration and analysis of the risks involved. It takes a significantly larger investment in scarce OEM resources to manage a tier 3 supplier versus managing tier 1 and tier 2 suppliers.

Many tier 2 and 3 suppliers outsource their wafer fabrication to dedicated foundries and their package assembly and test as well. Integrated device manufacturers, which are typically tier 1 suppliers, are beginning to outsource these operations as well. This removes the need for keeping pace with the latest technology developments and concomitant exponentially accelerating capital equipment expenditures. The major worldwide foundries are able to develop and maintain leading-edge processes and capability by spreading mounting process development and capital equipment costs across a large customer base. Many of these are "fabless" semiconductor suppliers who focus on their core competencies: IC circuit design (both hardware and software development), supply chain management, marketing, and customer service and support, for example.

Having said this, I need to also state that IC suppliers have not been standing still in a status quo mode. Like their OEM counterparts they are going through major changes in their structure for providing finished ICs. In the past (1960s and 1980s) IC suppliers were predominantly self-contained vertical entities. They designed and laid out the circuits. They manufactured the ICs in their own wafer fabs. They assembled the die into packages either in their own or in subcontracted assembly facilities (typically off-shore in Pacific Rim countries). They electrically tested the finished ICs in their own test labs and shipped the products from their own warehouses. This made qualification a rather simple (since they directly controlled all of the process and resources) yet time-consuming task. Mainline suppliers (now called integrated device manufacturers, or IDMs) such as AMD,

Component and Supplier Selection 191

IBM, Intel, National Semiconductor, Texas Instruments, and others had multiple design, wafer fab, assembly, electrical test, and warehouse locations, complicating the OEM qualification process.

Integrated circuit suppliers were probably the first link in the electronic product food chain to outsource some of their needs; the first being mask making and package assembly. Next, commodity IC (those with high volume and low average selling price) wafer fab requirements were outsourced to allow the IDM to concentrate on high-value-added, high-ASP parts in their own wafer fabs. Electrical testing of these products was also outsourced. Then came the fabless IC suppliers such as Altera, Lattice Semiconductor, and Xilinx who outsource all of their manufacturing needs. They use dedicated pure play foundries for their wafer fabrication needs and outsource their package assembly, electrical test, and logistics (warehouse and shipping) needs, allowing them to concentrate on their core competencies of hardware and software development, marketing, and supplier management. A newer concept still is that of chipless IC companies. These companies (Rambus, ARM, and DSP Group, for example) develop and own intellectual property (IP) and then license it to other suppliers for use in their products. Cores used to support system-on-a-chip (SOC) technology are an example.

Currently, IC suppliers are complex entities with multiple outsourcing strategies. They outsource various parts of the design-to-ship hierarchy based on the functional part category, served market conditions, and a given outsource provider's core competencies, design-manufacturing strategy, and served markets. Each of the functions necessary to deliver a completed integrated circuit can be outsourced: intellectual property, design and layout, mask making, wafer fabrication, package assembly, electrical test, and warehousing and shipping (logistics). In fact, for a specific component, there exists a matrix of all possible steps in the IC design-to-ship hierarchy (that the supplier invokes based on such things as cost, delivery, market needs, etc.), complicating supplier selection and qualification, part qualification, and IC quality and reliability. From the OEM's perspective the overarching questions are who is responsible for the quality and reliability of the finished IC, to whom do I go to resolve any problems discovered in my application, and with so many parties involved in producing a given IC, how can I be sure that permanent corrective action is implemented in a timely manner.

Another issue to consider is that of mergers and acquisitions. Larger IC suppliers are acquiring their smaller counterparts. This situation is occurring as well across the entire supply chain (material suppliers, EMS providers, and OEMs). In these situations much is at risk. What products are kept? Which are discontinued? What wafer fab, assembly, test, and ship facilities will be retained? If designs are ported to a different fab, are the process parameters the same? Is the same equipment used? How will the parts be requalified? All of this affects the qualification status, the ability of the part to function in the intended application, and ensuring that a continuous source of supply is maintained for the OEM's manufacturing needs. Then there is the issue of large companies spinning off their semiconductor (IC) divisions. Companies such as Siemens (to Infineon), Lucent (to Agere Systems), and Rockwell (to Conexant), for example, have already spun off their IC divisions.

In fact, it is predicted that the IC industry will be segmented into a number of inverted pyramids characterized by dramatic restructuring and change (Fig. 6). The inverted pyramid shows that for a given product category, digital signal processors (DSPs), in this example, four suppliers will eventually exist: one large tier 1 supplier and 2 or 3 smaller suppliers will account for the entire market for that circuit function. A reduction in the supply base will occur due to mergers and acquisitions and the process of natural selection. This model has proven to be valid for programmable logic devices (PLDs) and dynamic random access memories (DRAMs), with the number of major DRAM suppliers being reduced from 10 five years ago to five today. Other IC product types are likely to follow this path as well. This structure will prevent second tier or start-up companies from breaking into the top ranking. The barriers to entry will be too formidable for these companies to overcome and the inertia too great to displace the market leader. Such barriers include financial resources, process or design capabilities, patent protection, sheer company size, and the like. This will serve to further complicate the selection and qualification of IC suppliers and the parts they provide. OEMs will be limited in the number of suppliers for a given functional IC.

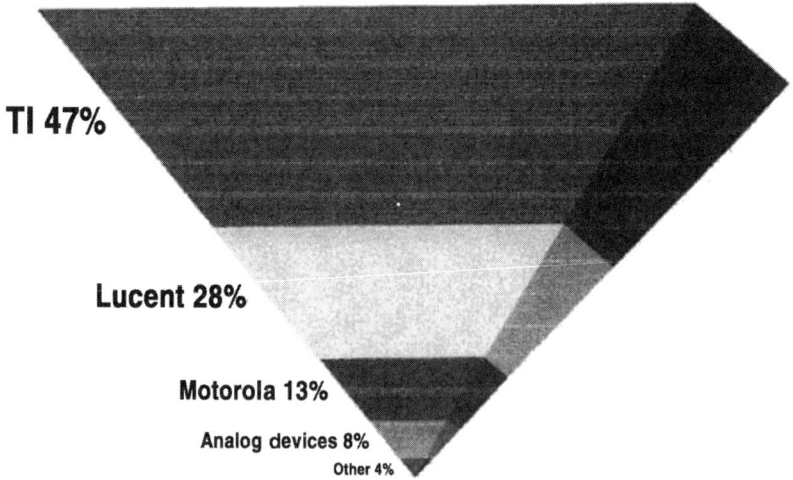

FIGURE 6 Inverted pyramid model for the 1998 DSP market. (Courtesy of IC Insights, Scottsdale, AZ.)

Component and Supplier Selection

4.3.6 Customer-Supplied Relationships

Partnerships (i.e., mutually beneficial relationships) are absolutely essential in today's marketplace. Both buyers and suppliers need the stability and security provided by establishing long-term relationships. Good supplier relationships don't just happen. They take a conscious and concerted effort and require a lot of hard work. It takes time to develop relationships based on mutual trust and respect. People issues, which are more critical than technical issues, must be allocated the proper resources for these relationships to succeed. Supplier–OEM relationships are all about people skills, mutual trust, and commitment to the relationship. These long-term relationships must be developed through the efforts of both customers and suppliers. They are mutual stakeholders in each other's success. Each must make concessions to the other in order to arrive at a relationship satisfactory to both. Continuing coordination and cooperation are important in retaining the relationship once it is established. Collaboration with suppliers, rather than an adversarial relationship, will result in more successful projects; leading to suppliers having more responsibility and influence during product design.

Why have many companies, which for so long emphasized price competition, changed direction to place greater emphasis on the long term relationship?

1. They have found that this relationship results in both fewer quality problems and fewer missed delivery schedules.
2. When there is a supply shortage problem, the firms with long-term relationships suffer less than do opportunistic buyers.
3. Frequent changes in suppliers take a lot of resources and energy and require renewed periods of learning to work together. Relationships can't be developed and nurtured if the supply base is constantly changing.
4. Product innovations require design changes. Implementation of changes in requirements is less costly and time consuming when a long-term relationship has been developed.
5. If either party runs into a financial crisis, concessions are more likely if a long-term relationship exists.
6. Cooperation helps minimize inventory carrying costs.
7. A company and its suppliers can work together to solve technical issues to achieve the desired product performance and quality improvement.

Commitment to these relationships is essential. But strong commitments to suppliers over the long term do not diminish the importance of other factors. The doors should be left open for new suppliers who have new and applicable technologies to the OEM's products. Suppliers in turn are expected to contribute both cost and quality improvement ideas and they are expected to understand the goals, products, and needs of their partner OEM.

Table 7 presents a high level supplier–OEM relationship model put forth in *Beyond Business Process Reengineering*. It progresses from the "conventional" (or historical) OEM–supplier relations that were built on incoming inspection and driven by price to what the authors call a holonic node relationship. This, in effect, represents the ultimate relationship: a truly integrated supplier–OEM relationship in which both are mutual stakeholders or comakers in each other's success.

In this ultimate relationship, there is cooperation in designing new products and technologies. Suppliers are integrated into the OEM's operations, and there is a feeling of mutual destiny—the supplier lives for the OEM and the OEM lives for the supplier's ability to live for its business. Increasingly, the OEM's product components are based on the supplier's technology.

There is also a constant exchange of information concerning both products and processes. The client company's marketing people feed back information directly to the supplier company. This allows the partners to make rapid, global decisions about any required product change.

4.3.7 Commodity Teams

There is no substitute for knowing (developing close relationships with) your key suppliers, especially those that provide components for critical parts of the design. An equipment manufacturer needs to understand an IC supplier's design practices, modeling techniques, in-line process monitors and feedback mechanisms, manufacturing (both wafer fabrication and package assembly) and electrical testing processes, yields, qualification and long-term life test methodology and results. The OEM also needs to understand the financial health and stability of a supplier and its long-term business outlook.

Industry-leading companies have recognized the multiplicity of skills required to truly know and understand, in depth, a supplier's capabilities. These companies have created and empowered commodity teams (also called supply base teams) composed of experts from multiple disciplines to have complete sourcing responsibility for designated strategic or critical components. I use the word *commodity* to denote critical or strategic components, not to denote widely available low-priced components.

The commodity team

"Owns" the supply base for a given strategic technology/commodity
Assesses product technology needs
Acts as a resource to new product teams and ensures that supply strategies are established early in new product development

Component and Supplier Selection

TABLE 7 Supplier–Customer Relationship Models

Supplier level	Quality	Logistics	Product and technology development	Node choice criteria
Class III—conventional	Supplier responsible to furnish in accordance with quality specifications Client makes incoming inspections and source inspections	Supplies "order by phone", with specific delivery times Reserve stocks are necessary	Product and component characteristics designed solely by client First supply verification	Price
Class II—associated	Self-certification (supplier) Free pass (client) Quality improvement programs (supplier and client)	Long-term contracts JIT/synchronized deliveries directly to production departments (no stock) Continuous reduction of stock and lead times (together)	Technical requirements of components and technology defined with supplier Supplier consulted in advance	Total cost
Class I—partner	Supplier responsible for conformity of components to final customer satisfaction	Supplier-integrated in the client's logistic process (same documents, same operative system)	Supplier involved in the development process, starting from product concept	Speed
Class 0—Holonic nodes	Mutual continuous improvement Codesign of quality requirements Codesign virtual company business processes	Shared information and planning systems (electronic data interchange network) Same information systems	Supplier involved in product planning process Supplier proactive Supplier commitment to product development and planning process	Market innovation and support Shared values Flexibility

Source: McHugh, Patrick, et al., Beyond Business Reengineering, John Wiley & Sons, 1995.

- Maintains knowledge of industry and technology developments and matches the technology to the available supply base
- Defines, evaluates, and ranks suppliers and implements the long-term supply base/commodity strategy
- Identifies benchmarks for the commodity
- Defines the criteria and manages supplier selection, qualification, development, and the qualified part list (QPL)
- Generates supplier report cards based on quality, on-time delivery, price, flexibility, service, continuous improvement, reliability, and other metrics
- Determines deficiencies, isolates their root cause, and drives corrective actions with suppliers
- Develops, implements, and monitors supplier performance target plans to ensure that continuous improvement is occurring.

In addition to these tasks, I present two issues for consideration: price and time to volume. In a previous section it was mentioned that price was no longer the primary focus of the purchasing equation. However, this is not entirely accurate. Manufacturers of products destined for high volume use applications such as consumer products (video games, DVD players, mobile communication devices, and the like) and personal computers select and trade suppliers solely on price. (Questions such as what is the best price I can get for a DRAM, a microprocessor, a disk drive, etc., are routine. In fact many OEMs for these products are so bold as to state, "I expect the price for your DRAM to be $W this quarter, $X the next quarter, $Y the following quarter, and $Z four quarters out. If you can't meet these prices we will take our business to someone who can.") When many millions of components are used in manufacturing one of these products, several cents shaved off the price of true commodity components affects the product cost and a company's gross margin. For more complex and less price-sensitive products/equipment, component price is not the primary focus, although it certainly is important.

In Chapter 1 I talked about time to market and customer-centered issues. Just as important as time to market is time to volume. Can an IC supplier meet IBM's, Cisco System's, Nokia's, Dell Computer's, or Sony's huge volume ramp reqirements? Being able to deliver samples and early production quantitites is well and good, but can volume ramp up and production be sustained at the levels required by the OEM to produce and ship the product? This is a serious consideration not to be taken lightly.

A typical commodity team might include representatives from Purchasing, Supplier Quality Engineering, Development Engineering, Reliability Engineering, Component Engineering, and Manufacturing Engineering. The technical requirements are so important that in many organizations, Purchasing contains an embedded Engineering organization. In the commodity team structure each

Component and Supplier Selection

functional discipline brings various strengths and perspectives to the team relationship. For example,

1. *Purchasing* provides the expertise for an assured source of supply, evaluates the suppliers financial viability, and helps architect a long-term sourcing strategy.
2. *System Engineering* evaluates and ensures that the part works in the product as intended.
3. *Manufacturing Engineering* ensures that the parts can be reliably and repeatedly attached to the PCB and conducts appropriate evaluation tests of new PCB materials and manufacturing techniques, packaging technology (BGA, CSP, etc.), and attachment methods (flip chip, lead free solder, etc.).
4. *Component Engineering* guides Develop Engineering in the selection and use of technologies, specific IC functions, and suppliers beginning at the conceptual design phase and continuing through to a fixed design; evaluates supplier-provided test data: conducts specific analyses of critical components; generates functional technology road maps; and conducts technology audits as required.
5. *Reliability Engineering* ensures that the product long-term reliability goals, in accordance with customer requirements, are established; that the supplier uses the proper design and derating guidelines; and that the design models are accurate and in place.

Each commodity team representative is responsible for building relationships with her/his counterparts at the selected component suppliers.

4.4 THE ROLE OF THE COMPONENT ENGINEER

The component engineer (CE) (see also Chapter 3) is a resource for product development and strategic planning. The focus of the CE is on matching the company's product direction with the direction of the supplier base. The most effective development support that can be provided by the CE is to implement fast and effective means of evaluating new products and technologies before they are needed, provide product road maps, participate in design reviews, and provide component application assistance. The CE has traditionally been a trailer in the product development cycle by qualifying components that are already designed into new products. Lets look at the evolution of the CE function.

In the 1960s and and 1970s, the job of the component engineer was to write specifications and issue component part numbers. The specifications were used by Receiving Inspection to identify and measure important parameters. The component engineer was not responsible for component quality; component quality was the responsibility of the quality organization and was controlled by test and measurement techniques with an objective of finding defective components. The average defect rate was 1 to 5%. For example, a 1% defect rate for a 1000-piece lot of components received translated into an average of 10 bad parts. Screening material at incoming was an effective and practical, although expensive, way of finding defective components, but it did nothing to institute/drive continuous improvement on the supplier's part. By this practice, the supplier was not held accountable for the quality of the components.

In the 1980s, the scope of responsibility of the component engineer expanded to include quality as well as specification and documentation of components. Component engineers worked with quality engineers to improve product quality. During the 1980s, the average component defect rate was reduced from 10,000 ppm to around 100 ppm. This reduction in defect rate occurred at the same time that devices became more complex. Electrically testing these complex devices became more and more expensive, and the number of defects found became less and less. A lot of 1000 parts received would yield 0.1 defects, or 100 ppm. Another way of looking at it is that it would now take 10,000 parts to find a single defect versus finding 1 defect in 100 parts in the 1970s. Electrical testing at receiving inspection became more and more expensive (next to impossible for VLSI circuits) and was an ineffective means of controlling component quality, and so was discontinued.

Component engineers decided that component quality was best controlled at the supplier's site. Thus the suppliers should be responsible for quality, they should regularly be assessed by means of on-site audits, and they should provide process capability indices (Cp and Cpk) for critical processes and electrical performance parameters. This provided an acceptable means of assessing a supplier's ability to consistently build good components. Component engineers also discovered that construction analysis was an effective means of evaluating the quality of execution of a component supplier's manufacturing process. Thus, both audits and construction analyses were added to the qualification process. Because of the increased cost and complexity of ICs, qualification tests per MIL-STD-883 method 5005 or EIA/JESD 47 (Stress-Test-Driven qualification of Integrated Circuits) were found to add little value, and their use by OEMs was terminated by the late 1980s. Wafer level reliability tests, in-line process monitors, and step stress-to-failure methods were used by the IC suppliers instead.

In the 1990s, component engineers de-emphasized the use of on-site audits and construction analyses, devices became even more costly and complex, and pin counts rose dramatically. New package types (such as BGAs and CSPs) added new complications to device qualification and approval. Many devices were custom or near custom (PLDs, FPGAs, and ASICs), and because of their associated tool sets it became easier to develop applications using these devices. With this change came still higher component quality requirements and reduced defect rate; the average defect rate was approximately 50 ppm. A lot of 1000 parts received would yield 0.05 defective components. That is, it would take 50,000 parts to find a single defect versus finding 1 defect in 100 parts in the 1970s.

A dramatic improvement in the quality of integrated circuits occurred during the past 30–40 years. Since the early 1970s, the failure rate for integrated circuits has decreased aproximately 50% every 3 years! A new phenomenon began to dominate the product quality area. A review of OEM manufacturing issues and field returns showed that most of the problems encountered were not the result of poor component quality. Higher clock rate and higher edge rate ICs as well as impedance mismatches cause signal integrity problems. Component application (fitness for use) and handling processes, software/hardware interaction, and PWA manufacturing quality (workmanship) and compatibility issues are now the drivers in product quality. Many misapplication and PWA manufacturing process problems can be discovered and corrected using HALT and STRIFE tests at the design phase. Furthermore, many component- and process-related defects can be discovered and controlled using production ESS until corrective actions are in place.

4.4.1 Beyond the Year 2000

The traditional tasks of the component engineer do not fit in the current business model; therefore a new paradigm is needed. To meet the challenges of the 21st century, the component engineer's role in generating specifications and issuing part numbers is minimal. In the years since the referenced paper was written much has changed. Readily available component and technology databases have been established across the industry on the Internet and are easily licensed and downloaded to the component and development engineer's desktops, easing much of the research and documentation required for new components and suppliers. The services offered by these Internet database providers include bill-of-materials optimization, component and sourcing strategies, component availability and lifecycle issues, and price tracking. The amount of information available on the Internet allows more efficient use of both the design and component engineers' time.

The role of the component engineer has changed from an emphasis on device specification and reactive quality control to a more global concept that includes active up-front design involvement, Alpha/Beta application testing, manufacturing, specification, and quality control. Component engineering has moved from a back-end reactionary and documentation mindset function to proactive involvement at the conceptual phase of a design. Technology and product road maps are an important part of the component engineer's tool set in providing source and component recommendations that satisfy both the circuit performance requirements and the issues of manufacturability and an assured source of supply. The CE must be a forward-looking resource for product development and strategic planning. The focus is on matching a company's product direction with the direction of the supply base (see Chapter 3, Fig. 2). Throughout the design process, the component engineer questions the need for and value of performing various tasks and activities, eliminating all that are non–value added and retaining the core competencies. As more IC functions are imbedded into ASICs via coreware, the coreware will have to be specified, modeled, and characterized, adding a new dimension to a CE's responsibility.

The component engineer is continuously reinventing the component qualification process (more about this in a subsequent section) in order to effectively support a fast time-to-market requirement and short product development and life cycles. The CE has traditionally been a trailer in the product development cycle by qualifying components that are designed into new products. Now components need to be qualified ahead of the need, both hardware and software—and their interactions.

A listing of the component engineer's responsibilities is as follows:

1. Eliminate non–value added activities.
   Reduce the time spent on clerical activities and documentation.
   Eliminate the development of SCDs for standard off-the-shelf components unless special testing is required.
   Create procurement specifications only for special/custom components.
2. Implement the use and placement of Internet-driven component databases on designer desktops or work stations. These databases will be expanded to include previous component/supplier useage history, field experience, qualification information, application information, and lessons learned.
3. Adopt the use of standard off-the-shelf components (such as those used in PCs, networking, telecommunication equipment, and mobile applications) as much as possible since they drive the marketplace (i.e., IC manufacturing) by their volume.
4. Use of ASICs and PLDs where they provide design flexibility, product performance, and advantage and thus competitive marketplace advantage.
5. Provide product definition and development support—early involvement in technology and functional component assessment and selection in the conceptual design phase.
   Work closely with design teams and development labs to mold technology needs, product direction, and component and supplier selection.
   Provide component application assistance.
   Disseminate basic technology and road map information to the com-

Component and Supplier Selection 201

pany via presentations and the use of the intracompany worldwide web. Examples of these technologies include PCI, HSTL, GTL, Fiber Channel, ATM, SSA, I2O, Optimal interconnect, etc.

Conduct ongoing component/supplier selection application reviews (i.e., bill of material reviews) throughout the design process to ensure that components and suppliers selected are fit for the application, won't go end-of-life for the projected product life, are manufacturable, and can meet delivery, quality, and reliability needs.

6. Coordinate component and technology qualification.

Develop a qualification plan for strategic components.

Begin early qualification of technologies, devices, and package styles that will be required for new designs when the parts become available such as low-voltage (1.2-V) logic—before they are needed in product design.

Ensure that the component is qualified and available just as the circuit design is complete, insert the new component in the product, and ship it.

Obtain device SPICE models for signal integrity analysis and use in system modeling and evaluation.

Determine the need for construction analyses, special studies, and design reviews and coordinate their conduct.

Define combinational "smorgy" test plans (mixing and matching suppliers on a PWA) for product test.

7. Manage supplier design usage.

Consider current suppliers first. Do the current suppliers' technology and component road maps match ours? Is there a good fit? Are there other suppliers we should be using?

Only approved suppliers will be used for new designs. Suppliers not on the approved suppliers list (ASL) need to be justified for use and pass the supplier approval process. If the business dictates (lower price or increased performance), new suppliers will be considered.

Reduce the number of supplier audits by eliminating the need to audit the top tier suppliers. Institute supplier self-audits in lieu of customer audits for top tier suppliers.

Focus resources on components that pose the greatest quality or design risks, greatest areas for improvement, and/or those that are strategic to a given product design.

8. Provide manufacturing support.

Work with suppliers on quality and availability issues. Ensure that the right devices are used for application requirements. Ensure that parts meet safety requirements and manufacturing requirements, such as PWA water wash and reducing the number of device families to support.

Work with Development, Manufacturing, and Purchasing on cost reduction programs for existing and new designs. Focus on Total cost of ownership in component/supplier selection.

9. Provide CAD model support. Component engineers MCAD, and ECAD designers should combine efforts to create singular and more accurate component models. The MCAD and ECAD models will become the specification control vehicle for components in replacement of SCDs.
10. Coordinate the establishment of quality (ppm) and reliability (FIT) goals for critical, core, or strategic commodites including a 3-year improvement plan. Establish a plan to monitor commodity performance to these quality and reliability goals on a quarterly basis.
11. Institute obsolescence BOM reviews for each production PWA every 2 years. Institute obsolescence BOM reviews when a portion of an existing board design will be used in a new design.

4.5 COMPONENT SELECTION

As discussed in Chapter 3, the right technology and functional implementation (i.e., specific component) must be selected based on the application requirements. The type of component selected is a double-edged sword and can significantly impact a product manufacturer's ability to compete in the marketplace. For example, in a low-volume product application market such as military/aerospace or high-end computer manufacturing a very viable component strategy is to use as many industry standard off-the-shelf components as possible. The advantages this provides are low price, high availability, low risk, manufacturing process stability due to high volumes produced, high quality and reliability, common building blocks, and fast product ramping. High-volume industry standard ICs tend to have sufficient design margins and minimum variability of critical parameters indicating good manufacturing (wafer fab) process control. However, when multiple sources are used for a given component there is significant variability between all sources of supply. Mixing and matching various suppliers' components can lead to timing violations due to variability and lack of margin, for example. Essentially the low-volume users "piggyback" on the high-volume users who drive the IC marketplace and who resolve any issues that occur. The disadvantage of this strategy is that anyone can use the components in their designs. Thus, all competing hardware designs could end up looking alike (true commodity products) with there being no real leverage, competitive performance advan-

Component and Supplier Selection 203

tage, or differentiating factors between the various manufacturers' product offerings, such as is the case for personal computers.

On the flip side of this issue is the competitive advantage and differentiation provided by the use of customizable application specific ICs (ASICs), PLDs, and systems on a chip (SOCs). They provide the circuit designer with flexibility, performance advantages, and reduced size. But for the low-volume product markets, the device useage is small; there is no benefit to be gained from the quality improvement inherent in large manufacturing (wafer fab) runs; and because of the small quantities purchased the component prices will tend to be high. Also, since many of these high-leverage parts are designed by fabless IC companies involving multiple outsource providers (wafer fab, assembly, electrical test), several questions arise: who, how, when, and how fast with regards to problem resolution and who has overall IC quality responsibility? Conversely, due to the long product lifetimes for these product markets, the use of ASICs and PLDs provides a solution to the escalating obsolescence of standard off-the-shelf ICs and the subsequent expensive and time-consuming system requalification efforts required for substitute parts.

Today, there is also the issue of leverage in bringing products to market rapidly. Printed wire assembles are being designed using ASICs, PLDs, and SOCs as placeholders before the logic details are determined, allowing a jumpstart on PWA design and layout. This is contrary to waiting until the circuit design (actual components used) is fixed using standard off-the-shelf components before PWA layout begins. Again the issues are market leverage, product performance, and time to market. Another problem is that incurred in using ASICs. Normally, several "spins" of a design are required before the IC design is finalized, resulting in high nonrecurring engineering (NRE) and mask charges being incurred by the product manufacturer. Many designers have turned to the use of programmable devices [PLDs and field programmable gate arrays (FPGAs)] because of their flexibility of use and the fact that device densities, performance, and price have approached those of ASICs. Another strategic decision needs to be made as well. For which components are multiple sources both available and acceptable for use in a given design? For which components will a single-source solution be used? These decisions are important from a risk perspective. Using a single-sourced component may provide a performance advantage, but at the same time may involve increased risk from a yield perspective and in being able to ensure a continuous source of components in the manufacturing flow. This last point is key because often (but not always) unique sole-sourced components (which provide product functional and performance advantages) are provided by smaller boutique suppliers that do not have the fab capacity arrangements and quality infrastructure in place to provide the level of support required by the OEM. Multiple suppliers for any component increase the number of variables in the product and manufacturing process, both of which make it difficult to ensure a consistent product.

4.6 INTEGRATED CIRCUIT RELIABILITY

The reliability of ICs critical to equipment operation (primarily microprocessors, memories, ASICs, and FPGAs) by and large determines (or drives) equipment (product) reliability. This section traces the historical perspective of IC quality and reliability and how it evolved to where we are today. I then discuss accelerated (life) testing and how it is used to estimate fielded reliability.

4.6.1 Historical Perspective: Integrated Circuit Test

Accelerated testing has been extensively used for product improvement, for product qualification, and for improving manufacturing yields for several decades. Starting with the lowest level in the electronics equipment food chain, accelerated testing was rigorously applied to the packaged integrated circuit, for economic reasons. Component screening has been a way of life for the military/aerospace community, led by NASA and the U.S. Air Force, since the inception of the integrated circuit industry in the late 1950s. Almost every major computer, telecommunications, and automotive manufacturer performed accelerated stress testing on all purchased integrated circuits until major quality improvements became widely evident.

The reason why OEMs performed these seemingly non–value added screening requirements was that U.S. IC manufacturers had an essentially complacent, almost arrogant, attitude toward quality. "We'll decide what products you need and we'll provide the quality levels that we feel are appropriate." They turned a deaf ear to the user's needs. In the 1960s and 1970s, IC suppliers often had a bone pile, stored in 50-gallon drums, of reject devices at final electrical test. Additionally, the ICs that were delivered to customers had a high field failure rate. There was no concerted effort on the part of the IC suppliers to find the root cause of these in-house rejects or the high field failure rates experienced and institute a corrective action feedback loop to design and manufacturing.

As a result, the automotive industry and the U.S. computer industry (led to Burroughs/Unysis, Honeywell, and others), due to usage problems encountered with the first DRAMs (notably the Intel 1103), came to the conclusion that it would be prudent for them to perform 100% incoming electrical inspection, burn-in, and environmental stress testing. Other industries followed suit.

The decision to perform accelerated stress testing (AST) resulted in the creation of an open forum in 1970 for the discussion of electrical testing issues between IC manufacturers and users; the International Test Conference. Testing required an expensive capital equipment and technical personnel overhead structure at most large users of ICs. Some of the smaller users chose to outsource their incoming electrical inspection needs to independent third-party testing laboratories, thus fueling that industry's growth.

Up to the mid-1980s most users of integrated circuits performed some level of screening, up to the LSI level of integration, for the following reasons:

> They lacked confidence in all suppliers' ability to ship high-quality, high-reliability ICs.
>
> They felt some screening was better than none (i.e., self-protection).
>
> They embraced the economic concept that the earlier in the manufacturing cycle you find and remove a defect, the lower the total cost.

The last item is known as the "law of 10" and is graphically depicted in Figure 7. From the figure, the lowest cost node where the user could make an impact was at incoming test and thus the rationale for implementing 100% electrical and environmental stress screening at this node.

The impact of failure in the field can be demonstrated by considering some of the effects that failure of a computer in commercial business applications can have. Such a failure can mean

> The Internet going down
> A bank unable to make any transactions
> The telephone system out of order
> A store unable to fill your order
> Airlines unable to find your reservation
> A slot machine not paying off

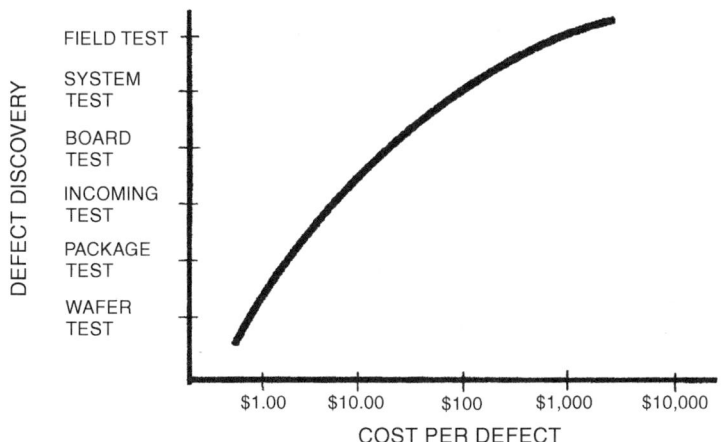

FIGURE 7 Cost of failure versus place of discovery for electronic devices.

TABLE 8 Cost of Failures

Type of business	Cost per hour of downtime
Retail brokerage	$6,450,000
Credit card sales authorization	$2,600,000
Home shopping channels	$113,700
Catalog sales center	$90,000
Airline reservation centers	$89,500
Cellular service activation	$41,000
Package shipping service	$28,250
Online network connect fees	$22,250
ATM service fees	$14,500

Source: IBM internal studies.

The dramatic increase in cost of failures when moving from identification at the component level to identification in the field is shown in Table 8. This is the overwhelming reason why accelerated stress testing was performed.

The arrogant independence of the IC suppliers came to a screeching halt in 1980 when Hewlett-Packard dropped a bombshell. They announced that Japanese 16K DRAMs exhibited one to two orders of magnitude fewer defects than did those same DRAMs produced by IC manufacturers based in the United States. The Japanese aggressively and insightfully implemented the quality teachings of such visionaries as Drs. W. Edwards Deming and Joseph Juran and consequently forever changed the way that integrated circuits are designed, manufactured, and tested. The Japanese sent a wake-up call to all of U.S. industry, not just the semiconductor industry. Shoddy quality products were no longer acceptable. They raised the bar for product quality and changed the focus from a domestic one to a global one.

The Hewlett-Packard announcement and subsequent loss of worldwide DRAM market share served to mobilize the U.S. IC industry to focus on quality. The result was a paradigm shift from an inspection mindset to a genuine concern for the quality of ICs produced. It was simply a matter of design and manufacture quality ICs or go out of business. Easy to say; longer and tougher to realize.

Also, in the 1980–1981 timeframe, the U.S. Navy found that over 50% of its fielded F14 aircraft were sitting on the flight line unavailable for use. Investigations traced the root cause of these unavailable aircraft to IC defects. As a result, Willis Willoughby instituted the requirement that all semiconductors used in Navy systems must be 100% rescreened (duplicating the screens that the manufacturer performs) until such time as each IC manufacturer could provide data showing that the outgoing defect level of parts from fab, assembly, and electrical test was less than 100 ppm. This brute force approach was necessary to show the semiconductor industry that the military was serious about quality.

Component and Supplier Selection

Since components (ICs) have historically been the major causes of field failures, screening was used

- To ensure that the ICs meet all the electrical performance limits in the supplier's data sheets (supplier and user).
- To ensure that the ICs meet the unspecified parameters required for system use of the selected ICs/suppliers (user).
- To eliminate infant mortalities (supplier and user).
- To monitor the manufacturing process and use the gathered data to institute appropriate corrective action measures to minimize the causes of variation (supplier).
- As a *temporary solution* until the appropriate design and/or process corrective actions could be implemented based on a root cause analysis of the problem (user or supplier) and until IC manufacturing yields improved (user).
- Because suppliers expect sophisticated users to find problems with their parts. It was true in 1970 for the Intel 1103 DRAM, it is true today for the Intel Pentium, and it will continue to be true in the future. No matter how many thousands of hours a supplier spends developing the test vectors for and testing a given IC, all the possible ways that a complex IC will be used cannot be anticipated. So early adopters are relied on to help identify the bugs.

As mentioned, this has changed—IC quality and reliability have improved dramatically as IC suppliers have made major quality improvements in design, wafer fabrication, packaging and electrical test. Since the early 1970s IC failure rates have fallen, typically 50% every 3 years. Quality has improved to such an extent that

1. Integrated circuits are not the primary cause of problems in the field and product failure. The major issues today deal with IC attachment to the printed circuit board, handling issues (mechanical damage and ESD), misapplication or misuse of the IC, and problems with other system components such as connectors and power supplies. Although when IC problems do occur, it is a big deal and requires a focused effort on the part of all stakeholders for timely resolution and implementation of permanent corrective action.
2. Virtually no user performs component screening. There is no value to be gained. With the complexity of today's ICs no one but the IC supplier is in a position to do an effective job of electrically testing the ICs. The supplier has the design knowledge (architectural, topographical, and functional databases), resources (people) who understand the device operation and idiosyncracies, and the simulation and the test

tools to develop the most effective test vector set for a given device and thus assure high test coverage.
3. United States IC suppliers have been continually regaining lost market share throughout the 1990s.
4. Many failures today are system failures, involving timing, worst case combinations, or software–hardware interactions. Increased system complexity and component quality have resulted in a shift of system failure causes away from components to more system-level factors, including manufacturing, design, system-level requirements, interface, and software.

4.6.2 Failure Mechanisms and Acceleration Factors

Integrated circuits are not the primary cause of product failure, as they were in the 1960s through the 1980s. The reason for this improvement is that IC suppliers have focused on quality. They use sophisticated design tools and models that accurately match the circuit design with the process. They have a better understanding and control of process parameters and device properties. There is a lower incidence of localized material defects; greater attention is paid to detail based on data gathered to drive continuous improvement; and when a problem occurs the true root cause is determined and permanent corrective action is implemented. Figure 8 shows this improvement graphically using the bathtub failure rate curve. Notice the improved infant mortality and lower useful life failure rates due to

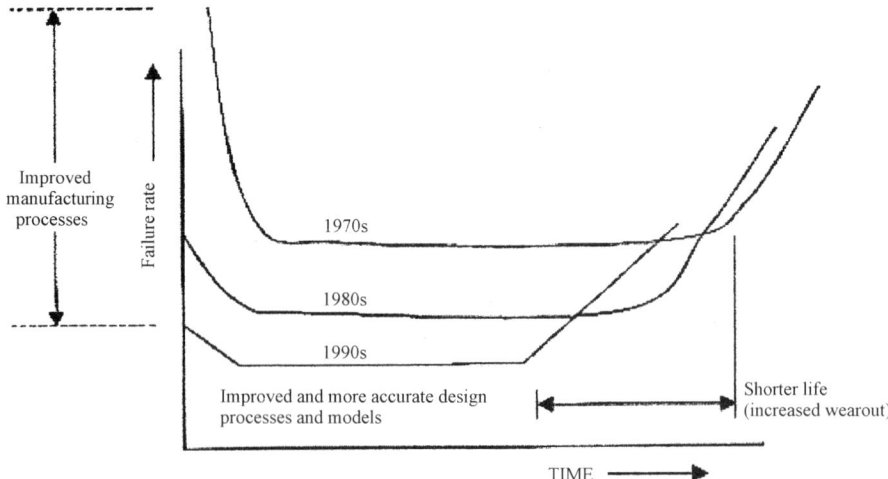

FIGURE 8 The impact of advanced technology and improved manufacturing processes on the failure rate (bathtub) curve.

Component and Supplier Selection

improved manufacturing processes and more accurate design processes (design for quality) and models. Also notice the potentially shorter life due to smaller feature size ramifications (electromigration and hot carrier injection, for example).

Nonetheless failures still do occur. Table 9 lists some of the possible failure mechanisms that impact reliability. It is important to understand these mechanisms and what means, if any, can be used to accelerate them in a short period of time so that ICs containing these defects will be separated and not shipped to customers. The accelerated tests continue only until corrective actions have been implemented by means of design, process, material, or equipment change. I would like to point out that OEMs using leading edge ICs expect problems to occur. However, when problems occur they expect a focused effort to understand the problem and the risk, contain the problem parts, and implement permanent corrective action. How issues are addressed when they happen differentiates strategic suppliers from the rest of the pack.

TABLE 9 Potential Reliability-Limiting Failure Mechanisms

Mobile ion contamination
   Impure metals and targets
   Manufacturing equipment
Metals
   Electromigration (via and contact)
   Stress voiding
   Contact spiking
   Via integrity
   Step coverage
   Copper issues
Oxide and dielectric layers
   TDDB and wearout
   Hot carrier degradation
   Passivation integrity
   Gate oxide integrity (GOI)
   Interlayer dielectric integrity/delamination
EOS/ESD
Cracked die
Package/assembly defects
   Wire bonding
   Die attach
   Delamination
   Solder ball issues
Single event upset (SEU) or soft errors
   Alpha particles
   Cosmic rays

TABLE 10 Example of Acceleration Stresses

Voltage acceleration
   Dielectric breakdown
   Surface state generation
   Gate-induced drain leakage current
   Hot carrier generation, injection
   Corrosion
Current density acceleration
   Electromigration in metals
   Hot-trapping in MOSFETs
Humidity/temperature acceleration
   Water permeation
   Electrochemical corrosion

Integrated circuit failure mechanisms may be accelerated by temperature, temperature change or gradient, voltage (electric field strength), current density, and humidity, as shown in Tables 10 and 11.

For more information on IC failure mechanisms, see Ref. 2.

Temperature Acceleration

Most IC failure mechanisms involve one or more chemical processes, each of which occurs at a rate that is highly dependent on temperature; chemical reactions and diffusion rates are examples of this. Because of this strong temperature dependency, several mathematical models have been developed to predict temperature dependency of various chemical reactions and determine the acceleration factor of temperature for various failure mechanisms.

The most widely used model is the Arrhenius reaction rate model, determined empirically by Svante Arrhenius in 1889 to describe the effect of temperature on the rate of inversion of sucrose. Arrhenius postulated that chemical reactions can be made to occur faster by increasing the temperature at which the reaction occurs. The Arrhenius equation is a method of calculating the speed of reaction at a specified higher temperature and is given by

$$r(T) = r_0 \, e^{-E_a/kT} \qquad (4.1)$$

where
    A = constant
    e = natural logarithm, base 2.7182818
    E_a = activation energy for the reaction (eV)
    k = Boltzmann's constant = 8.617×10^{-5} (eV/K)
    T = temperature (K)

TABLE 11 Reliability Acceleration Means

Concerns for the performance of product elements	Examples of mechanisms	Stressors
Electrical parameter stability of active devices	Trapping of injected charges (hot carrier effects)	Electrical field strength Temperature
	Drift of ionic contaminants	Electrical field strength Temperature
Time-dependent breakdown of dielectrics (thin oxides)	Trapping of charges	Electrical field strength Temperature
Endurance of conductors, contacts	Interdiffusion of different metals with growth of voids	Electrical field strength Temperature
	Electromigration	Current density Temperature
Robustness of product construction		
Thermomechanical mismatch of product elements	Cracking Mechanical fatigue at material interfaces	Temperature interval Number cycles at temperature interval
Hermeticity	Corrosion	Relative humidity

Today the Arrhenius equation is widely used to predict how IC failure rates vary with different temperatures and is given by the equation

$$R_1 = R_2 \, e^{Ea/k(1/T_1 - 1/T_2)} \tag{4.2}$$

Acceleration is then

$$A_T = e^{Ea/k(1/T_1 - 1/T_2)} \tag{4.3}$$

where

T_1 and T_2 = temperature (K)
k = Boltzmann's constant = 86.17 μeV/K

These failure mechanisms are primarily chemical in nature. Other acceleration models are used for nonchemical failure mechanisms, as we shall shortly see. The activation energy Ea is a factor that describes the accelerating effect that temperature has on the rate of a reaction and is expressed in electron volts (eV). A low value of Ea indicates a reaction that has a small dependence on temperature. A high value of Ea indicates a high degree of temperature depen-

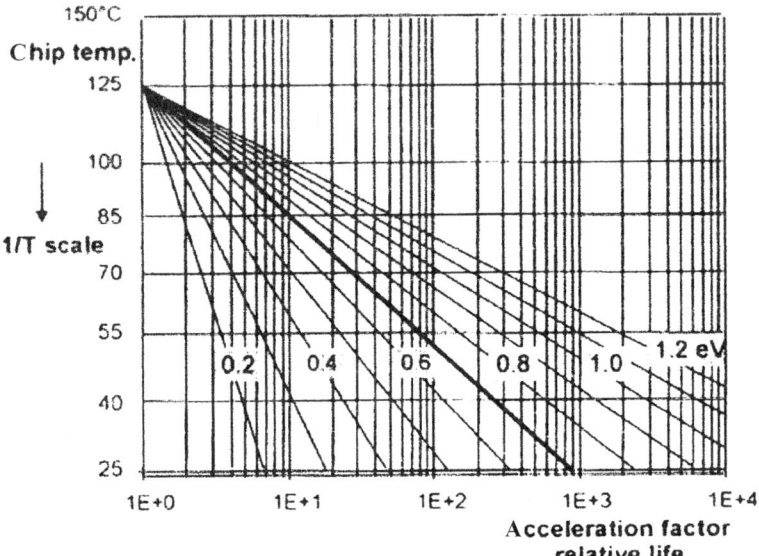

FIGURE 9 The Arrhenius model showing relationship between chip temperature and acceleration factor as a function of Ea.

dence and thus represents a high acceleration factor. Figure 9 shows the relationship between temperature, activation energy, and acceleration factor.

Voltage Acceleration

An electric field acceleration factor (voltage or current) is used to accelerate the time required to stress the IC at different electric field levels. A higher electric field requires less time. Since the advent of VLSICs, voltage has been used to accelerate oxide defects in these CMOS ICs such as pinholes and contamination. Since the gate oxide of a CMOS transistor is extremely critical to its proper functioning, the purity and cleanliness of the oxide is very important, thus the need to identify potential early life failures. The IC is operated at higher than normal operating V_{DD} for a period of time; the result is an assigned acceleration factor A_V to find equivalent real time. Data show an exponential relation to most defects according to the formula

$$A_V = e^{\gamma(V_S - V_N)} \tag{4.4}$$

where

V_S = stress voltage on thin oxide
V_N = thin oxide voltage at normal conditions
γ = 4–6 volt^{-1}

Humidity Acceleration

The commonly used humidity accelerated test consists of 85°C at 85% RH. The humidity acceleration formula is

$$A_H = e^{0.08(RH_s - RH_n)} \tag{4.5}$$

where
RH_s = relative humidity of stress
RH_n = normal relative humidity

For both temperature and humidity accelerated failure mechanisms, the acceleration factor becomes

$$A_{T\&H} = A_H A_T \tag{4.6}$$

Temperature Cycling

Temperature cycling, which simulates power on/off for an IC with associated field and temperature stressing is useful in identifying die bond, wire bond, and metallization defects and accelerates delamination. The Coffin–Manson model for thermal cycling acceleration is given by

$$A_{TC} = \frac{[\Delta T_{stress}]^c}{\Delta T_{use}} \tag{4.7}$$

where c = 2–7 and depends on the defect mechanism. Figure 10 shows the number of cycles to failure as a function of temperature for various values of c.

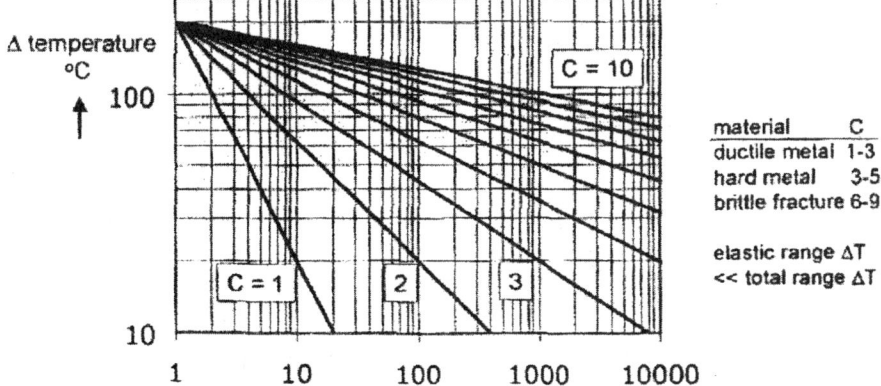

FIGURE 10 Coffin-Manson model showing the number of stress cycles as a function of temperature change or gradient.

The total failure rate for an IC is simply the mathematical sum of the individual failure rates obtained during the various acceleration stress tests, or

$$F_{total} = F_T + F_{VDD} + F_{T\&H} + F_{TC} \tag{4.8}$$

where
- F_T = failure rate due to elevated temperature
- F_{VDD} = failure rate due to accelerated voltage (or electric field)
- $F_{T\&H}$ = failure rate due to temperature humidity acceleration
- F_{TC} = failure rate due to temperature cycling

Temperature Acceleration Calculation Example

Reliability defects are treated as chemical reactions accelerated by temperature. Refining Eq. (4.3) for temperature acceleration factor, we get

$$T_{AF} = \frac{r_2}{r_1} = e^{Ea/k[1/T1-1/T2]} \tag{4.9}$$

Conditions: T_{BI} = 125°C, T_N = 55°C, k = 86.17 μeV/K.
For Ea = 0.3 eV,

$$T_{AF} = e^{0.3/86.17[1/328-1/398]} = 6.47$$

For Ea = 1.0 eV,

$$T_{AF} = e^{1.0/86.17[1/328-1/398]} = 504.1$$

What does this mean? This shows that a failure mechanism with a high activation energy is greatly accelerated by high temperature compared to a failure mechanism with a low activation energy.

Voltage Acceleration Calculation Example

Using the electric field form of Eq. (4.4), we get

$$V_{AF} = e^{(E_S-E_N)/E_{EF}}$$

where
- E_S = stress field on thin oxide (mV/cm)
- E_N = stress field at thin oxide at normal conditions
- E_{EF} = experimental or calculate (Suyko, IRPS'91)(MV/cm)

and

$$V_{AF} = f(t_{OX,\ process}) = \frac{1}{\ln(10)^\gamma} \quad \text{and} \quad \gamma = 0.4\exp(0.07/kT) \tag{4.10}$$

Component and Supplier Selection

Conditions: $t_{OX} = 60$ Å, $V_N = 3.3$ V, $V_s = 4.0$ V, $E_S = 6.67$ MV/cm, $E_N = 5.5$ MV/cm, $E_{EF} = 0.141$ MV/cm. Then

$$V_{AF} = e^{(6.67-5.5)/0.141} = 3920$$

If $V_s = 4.5$ V and the other conditions remain the same, then

$$V_{AF} = e^{[7.5-5.5]/0.141} = 1.45 \times 10^6$$

This shows the greater acceleration provided by using a higher voltage (E field) for a failure mechanism with a high voltage activation energy.

Integrated Circuit FIT Calculation Example

A practical example is now presented using temperature and voltage acceleration stresses and the activation energies for the encountered failure mechanisms to calculate the reliability of a specific IC. (This example provided by courtesy of Professor Charles Hawkins, University of New Mexico.)

Three different lots of a given IC were subjected to an accelerated stress life test (two lots) and a high voltage extended life test (one lot). The results of these tests are listed in the following table.

Stress: 125°C and 5.5V

	48 hr	168 hr	500 hr	1000 hr	2000 hr
Lot 1	1/1000	1/999	1/998	0/998	4/994
Lot 2	0/221	0/201	2/201	0/100	1/99
Total	1/1221	1/1200	3/1199	0/1098	5/1093
Failure code:		A	B	C	D

Stress: high voltage extended life test (HVELT): 125°C and 6.5 V

	48 hr	168 hr	500 hr
Lot 3	0/800	1/800 B (oxide)	0/799

The way you read the information is as follows: for lot 1, 1/999 at the 168-hr electrical measurement point means that one device failed. An investigation of the failure pointed to contamination as the root cause (see Step 2).

A step-by-step approach is used to calculate the FIT rate from the experimental results of the life tests.

Step 1. Organize the data by failure mechanism and activation energy.

Code	Failure analysis	Ea (eV)
A	1 contamination failure	1.0
B	1 charge loss failure	0.6
	1 oxide breakdown failure	0.3
C	1 input leakage (contamination) failure	1.0
D	3 charge loss failures	0.6
	2 output level failures	1.0

Step 2. Calculate the total device hours for Lots 1 and 2 excluding infant mortality failures, which are defined as those failures occurring in the first 48 hr.

Total device hours = $1200(168) + 1199(500-168) + 1098(1000 - 500) + 1093(2000-1000) = 2.24167 \times 10^{-6}$ hr

For Lot 3,

$800(168) + 799(500-168) = 3.997 \times 10^5$ hr

Calculate T_{AF} for Ea = 0.3 eV, $T_2 = 398.15$ K, and $T_1 = 328.15$ K.

$R_2 = R_1 e^{(0.3/86.17 \times 10^{-6}(1/328.15 - 1/398.15)}$
$R_2 = R_1 e^{1.8636} = R_1(6.46)$
$T_{AF} = \dfrac{R_2}{R_1} = 6.46$

The T_{AF} values for Ea = 0.6 eV and 1.0 eV are calculated in a similar manner. The results are as follows:

Ea(eV)	T_{AF}
0.3	6.46
0.6	41.70
1.0	501.50

Step 3. Arrange the data:

Ea	Device hours at 125°C	T_{AF} at 55°C	Equivalent at 55°C
0.3	2.24167×10^6	6.46	1.44811×10^7
0.6	2.24167×10^6	41.70	9.34778×10^7
1.0	2.24167×10^6	501.50	1.12420×10^9

Component and Supplier Selection

Step 4. Divide the number of failures for each Ea by the total equivalent device hours:

Ea	Device hours	Fails	% fail/1000 device hours
0.3	1.44811×10^7	1	0.0069
0.6	9.34778×10^7	4	0.0042
1.0	1.12420×10^9	4	0.00036
			0.01146, or 114.6 FITs

HVELT data, Lot 3: 125°C, 6.5 V, $T_{OX} = 150$ Å.

48 hr	168 hr	500 hr
0/800	1/800 B	0.799

Note: B is oxide failure at 0.3 eV thermal activation.

Total device hours = 3.997×10^5 ($T_{AF} = 6.46$)

V_{AF} must also be calculated since the accelerated voltage was 6.5 V versus the normal 5.5 V.

Thus,

$$V_{AF} = e^{(4.33-3.67)/0.141} = 113$$

Include Lot #3 failure data with Lots 1 and 2 data.

Device hours	Fails	Acceleration factor
3.997×10^5	1	$6.46 \times 113 = 2.9175 \times 10^8$ equivalent device hours

Calculate the number of total oxide failures in their time: two fails gives 6.5 FITs [$2 \div (1.44811 \times 10^7 + 2.9175 \times 10^8$ hours)].
Therefore, the total failure rate is $6.5 + 42 + 3.6 = 52$ FITs.

4.7 COMPONENT QUALIFICATION

Qualification tests are conducted to make a prediction about the fielded reliability of a product and its chance of success in fulfilling its intended application. Component qualification was established as an experimental means to answer the following two questions.

1. Will the component function in the application as designed?
2. Will the component function as intended for the life of the product?

Since specific ICs are typically the most critical components in a given product design (providing its market differentiation), the discussion on qualification will be focused specifically on ICs. However, the underlying points made and concepts presented are applicable to the qualification of all components.

We look first at what was done in the past in qualifying components to give us a perspective on where we've come from. Next I examine current qualification methodologies and follow this with a discussion of what will likely be required in the future. Finally, I provide a list of items to consider in developing a component qualification strategy.

4.7.1 Qualification Testing in the Past

Integrated circuit qualification tests prior to the 1990s were one-time events conducted by the user (OEM) or by the supplier for a specific customer under the customer's direction. They used a one-size-fits-all stress-driven methodology based on MIL-STD-883, rather than on specific application requirements. Table 12 lists the typically conducted qualification tests. Notice that 1000-hr operating life and storage tests are included to make an assessment of the long-term reliability of the IC.

Qualification testing was typically conducted just prior to release to production; it was long (the typical time to complete device qualification testing was 4–6 months or longer), costly (direct costs varied from $10,000 to $50,000 or more per device/package type and costs associated with system testing and product introduction delays added several orders of magnitude to this number), and

TABLE 12 Typical IC Qualification Tests

Test	Condition
Operating life	125°C, 1000 hr
High-temperature storage	T_{max}, 1000 hr
Moisture resistance	85°C, 85% RH, 1000 hr
Temperature cycling	−55°C to 125°C, 2000 cycles
Thermal shock	100°C to 0°C, 15 cycles
Lead integrity	225 g, 90°, 3 times
Mechanical shock	1500G, 0.5 ms, 3 times each axis (x,y,z)
Constant acceleration	20,000G, 1 min each axis (x,y,z)

Component and Supplier Selection

TABLE 13 Traditional IC Qualification Test Practices

Stress test response–driven qualification uses a one-size-fits-all test mindset.
The wrong tests are performed.
The stresses aren't compatible with nor do they provide an assessment of current technology.
The tests can be performed incorrectly.
The tests can be interpreted incorrectly.
The tests are statistically insignificant and do not reflect process variations over the IC's life. The small sample size drawn from one or two production runs yielded results that were statistically meaningless and incapable of identifying any problems associated with wafer fabrication or assembly process variations. If the qualification tests were conducted on preproduction or "learning lots," no valid interpretation of the data gathered could be made.
The tests are too expensive to perform.
The tests take too long to perform, negatively impacting time to market.
The standards require using a large sample size resulting in too many ICs being destroyed.
The tests don't include application requirements.
The test structure doesn't take into account the impact of change (design, wafer fab, assembly, and test processes).

destroyed many expensive ICs in the process (an ASIC, FPGA, or microprocessor can cost between $500 and $3,000 each). For an ASIC, one could use (destroy) the entire production run (1–3 wafers) for qualification testing—truly unrealistic. Table 13 lists the issues with the traditionally used stress test–driven qualification test practices.

When the leadership in the military/aerospace community decided that the best way for inserting the latest technological component developments into their equipment was to use commercial off-the-shelf (COTS) components, the commercial and semiconductor industries looked to EIA/JESD 47 and JESD 34 to replace MIL-STD-883 for IC qualification testing. The following is a brief discussion of these standards.

Stress Test–Driven Qualification

EIA/JESD Standard 47 (stress test–driven qualification of integrated circuits) contains a set of the most frequently encountered and industry-accepted reliability stress tests. These tests (like those of MIL-STD-883), which have been found capable of stimulating and precipitating IC failures in an accelerated manner, are used by the IC supplier for qualifying new and changed technologies, processes, product families, as well as individual products.

The standard is predicated on the supplier performing the appropriate tests as determined by the IC's major characteristics and manufacturing processes. It states that each qualification project should be examined for

1. Any potential new and unique failure mechanisms
2. Any situations where these tests/conditions may induce falures

In either case the set of reliability requirements and tests should be appropriately modified to properly address the new situations in accordance with JESD 34.

Failure Mechanism–Driven Reliability Qualification

JESD 34 (failure mechanism–driven qualification of silicon devices) provides an alternative to traditional stress-driven qualification for mature products/manufacturing processes. (As a side note, both JESD 34 and EIA/JESD 47 were developed by an industry consortium consisting mainly of IC suppliers and thus have their buy-in.) This standard is predicated on accepting the qualification process performed by the supplier using qualification vehicles rather than actual product. Thus, JESD 34 does not address qualification of product, quality, or functionality. The standard is based on the fact that as failure rates (and thus the number of detectable defects) become smaller, the practice of demonstrating a desired reliability through the use of traditional stress-driven qualification tests of final product becomes impractical. The burden for this method of qualification falls on the shoulders of the IC supplier (as it should) through an understanding of the wafer fab and assembly processes, potential failure mechanisms of familiar products/processes; and the implementation of an effective in-line monitoring process of critical parameters. The standard provides a typical baseline set of reliability qualification tests.

From an OEM perspective there are several problems with this approach, including

> The supplier is not in a position to determine the impact that various failure mechanisms have on system performance.
> The method does not take into account functional application testing at the system level, which is really the last and perhaps most important step in component qualification.

Also, OEMs are active participants of failure mechanism–driven qualification by committing to collect, analyze, and share field failure data with the supplier. Then, from this data, the supplier identifies those failure mechanisms that may be actuated through a given product/process change and develops and implements reliability tests that are adequate to assess the impact of those failure mechanisms on system reliability.

4.7.2 Current Qualification Methodology

The component world has changed, as has the method of achieving component qualification. Stress test–driven qualification conducted by the supplier is an important and necessary part of component qualification, but it is not sufficient for today's (and future) technologies and components. Much more is needed to qualify an IC for use than is covered by either stress test–driven qualification or failure mechanism–driven qualification. The qualification test process must be compatible with the realities of both the technology and business requirements. It must leverage the available tools and data sources other than traditional IC life tests. The methodology focuses on reliability issues early in the design phase, ensuring that product reliability goals and customer expectations are met; it optimizes cost and cycle time; and it stimulates out-of-box thinking.

Today, IC qualification

Is a dynamic multifaceted process
Is application specific and determined, in the final analysis, by various system tests
Is based on an understanding of the specific potential failure mechanisms for the device and technology proposed for use in a specific application
Requires a clear understanding of the roles and responsibilities of the IC supplier and IC user (OEM) in qualifying components for use

The current qualification methodology recognizes that there are two distinct partners responsible for qualification testing of a given component. It is a cooperative effort between IC suppliers and OEMs, with each party focusing their time and energy on the attributes of quality and reliability that are under their respective spheres of control. The industry has moved from the OEMs essentially conducting all of the environmental and mechanical qualification tests to the IC suppliers performing them to ensure the reliability of their components.

The IC suppliers have the sole responsibility for qualifying and ensuring the reliability, using the appropriate simulations and tests, for the components they produce. Identification of reliability risks just prior to component introduction is too late; qualification must be concurrent with design. By implementing a design-for-reliability approach, IC suppliers minimize reliance on conventional life tests. Instead, they focus on reliability when it counts—during the IC circuit, package design, and process development stages using various validation vehicles and specially designed test structures to gather data in a short period of time. This IC qualification approach has been validated through the successful field performance of millions of parts in thousands of different fielded computer installations.

The component suppliers are also asked to define and verify the performance envelope that characterizes a technology or product family. They do this by conducting a battery of accelerated testing (such as 1000-hr life tests) for the electrical function and technology being used in the design: wafer fab process, package and complete IC. From these tests they draw inferences (projections) about the field survivability (reliability) of the component. Additionally, a commitment in supplying high-quality and high-reliability ICs rquires a robust quality system and a rigorous self-audit process to ensure that all design, manufacturing, electrical test, continuous improvement, and customer requirement issues are being addressed.

Table 14 details the tasks that the IC supplier must perform to qualify the products it produces and thus supply reliable integrated circuits. Notice the focus (as it must be) is on qualifying new wafer fabrication technologies. Equally important is the effort required to conduct the appropriate analyses and qualification and reliability tests for new package types and methods of interconnection, since they interconnect the die to the PWA, determine IC performance, and have an impact on product reliability.

The user (OEM) requires a stable of known good suppliers with known good design and manufacturing processes. This begins with selecting the right technology, part, package, and supplier for a given application early in the design

TABLE 14 Example of an IC Supplier Qualification Process: Elements and Responsibilities

Process technology qualification through advanced process validation vehicles, test structures (WLR), and packaged devices
    FEOL wearout: Vt stability, gate dielectric integrity, HCI, TDDB
    BEOL wearout: EM (in both Al and Cu metal lines and vias), stress voiding/migration, ILD integrity

Manages potential wearout mechanisms through robust design rules and process control
    Verifies that design simulations using sophisticated and accurate CAD, SPICE, and reliability models correlate with physical failure mechanisms and reliability design rules and validates that they match the wafer fab process.
    Controls process reliability interactions by management of critical process steps.
    Reliability monitors, WLR test structures, and qualification test data are used to verify reliability projections.

Conducts package reliability tests
    Computer simulations and tests are conducted for new package types and interconnect technologies (e.g., flip chip), without the die in the package, for assembly compatibility/manufacturability, materials compatibility, thermal characteristics, and electrical performance (parasitic effects)
    Wire bonding
    Die attach
    Solderability
    Moisture performance

Component and Supplier Selection

TABLE 14 Continued

Prerequisites for conducting complete IC qualification testing
    Topological and electrical design rules established and verified. There is a good correlation between design rules and wafer fab process models.
    Verifies that die package modeling (simulation) combination for electrical and thermal effects matches the results obtained with experimenting testing.
    Comprehensive characterization testing is complete including four-corner (process) and margin tests at both NPI and prior to each time a PCN is generated, as appropriate.
    Electrical test program complete and released.
    Data sheet developed and released.
    Manufacturing processes are stable and under SPC.
    In-line process monitors are used for real-time assessment and corrective action of critical process or product parameters with established Cpk metrics and to identify escapes from standard manufacturing, test, and screening procedures (maverick lot).

Conducts complete IC qualification tests
    Conducts electrical, mechanical and environmental stress tests that are appropriate for the die, package, and interconnection technologies used and potential failure mechanisms encountered and to assess time-dependent reliability drift and wear-out.

Product qualification	Medium risk	Low risk
	ESD	ESD
	Latch-up	Latch-up
	Soft error rate testing	Soft error rate testing
THB/HAST	500 hr	1000 hr
Temperature cycle	250 cycles	500 cycles
Steam test	50 hr	100 hr
Infant mortality	<1000 ppm	<500 ppm
Life test (HTOL)	<200 FIT	<100 FIT
Field reliability	<400 FIT	<200 FIT

Both categories of risk require three lots from three different time points (variability).

FEOL: Front end of line (IC transistor structure formation); BEOL: back end of line (transistor interconnect structures); EM = electromigration; FIT = failures in time (failures in 10^9 hr); HCI = hot carrier injection; HTOL = high temperature operating life; ILD = interlevel dielectric; NPI = new product introduction; PCN = product change notice; SPC = statistical process control; TDDB = time-dependent dielectric breakdown; THB/HAST = temperature-humidity bias/highly accelerated stress test; WLR = wafer level reliability.

phase, within an organizational structure that is focused on supplier management. The use of standard off-the-shelf components (such as those used in personal computers or mobile applications) should be encouraged since they drive the marketplace by their volume, process improvements, and process stability. But it must also be realized that product leverage often comes from using sole-sourced "bleeding-edge" components. The use of sole-sourced components presents unique product qualification issues due to the dependence of the product design on a single part and the need for continued availability of that part throughout the product manufacturing cycle. If the part becomes unavailable for any reason and the PWA must be redesigned to accommodate another part or requires a mezzanine card (which still entails somewhat of a PWA relayout), portions or all of the qualification process must be repeated. The price paid for this is added cost and delay of the product's market introduction. The critical element of successful component qualification, for both the present and the future, is OEM-conducted application-based (also called functional application) testing.

Different market segments/system applications have different component qualification requirements. Table 15 lists examples of some of the different requirements of various market segments. The OEM must first determine what components are critical to a design for a given end market (application) and then tailor the tests for these critical components and their associated PWAs to be compatible with the specific application needs. The OEM makes an assessment of the survivability of the product/equipment in the field with all components, modules, and subassemblies integrated into the design by conducting a battery of end use–specific tests.

Application testing has traditionally been the OEM's responsibility, but in the PC world suppliers are often given machines (computers) to run the application testing per OEM-provided documentation. This allows the OEMs to potentially qualify a number of suppliers simultaneously, ensuring a plentiful supply of low-priced parts.

The high-end PC server OEMs look at all components, but instead of qualifying each component, they qualify a family of devices (like ASICs and other commodities) with the same manufacturing process. The automotive industry essentially tells the supplier what tests to conduct and then they run the same tests on varying sample sizes. Automotive OEMs don't have the same level of trust of their suppliers, and they are always looking for lowest price rather than lowest overall cost (total cost of ownership).

Personal computer OEMs perform detailed specification reviews and conduct application tests. In some specific instances they may have the supplier do some special analyses and tests. Consumer product providers and low-end PC companies just rely on normal supplier qualification procedures and application tests that just look at the ability of different suppliers' parts to "play well" together.

Component and Supplier Selection

TABLE 15 Examples of End Market Qualification Test Needs

End use application/ market	Type of qualification	Typical qualification tests
Aerospace	Extremely comprehensive	Life test, mechanical stress, environmental stress, parametric characterization, destructive physical analysis, FMEA, HALT, STRIFE
Mainframe computer	Comprehensive	Application test, parametric characterization
High-end PC server and telecommunications equipment	Comprehensive (core critical components only)	Application test, parametric characterization
Automotive	Comprehensive	Life test, mechanical stress, environmental stress, parametric characterization, FMEA, HALT, STRIFE
PC	Comprehensive (core critical components only)	Application test, limited parametric characterization
Consumer	Minimal	Application test

Table 16 is a detailed list of the steps that the OEM of a complex electronic system takes in qualifying critical components for use in that system, both currently and looking to the future.

4.7.3 Forward-Thinking Qualification

One thing is for certain, technology improvements and market conditions will require suppliers and OEMs alike to continually review, evaluate, refine, and update the techniques and processes used to qualify both ICs (components) and products. New methods will need to be developed to keep pace with the following expected component technology improvements and market conditions.

1. *Shorter product design and life cycles.* The PC and consumer (mobile appliance) markets are spearheading a drive to shorter product design cycles. The pressure to keep up with technology changes is causing decreased design cycles for all electronic products (PCs, servers, mainframe computers, telecommunication equipment, and consumer products). Time to market is the key to market

TABLE 16 Example of Complex Equipment OEM's Qualification Model for Critical Components

1. Supplier business qualification. Is the company stable, financially viable, a good business and technology partner? (Responsibility: Purchasing)
2. Supplier technology qualification. Is this a viable/producible technology? (Responsibility: Technology Development and Component Engineering)
3. Supplier line qualification [wafer fabrication and assembly (ICs) and manufacturing (components)]. Can the parts be built reliably, consistently, and to OEM requirements? (Responsibility: Component Engineering and Supplier Quality Engineering)
4. Review supplier in-line process monitors, die and assembly qualification test data, and wafer level reliability test data (for ICs). (Responsibility: Component Engineering and Supplier Quality Engineering).
5. In-product application qualification tests and evaluations:
    a. Manufacturability test (on PWA for new package type—wash, solderability, etc.). Is the part compatible with our manufacturing process? (Responsibility: Manufacturing Engineering)
    b. SPICE and CAD models and supplier provided tools (for ICs). Are the models and tools available and ready for use? Can the component operate to our requirements? (Responsibility: Product Technology Modeling and Simulation)
    c. Test vector qualification (for ICs). Is there a high (>95%) level of test/fault coverage? Are a combination of AC, DC, I_{DDQ}, functional (stuck at), delay, and at-speed tests utilized to ensure the performance of the component? Is there a high correlation between electrical test yield, test coverage, and average outgoing quality level? Does wafer test yield correlate with final electrical test yield? (Responsibility: Technology Development)
    d. Smorgasbord (Smorgy) testing (mix and match all suppliers of multisourced components on the PWA. Can the different suppliers for a given part work in the application together? (Responsibility: Product Design and Product Quality Engineering)
    e. Matrix lot testing. Can the component be manufactured to consistently meet requirements even in a worst case/worst use environment? What are the operating margins/boundaries and safe design space (parts are manufactured at the process corners and tested on the PWA)? Will the components operate in the PWA to requirements (timing and signal integrity)? (Responsibility: Product Design for conducting tests/evaluations; Component Engineering for communicating requirements to suppliers and tracking completion of manufacturing parts at corners)
    f. Special studies. What are the nonspecified parameters that are critical to the application? The OEM performs reliability assessments, package studies (e.g., PWA solderability, lead-free evaluation), thermal studies, signal integrity testing, and electrical tests of critical IC parameters, often unspecified by the supplier, for proper system functioning. The latter includes timing analysis, cross-talk, ground bounce, simultaneous switching outputs, undershoot, power-on ramp/surge, and the like. (Responsibility: Component Engineering and Design Engineering)

TABLE 16 Continued

g. HALT testing at PWA or module (subassembly) level. Highly accelerated stress tests in which power cycling, temperature, temperature cycling, and random vibration testing are used simultaneously to take a product beyond its design requirements to determine its robustness (more about this in Chapter 7). (Responsibility: Product Design and Product Quality Engineering)
h. System test. Do the components work in the system as designed and intended? (Responsibility: Product Design)
6. PWA manufacturing ICT and manufacturing ATE functional qualification. Has the component been degraded due to test overdrive? (Responsibility: Manufacturing Engineering)
7. Supplier process change monitoring (as required).
   a. Destructive physical analysis. Has anything in the physical assembly of the component changed? (Responsibility: Component Engineering and Supplier Quality Engineering)
   b. Periodic manufacturing line auditing. Using SPC data it answers the question has anything changed (drifted) in wafer fabrication. (Responsibility: Component Engineering and Supplier Quality Engineering)
   c. Requalification of supplier processes based on process change notification being received. (Responsibility: Component Engineering and Supplier Quality Engineering)
8. Product requalification due to process change notification impacting product performance. (Responsibility: Product Design Engineering)

---

share and profitability; therefore, short product design cycles are here to stay. Newer designs face greater time-to-market pressures than previous designs. For example, new PCs are being released to production every 4 months versus every 6–9 months previously. In order to support these short design cycles and increased design requirements, component qualification processes must be relevant and effective for this new design environment. Figure 11 shows the dramatically shortened product design and qualification timeframes.

Traditional back-end stress-based qualification test methods will not meet the short cycle times for today's market. Integrated circuit suppliers need to develop faster and more effective technology and process-qualification methods (test vehicles and structures) that give an indication of reliability before the IC design is complete.

2. *Shorter component life cycles.* Component production life cycles have significantly been reduced over the last 5–10 years. A typical component has a 2- to 4-year production life cycle (time to major change or obsolescence). The shortened component life cycle is due to the effect of large PC, telecommunication equipment, and consumer product manufacturers pressuring component suppliers for cost reductions until a part becomes unprofitable to make. Often, a

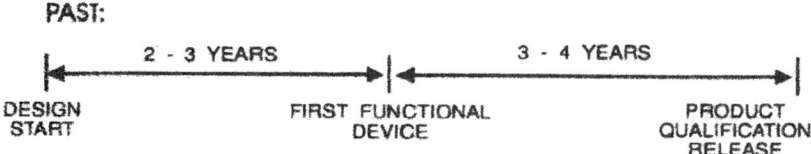

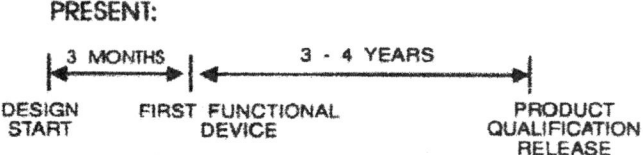

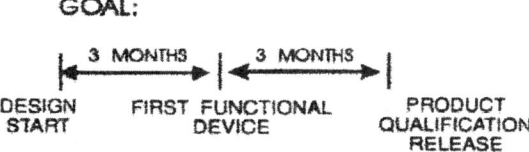

FIGURE 11 The changing process and product qualification timeframe.

reduction in demand causes component manufacturers to make obsolete devices that are suddenly unprofitable to produce. A second factor reducing component life cycles is large equipment manufacturers discontinuing products due to technological obsolescence and market pressure to provide the latest technologies in their products.

3. *Complications of high-speed designs.* The design of high-speed circuits adds many complications that affect component qualification, such as high-frequency operation, timing variances, and signal quality. Maximum IC operating frequency specifications are being driven by newer and faster designs. In order to develop components that are faster, suppliers are reducing rise and fall times, which affect functionality in high-speed applications. The resultant decreasing setup and hold times cause minimal propagation delays to become critical as well. The effect of these shorter rise and fall times must be considered in IC qualification. Designing for high-speed logic applications is very difficult and requires the use of transmission line design techniques to ensure good signal quality. Both IC and product designers are responsible for signal quality. The increased power dissipation and resultant thermal issues generated by these high operating speeds must be addressed as well (See Chapters 3 and 5 for more details).

Component and Supplier Selection

4. *Highly integrated component designs.* The highly integrated component designs such as system-on-a-chip, high-density ASICs, and field-programmable gate arrays add special challenges. How do you qualify the RAM embedded within a larger device? How do you characterize the performance of a component that is configured prior to use while embedded in the application? Highly integrated component designs add the failure mechanism of different component types onto one chip. Component qualification must consider the failure mechanisms of all of the parts that make up a highly integrated device (i.e., microprocessors, RAM, PLLs, logic, mixed signal, etc.).

Also added to these are

- More complex electrical test programs and shorter allowed test program development time.
- More complex package qualification.
    - The package is an integral part of the complete IC solution from electrical performance, thermal management, and mechanical protection perspectives.
    - Stacked packages present new and unique issues.
- Faster time to market and shorter component qualification times.
- Less components to qualify.
- More knowledge of component required.
- More application (product) knowledge required.
- Shorter product change notice review. IC suppliers make a myriad planned and unplanned die shrinks to reduce wafer costs and improve IC performance. This requires that they conduct the appropriate tests and analyses prior to issuing a product change notice (PCN). They want to have a shorter PCN review cycle so they can quickly get into production with the proposed changes. But the OEM needs time to evaluate these changes and their impact on the circuit design, often requiring expensive and time-consuming system requalification testing.
- Greater quality and reliability expectations.
- High-speed highly integrated and complex product designs.

4.7.4 Application-Based Qualification: The Future

The environmental and mechanical "shake-and-bake" testing philosophy of component qualification as espoused by MIL-STD-883 has become entrenched into our component qualification philosophy. For some product environments stress-based qualification testing is still a critical aspect of component qualification that cannot be discounted. Stress test–based component qualification testing addresses the physical attributes that cause early-life failures, but does not address all aspects of electrical or functional fitness for the application that uses the component.

For product designs of the new millennium component qualification focuses on two critically important areas from an OEM perspective: a consideration of the application requirements and qualifying the technology that creates the components (rather than qualify the resulting output of the technology; i.e., the specific component). Application-based qualification allows a family of components to be qualified and maintained for its lifetime with review of technology data and the results of changes in a product family. This approach results in reduced component qualification cycle times and provides the means to perform an engineering analysis of supplier-provided data to evaluate a component product change versus the "try one and see if it fails" approach.

The fundamental difference between application-based qualification and traditional component qualification is understanding what a technology can and cannot do; how it behaves with various loading conditions versus testing a component to a data sheet; and the specific application requirements. However, not all is new. Component qualification is an evolutionary process that meets the current time period technology realities and market needs. Many aspects of application-based qualification are already in place and operating out of necessity. As mentioned in Chapter 3, the OEM's technology needs must be matched with the supplier's technology availability (see Fig. 2 of Chapter 3). Passing qualification is an assessment that there is a low level of risk that a technology will have issues in user applications.

Application-Based Qualification Details

1. *Identifying device technology and system requirements.* The engineer responsible for qualifying a device needs to understand both the device technology requirements and the application requirements. This knowledge is gained in part by working closely with the responsible product design engineer. Thus, the qualifying engineer is an expert consultant for new application uses of previously qualified devices and technologies.

2. *Quality assessment.* Supplier qualification is a prerequisite to application-based qualification. After a supplier is qualified, a quality assessment of the technology being qualified should begin. Process quality attributes that affect quality and reliability should be evaluated. These include

> Quality. Conformance to data sheet specifications (see following paragraph on self-qualification) and process control (accuracy, stability, repeatability, and reproducibility)
> Reliability. Wearout (i.e., high-temperature operating life, gate oxide dielectric integrity, hot carrier degradation, electromigration, etc.)
> Susceptibility to stress. Electrostatic discharge, latch-up, electromigration, delamination, etc.
> Soft error rate (SER). Transient errors caused by alpha particles and cosmic rays (devices with memory cells)

3. *Self-qualification process.* The majority of a technology qualification's requirements can be satisfied by the supplier's submission of information on a device technology via a self-qualification package. The self-qualification package should address the technology attributes that define the technology limits and process capability. Device characterization data showing minimum, maximum, mean, and standard deviation of device parametrics should be evaluated. Many large OEMs (such as Nortel and Lucent, to name two) are allowing some of their top suppliers to self-qualify by filling out a preestablished template attesting to the fact that various tests and analyses were conducted and that a robust quality infrastructure is in place and functioning. Appropriate documentation is referenced and available to the OEM on demand. Appendix B of Chapter 4 is an example of a form used by Nortel Networks for self-qualification of programmable logic ICs.

4. *Special studies.* Special studies are performed for application-specific requirements that the supplier does not specify but are key to successful operation in an application. They include SPICE modeling of I/O to determine signal quality under different loading conditions and varying trace lengths. Special studies provide two benefits:

1. The information gathered allows effective design using a supplier's components.
2. The designer can evaluate second source components for the same application as the primary source.

Non-specified parameter studies/tests should be performed to determine the effects on design requirements. Examples of some nonspecified parameters that are critical to digital designs are as follows:

Hot-plot characteristics define the behavior of a device during live insertion/withdrawal applications.
Bus-hold maximum current is rarely specified. The maximum bus-hold current defines the highest value of pull-up/pull down resistors for a design.
It is important to understand transition thresholds, especially when interfacing with different voltage devices. Designers assume 1.5-V transition levels, but the actual range (i.e., 1.3–1.7 V) is useful for signal quality analysis.
Simultaneous switching effect characterization data with 1, 8, and 16 or more outputs switching at the same time allow a designer to manage signal quality as well as current surges in the design.
Pin-to-pin skew defines the variance in simultaneously launched output signals from package extremes.
Group launch delay is the additional propagation delay associated with simultaneous switching of multiple outputs.

5. *Package qualification.* Component packaging constitutes a technol-

ogy that often requires special attention to assure a good design fit. Some package characteristics for evaluation are

> Thermal characteristics in still air and with various air flow rates.
> Package parasitics (i.e., resistance, capacitance, and inductance) vary with package type and style. Some packages have more desirable characteristics for some designs.
> Manufacturing factors such as solderability, handling requirements, mechanical fatigue, etc.

Advanced packaging innovations such as 3D packages are being used to provide increased volumetric density solutions through vertical stacking of die. Vertical stacking provides higher levels of silicon efficiency than those achievable through conventional multichip or wafer-level packaging (WLP) technologies.

Through 3D packaging innovations, a product designer can realize a 30 to 50% PWA area reduction versus bare die or WLP solutions. Stacked chip-scale packaging (CSP) enables both a reduction in wiring density required in the PWA and a significant reduction in PWA area. A 60% reduction in area and weight are possible by migrating from two separate thin small outline package (TSOPs) to a stacked CSP. Nowhere is this more important than in the mobile communications industry where aggressive innovations in packaging (smaller products) are required. Examples of several stacked die chip scale packages are shown in Figure 12.

As the packaging industry migrates to increased miniaturization by employing higher levels of integration, such as stacked die, reliability issues must be recognized at the product development stage. Robust design, appropriate materials, optimized assembly, and efficient accelerated test methods will ensure that reliable products are built. The functionality and portability demands for mobile electronics require extensive use of chip scale packaging in their design. From a field use (reliability) perspective portable electronics are much more subject to bend, torque, and mechanical drops than other electronic products used in business and laboratory environments. As a result traditional reliability thinking, which focuses on having electronic assemblies meet certain thermal cycling reliability requirements, has changed. There is real concern that these products may not meet the mechanical reliability requirements of the application. For stacked packages the combined effects of the coefficient of thermal expansion (C_{TE}) and elastic modulus determine performance. In stacked packages there is a greater C_{TE} mismatch between the laminate and the package. The failure mechanism may shift to IC damage (cracked die, for example) instead of solder joint damage. Failures occur along the intermetallic boundaries. Drop dependent failures depend on the nature of the intermetallics that constitute the metallurgical bond. During thermal cycling, alternating compressive and tensile stresses are operative. Complex structural changes in solder joints, such as intermetallic growths,

Component and Supplier Selection

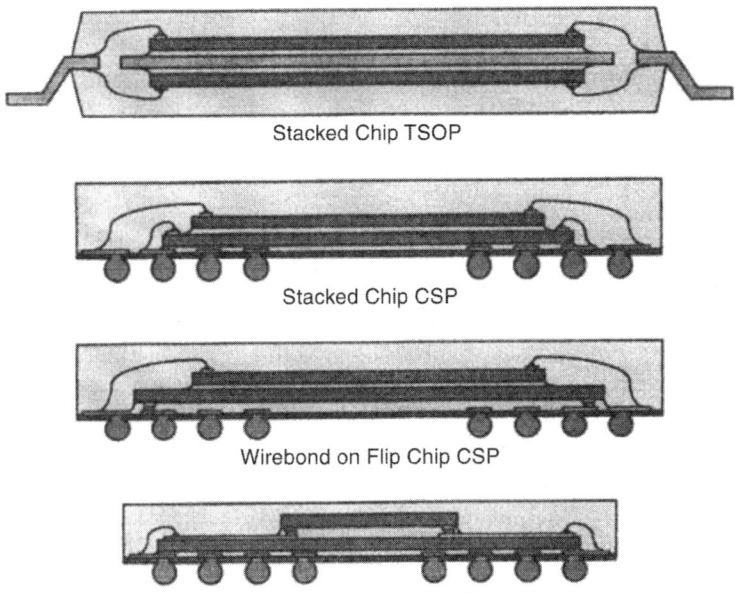

FIGURE 12 Several stacked die CSPs. (Courtesy of Chip Scale Review.)

grain structure modifications (such as grain coarsening and elastic and plastic deformations due to creep) are operative. The different C_{TE} values of the die that make up the stacked package could lead to the development of both delamination and thermal issues. The surface finish of the PWA also plays a significant role in the reliability of the PWA.

Thus, new package types, such as stacked packages, provide a greater challenge for both the IC supplier and the OEM user in qualifying them for use.

6. *Functional application testing.* Functional application testing (FAT) is the most effective part of component qualification, since it is in essence proof of the design adequacy. It validates that the component and the product design work together by verifying the timing accuracy and margins; testing for possible interactions between the design and components and between hardware, software, and microcode; and testing for operation over temperature and voltage extremes. The following are examples of functional requirements that are critical to designs and usually are not tested by the component supplier:

> Determinism is the characteristic of being predictable. Complex components such as microprocessors, ASICs, FPGAs, and multichip modules should provide the same output in the same cycle time for the same instructions consistently.

Mixed voltage applications, i.e., interfacing devices that operate at different voltages.

Low-frequency noise effects should be assessed for designs that contain devices with phase-locked loops (PLLs). Phase-locked loops are susceptible to low-frequency noise, which could cause intermittent problems.

Hot-plug effects on the system that a module is being hot-plugged into. Hot-plugging may cause intermittent functional problems.

Figure 13 is a fishbone diagram listing the various items involved in application-based qualification. Manufacturers of consumer products with short design and manufacturing life (30–90 days), and short product cycles (6–12 months) require a fast time-to-market mindset. They cannot afford the time necessary to conduct any but the most essential testing. As a result some manufacturers of consumer products are implementing a radical concept: forget about conducting any formal qualification testing and go straight to functional application testing. In fact, they're going one step beyond and letting FAT decide what components and suppliers are right for the application. Thus, in a nutshell, FAT becomes the entire component/supplier selection and qualification process.

Customer and Supplier Partnership in Application-Based Qualification

Here is that phrase again: customer and supplier partnerships. These relationships are necessary to develop timely and effective qualification of state-of-the-art components and the products that use them. Two specific points are made here.

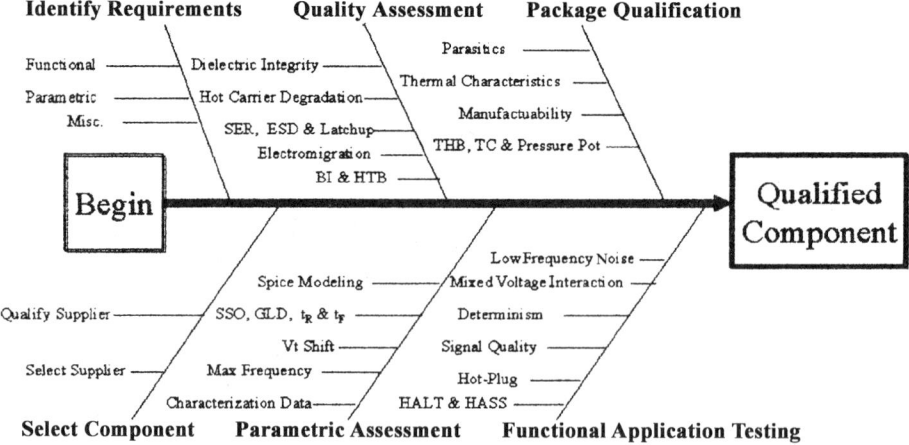

FIGURE 13 Example of application-based qualification process flow.

Component and Supplier Selection

1. Sharing information is critical to the use of new devices. Suppliers and customers need to identify their needs and explain the rationale for specified requirements. This interchange of information allows both parties to benefit from the experience and knowledge base of the other to create a product that meets the needs of the customer and the supplier.
2. Product co-development will become more common in the new millennium as custom and semicustom devices become easier to create. Suppliers and customers will jointly develop new devices to meet specific needs.

Benefits of Application-Based Qualification

1. Lower cost component qualification is a benefit of application-based qualification versus traditional component qualification. Once a technology (wafer process and package type) and/or part family is qualified for a given application there is no need to perform device-by-device qualifications of products from the same family. Only FAT is required. The end result of the application-based qualification is reduced component qualification cycle times.

2. Higher quality products are expected in the new millennium and to achieve product quality improvements there is a need to address an increasing number of no-defect-found industry problems. Application-based qualification uses component application studies to verify that a device is a good fit for a design, placing a greater emphasis on learning about device capability versus testing it to a data sheet. Better understanding of device capability and characteristics enhance a designer's ability to develop a robust design.

3. And, finally, application-based qualification is a process that can grow with changing requirements and technology. The specific attributes evaluated may change with time, but the method and objective of understanding and assessing device capability remains the constant objective.

4.7.5 Developing A Component Qualification Strategy

Now that I have talked about the past, present and future as regards component qualification, I want to list some of the items that must be considered in developing a component qualification strategy: a standard methodology or thought process that one uses to develop the qualification requirements for a specific application. Every company (OEM) needs to develop a component qualification strategy (typically for their critical components) that is appropriate for the product it makes and the end markets it serves.

Developing a component qualification test strategy involves using the engineering decisionmaking process in a logical step-by-step situation analysis and applying it to the die, package, technology, and application under consideration. It involves gathering all available data: market trends, technology trends, packaging trends, fab and assembly process data, what has and what hasn't worked in the past and why. It involves an intimate knowledge of both the component and the application, thus necessitating a close working relationship between supplier and OEM and a division of responsibilities. A qualification strategy addresses the application-specific tradeoffs between IC performance, reliability, risk, product performance, and cost and has the following attributes.

- It is based on an understanding of the technology and business trends detailed earlier.
- It is a multistep process that begins early in both the component and system design cycles while the designs are in the embryonic stages. From a system perspective it involves the appropriate technology, part, package, and supplier selection.
- It is a simple, standard, yet flexible, methodology based on all available data sources—not necessarily a battery of standard tests.
- It is concurrent with other engineering activities, not a pass/fail gate.
- It is best managed through adherence to a rigorous set of design rules, design reviews, and process controls.
- It is based on a detailed understanding of the IC design, physical layout, materials used, material interfaces, packaging details, and potential failure mechanisms (physics of failure concept) to which the device is sensitive by virtue of its wafer fab and assembly processes.
- It is application dependent rather than being a standard test or series of tests.
- It is easily portable across processes and designs, not requiring reinitialization or even use of all steps taken to date. The caveat here is that the strategy may not be easily ported from one IC technology node to the next (e.g., 0.07-μm technology versus 0.15-μm technology).
- It is fast, accurate, and low in cost, not requiring months and tens of thousands of dollars or more in the design process.
- It is specific-component independent, but technology dependent, allowing the rapid qualification of many device types of a given technology.
- It is based on a real understanding of reliability
    - Reliability rules are best verified through design simulations using accurate design-to-process models rather than by life testing an IC.
    - Reliability issues are an attribute of both the wafer fab processes and the application (e.g., logic states that cause floating buses) and should be addressed through process management techniques (in-line process monitors of critical parameters and defect control).

Life tests should be performed to characterize the time dependency of reliability defects, an attribute of the process, not specific ICs. Reliability phenomena need to be checked and verified throughout the design cycle. 70–80% of the operation and maintenance costs are due to choices made in design.

Formal reliability verification must be defined at a higher level and then performed at every level of the design hierarchy, starting at the transistor level and working through to the full chip (IC).

The reality of component qualification is that it is a constantly evolving process, requiring faster throughput and lower cost. Each component (die, package, and IC) must be qualified individually by the supplier and then in the product application by the OEM. Each of these requires a complex subset of various evaluations, tests, and analyses. Beyond the year 2000, as the rate of change is accelerating we need to adopt new methods for meeting our overall objectives—fielding reliable products. This means the issue of component qualification needs to be approached with a mind open to all possibilities in meeting end customer needs.

ACKNOWLEDGMENTS

Portions of Sections 4.3.2 and 4.3.4 excerpted by permission from the short course "Supplier Management" at the courtesy of Ken Stork and Associates, Inc. California Institute of Technology. Portions of Section 4.4 were excerpted from Ref. 3. Portions of Section 4.6.1 were excerpted from Ref. 4. Sections 4.7.3 and 4.7.4 excerpted from Ref. 5.

REFERENCES

1. Moore GA. Living on the Fault Line, Harper Business, 2000.
2. Hnatek ER. Integrated Circuit Quality and Reliability, 2nd ed., Marcel Dekker, 1995.
3. Hnatek ER, Russeau JB. Component engineering: the new paradigm. Advanced Electronic Acquisition, Qualification and Reliability Workshop, August 21–23, 1996, pp 297–308.
4. Hnatek ER, Kyser EL. Practical lessons learned from overstress testing: a historical perspective. EEP Vol. 26-2, Advances in Electronic Packaging, ASME, 1999.
5. Russeau JB, Hnatek ER. Technology qualification versus part qualification beyond the year 2001. Military/Aerospace COTS Conference Proceedings, Berkeley, CA, August 25–27, 1999.

FURTHER READING

1. Carbone J. HP buyers get hands on design. Purchasing, July 19, 2001.
2. Carbone J. Strategic purchasing cuts costs 25% at Siemens. Purchasing, September 20, 2001.
3. Greico PL, Gozzo MW. Supplier Certification II, Handbook for Achieving Excellence Through Continuous Improvement. PT Publications Inc., 1992.
4. Kuglin FA. Customer-centered supply chain management. AMACOM, 1998.
5. Morgan JP, Momczka RM. Strategic Supply Chain Management. Cahners, 2001.
6. Poirier CC. Advanced Supply Chain Management. Berrett-Koehler, San Francisco, CA, 1999.
7. Supply Chain Management Review magazine, Cahners business information publication.

APPENDIX A: SUPPLIER SCORECARD PROCESS OVERVIEW

4.A.1 Introduction to Scorecard Process

What Is a Supplier Scorecard?

A quick and practical approximation of total cost of ownership (TCOO) by measuring major business and performance parameters.
A measure of supplier performance, not just cost.
An evaluation of a supplier's competitiveness (shown by trend analysis).

Why Is a Supplier Scorecard Needed?

Gives suppliers status on performance issues and accomplishments.
Provides a clear set of actions needed for continuous improvement, i.e., it sets the baseline.
Provides a regularly scheduled forum between supplier's and customer's top management to discuss overall business relationships and future strategies for moving forward.

Example of the Scorecard Process

Top 40 Suppliers
Quarterly process: management reviews for top six to eight suppliers

Measurement attributes	Possible points
Quality	25
On-time delivery (OTD)	25
Price	25
Support	15
Technology	10

Component and Supplier Selection

Scoring methodology:

1. Add up points earned by supplier for each metric.
2. Apply following formula to get a "cost" score:

$$\frac{100 - \text{performance score}}{100} + 1.00$$

For example, a supplier's score is 83 points upon measurement; thus

$$\text{Cost score} = \frac{100 - 83}{100} + 1.00 = 1.17$$

3. The TCOO score is then compared to $1.00. In the example above, the supplier's TCOO rating would indicate that for every dollar spent with the supplier, the OEM's relative cost was $1.17.

Who Generates the Score?

Attribute	Responsibility
Quality	Supplier Quality Engineering and Component Engineering
OTD	Purchasing at both OEM and EMS providers
Price	OEM commodity manager and Purchasing and EMS Purchasing
Support	OEM commodity manager and Purchasing and EMS Purchasing
Technology	Product Design Engineering and Component Engineering

General Guidelines

OEM purchasing:
- Maintains all historical files and data
- Issues blank scorecards to appropriate parties
- Coordinates roll-up scores and emailing of scorecards
- Publishes management summaries (trends/overview reports)

If a supplier provides multiple commodities, then each commodity manager prepares a scorecard and a prorated corporate scoreboard is generated. Each commodity manager can then show a supplier two cards—a divisional and an overall corporate scorecard.

The same process will be applied (prorated by dollars spent) for an OEM buyer and a EMS buyer.

Scorecard Process Timetable Example

Week one of quarter: gather data.
Week two of quarter: roll-up scorecard and review results.
Goal: scorecards will be emailed to suppliers by the 15th of the first month of the quarter.
Executive meetings will be scheduled over the quarter.

4.A.2 Scorecard Metrics

Product and process quality might be worth up to 20 points, for example. Points are deducted for each quality problem resulting in the supplier's product being returned for rework or requiring rework by the OEM in order to be usable in the OEM's product. The number of points deducted is determined by the supplier quality engineer based on the severity and impact of the problem, and by the age of the items involved:

    Catastrophic   Minus 20 points
    Serious   Minus 5 points
    Minor   Minus 1 point

Management system quality might be rated for up to 5 points, as follows:

1 point for an internal continuous process improvement program with defined goals and documented improvement plans that are shared with the OEM on a quarterly basis.

1 point for improved first pass yield and/or other key quality indicator reports that are provided to the OEM monthly.

1 point for no open supplier corrective action requests older than 30 calendar days.

1 point for documented quality requirements imposed upon supplier's purchased material; monitored material quality reports provided to the OEM.

1 point if the supplier is certified to ISO 9000.

On-time delivery points might be allocated as follows:

On-time delivery	Points
75% (or below)	0
80%	3
85%	5
90%	10
95%	15
100%	25

Requested due date is usually the suggested due date.
Original due date is the supplier's original commit date.
Latest due date is the supplier's latest commit date.
Supplier performance: 5 days early/0 days late (measured as difference between the OEM receipt date and original due date).
Supplier flexibility: 5 days early/0 days late (measured as difference between receipt date and requested due date).

Component and Supplier Selection

Other scorecard metrics.

	# Points
Price	
Meets OEM price goals	25
Average price performance	15
Pricing not competitive	0
Support	
Superior support/service	15
Acceptable support/service	10
Needs corrective action	0

Technology[a]	Points
Meets OEM technology requirements	10
Average, needs some improvement	5
Does not meet OEM requirements	0

[a] Technology for OEM's current manufacturing processes.

An overall supplier scorecard may look like the following:

Supplier scorecard for _____

Performance attributes	Data/issues	Max. points	Score
Quality			
Product and process		20.0	
Management system		5.0	
On-time delivery		25.0	
Price			
Meets OEM goals		25.0	
Average price performance		15.0	
Not competitive		0.0	
Support service			
Superior support and service		15.0	
Acceptable support		10.0	
Needs corrective action		0.0	
Technology			
Meets OEM's reqirements		10.0	
Does not meet requirements		0.0	
Total			

Total cost of ownership $= \dfrac{100 - \text{SCORE}}{100} + 1.00$ (Goal: 1.0)

APPENDIX B: SELF-QUALIFICATION FORM FOR PROGRAMMABLE LOGIC ICs

The following documentation reproduces the self-qualification form for programmable Logic ICs used by Nortel Networks Inc.

Courtesy of Nortel Networks Inc., Ottawa, Canada.

Component and Supplier Selection

PROGRAMMABLE LOGIC PRODUCT QUALIFICATION REPORT/FORM

Qualification of Programmable Logic components for Nortel applications requires completion of the following form assuring compliance to the Applicable 'Nortel Procurement Specifications (NPS)' i.e. NPS00018 (general 'Nortel Procurement Specification' for Product Assurance requirements for the procurement of microcircuits) and the detailed technical requirements as defined in the 'Detailed Nortel Procurement Specification (Detailed NPS)' for the device. This form needs to be completed for each Nortel part number (and the corresponding supplier part number). The detailed NPS may be a supplier datasheet used as the prime design intent or Nortel design requirements document. Completion of this summarized form does not exempt supplier from compliance to all requirements as specified in NPS00018 and the detailed NPS for the device being qualified. A waiver to any non-compliance of these requirements can only be granted by the appropriate Nortel Network's authority and No approval or Qualification of a part will be granted without such approval.
It is the 'Supplier' responsibility to complete this checklist for each Nortel Networks part# and the corresponding supplier part#, and submit it to the Nortel Networks Qualification Checklist repository.
Please place an 'X' in the appropriate 'Yes' or 'No' box when answering Yes/No questions and answer all the questions in the checklist.

Northern Networks Part Number: _____
Programmable Logic Device Description: _____
Detailed NPS of Device (Controlling Document): _____

Supplier and Supplier-Part Identification
Supplier Name: _____ / Manufacturer Name: _____
Manufacturer Part # (Ordering Code) :_____
Die/Mask Revision Level (required for qual) :_____ What ID markings on Die (required for qual): _____
Name Generic Process Technology / Design Rule (u-drawn):_____/_____
Die Size :_____ Number of Metal layers : _____
Name / Location of Wafer Fab Facility (the facility being qualified):_____
Is ISO9000 Certification achieved for Fab Location? Yes[ ] No[ ]
Package Outline, Type or Name: _____ Number of Contacts :_____
Lead Frame Material: _____Lead Finish composition:_____
Die Attach material and method:_____ / _____
Wire Bond Method /Material /Size :_____/ _____ /_____
Name / Location of Assembly Facility (the facility being qualified):_____
Mould Compound Supplier and Name of Compound :_____/_____
Flammability Classification: UL-94 V0? Yes[ ] UL-94 V1 Yes[ ] No[ ]
provide Oxygen Index (% per ASTMD 2863-87):_____
Is ISO9000 Certification achieved for Assembly Location? Yes[ ] No[ ]

Compliance Assurance to Nortel Procurement Specifications
- Indicate if the supplier has reviewed and assures compliance to NPS00018 ?: Yes[ ] No[ ]
If 'No', provide details: _____
- Indicate if the supplier has reviewed and assures compliance to 'Detailed NPS' for the device(s):
Yes[ ] No[ ]
If 'No', provide details: _____
Compliance to 'Detailed NPS' means drop-in (mechanical / electrical) compliance with part specifications in the 'Detailed NPS'. Supplier shall identify non-compliant attributes. A waiver to any specified requirements must be received in writing from Nortel prior to qualification.

Quality Assurance Check

Is the supplier ISO9001 / 9002 Certified?	Yes[ ]	Date?_____		No[ ]
Is the supplier TL9000 Certified?	Yes[ ]	Date?_____		No[ ]
Is the supplier Stack Registered?	Yes[ ]	Date?_____		No[ ]
Is the supplier Stack Level 1 Certified?	Yes[ ]	Date?_____	No[ ]	
Is the supplier Stack Level 2 Certified?	Yes[ ]	Date?_____	No[ ]	

Reliability Assurance Check

Is Mass Production started (for current Rev)? :   Yes [ ]   No[ ]
Is Manufacturer's Internal Qual Completed (for current Rev)? :   Yes [ ]   No[ ]
What is the name of the Qualification Report document & Date completed :_____
Has the above report been added to Nortel Networks qualification reports repository: Yes [ ]   No[ ]
- Is the above 'Qualification Report' based on qualification tests on samples with actual die-package combination for the device listed above? The answer is 'No' if the 'Qualification Report' is applied to the listed device by similarity) using other test vehicles: Yes [ ]   No[ ]
  - If 'No' above, and die or package similarity rules are applied, can the supplier assures that due diligence and sound engineering judgments have been applied in identifying the relevancy of the die and package related test on 'test vehicles' to the actual devices being qualified?:   Yes [ ]   No[ ]

Has the supplier completed reliability test for the device (or by similarity) and included in the qualification report all applicable tests in the following list of standard reliability assurance test with the test conditions, sample size and accept/reject criteria meeting tests and acceptability requirements specified in NPS00018 and Detailed NPS?

Die Related Tests

• High Temp Operating Life Test	(per JESD 22-A108)	Yes [ ]	No[ ]
• ESD - Human Body Model	(per JESD 22-A114)	Yes [ ]	No[ ]
• Latch-Up Sensitivity Test	(per JESD78)	Yes [ ]	No[ ]
• Input / Output Capacitance	(per MIL 883-3012)	Yes [ ]	No[ ]
• Low Temp Op Life (LTOL) Test	('Hot Electron Effect' evaluation)	Yes [ ]	No[ ]

List actual Test Temperature and Voltage used in LTOL Test : _____ °C / _____ Volts

Package Related Tests

• Biased Moisture Endurance Test	(per JESD 22-A101 or 110)	Yes [ ]	No[ ]
• Temperature Cycling Test	(per JESD 22-A104)	Yes [ ]	No[ ]
• Autoclave (Pressure Cooker) Test	(per JESD 22-A102)	Yes [ ]	No[ ]
• Solderability Test	(per JESD 22-B102)	Yes [ ]	No[ ]

• Is there any test in 'Qualification Test Schedule, Table 4 per NPS00018' which the supplier will not assure compliance to?   No[ ]   Yes[ ]
  If 'Yes', provide details: _____

Extended (Industrial) Temperature Range Devices (-40 °C to 85 °C rated)

• Which of the following temp range parts are currently marketed (test screened) from the same base die (mark Y/N)?
  - Commercial Temp Range   (0 °C to 70 °C)   Yes [ ]   No[ ]
  - Extended (Industrial) Temp Range   (-40 °C to 85 °C) .   Yes [ ]   No[ ]
  - Military Temp Range   (-55 °C to 125 °C)   Yes [ ]   No[ ]
• Are the Extended (Industrial) Temp Range devices actually screened at the specified -40 °C to 85 °C
                                                                Yes [ ]   No[ ]
  - If 'No', Indicate lower and upper temp (in °C) used for 100% production screen: Lower[   ]°C, Upper[   ]°C

Component and Supplier Selection

General
• Can the supplier assure that the quality level in Defects-Per-Million (DPM) shall not exceed 100 DPM for Devices that are supplied Programmed and tested, for the shipped parts for all defects (i.e., not performing to the specified requirement)?: Yes [ ] No[ ]
• Can the supplier assure that the system level failure rate, when operated within the specified conditions at an ambient temperature of +50°C, shall not exceed 35 FITs @ 60 UCL for FPGA's (Field Programmable Logic Array) and 15 FIT's @ 60 UCL for CPLD (Complex Programmable Logic Devices after Programming)?: Yes [ ] No[ ]
• Is there a reliability monitor program for the device's ongoing assurance?: Yes [ ] No[ ]
  If 'Yes' what is the test vehicle. if 'Not', explain why: _____

• Has there been any reliability/application problem with any component derived from the die used by the above device, either found by the supplier or its customers, that may become a potential problem for Nortel applications?:
  No [ ]   Yes[ ]
If yes, please provide details including corrective action taken and its results: _____

• Is this supplier part recommended for new designs? Yes [ ]   No[ ]
  If 'No', do you recommend design out of existing products? No [ ]   Yes[ ]
• What is the projected 'End of Life' for (the very last die revision) of this product? _____
• Has the device Data Sheet been declared Stable? Stability implies that the Supplier has completed and locked the timing analysis, Parametric's, package and Pin out assignment, etc. The supplier will not change the Data Sheet without informing Nortel of proposed change. Yes [ ]   No[ ]

Tools
• Are the suppliers/manufacturer's tools to be used by the Nortel Design authority, to create Nortel products with the PLD Device, Complete/Stable? Complete/Stable, implies that the Tools have been updated to reflect the final device functionality, specifications, and timing parameters. Yes [ ]   No[ ]
• Have the programming algorithms been created and are they commercially available? Yes [ ]   No[ ]

Name of Supplier employee providing this information: _____ *Phone #:* _____
Function/Title: _____ *Email:* _____
Name of supplier Quality/Reliability Director Authorizing this report: _____
Date Completed: _____
==(010820) =====

5
Thermal Management

5.1 INTRODUCTION

The objective of thermal management is the removal of unwanted heat from sources such as semiconductors without negatively affecting the performance or reliability of adjacent components. Thermal management addresses heat removal by considering the ambient temperature (and temperature gradients) throughout the entire product from an overall system perspective.

Thermal removal solutions cover a wide range of options. The simplest form of heat removal is the movement of ambient air over the device. In any enclosure, adding strategically placed vents will enhance air movement. The cooling of a critical device can be improved by placing it in the coolest location in the enclosure. When these simple thermal solutions cannot remove enough heat to maintain component reliability, the system designer must look to more sophisticated measures, such as heat sinks, fans, heat pipes, or even liquid-cooled plates. Thermal modeling using computational fluid dynamics (CFD) helps demonstrate the effectiveness of a particular solution.

The thermal management process can be separated into three major phases:

1. Heat transfer within a semiconductor or module (such as a DC/DC converter) package
2. Heat transfer from the package to a heat dissipater
3. Heat transfer from the heat dissipater to the ambient environment

The first phase is generally beyond the control of the system level thermal engineer because the package type defines the internal heat transfer processes. In the second and third phases, the system engineer's goal is to design a reliable, efficient thermal connection from the package surface to the initial heat spreader and on to the ambient environment. Achieving this goal requires a thorough understanding of heat transfer fundamentals as well as knowledge of available interface and heat sinking materials and how their key physical properties affect the heat transfer process.

5.2 THERMAL ANALYSIS MODELS AND TOOLS

Thermal analysis consists of calculating or measuring the temperatures at each component within a circuit or an assembly. Thermal analysis, which is closely related to stress derating analysis, concentrates on assuring both freedom from hot spots within equipment and that the internal temperature is as uniform and low as feasible and is well within the capabilities of the individual components. Arrhenius theory states that chemical reaction rates double with each 10°C increase in temperature. Conversely, reducing the temperature 10°C will reduce the chemical reaction rate to one-half. Many part failures are attributable to chemical activity or chemical contamination and degradation. Therefore, each 10°C reduction of the temperature within a unit can effectively double the reliability of the unit.

As printed wire assemblies (PWAs) continue to increase in complexity, the risk of field failures due to unforeseen thermal problems also increases. By performing thermal analysis early in the design process, it becomes possible to ensure optimal placement of components to protect against thermal problems. This in turn minimizes or eliminates costly rework later.

The sweeping changes taking place in the electronics and software industries are resulting in dramatic improvements in the functionality, speed, and compatibility of the computer-aided design (CAD) tools that are available. As a result, thermal modeling software is gaining widespread use today and is now part of the standard design process at most major electronics manufacturers around the world.

Modern electronic systems incorporate multitudes of components and subassemblies, including circuit boards, fans, vents, baffles, porous plates [such as electromagnetic interference (EMI) shields], filters, cabling, power supplies, disk drives, and more. To help designers cope with this complexity, the most advanced thermal modeling solutions provide a comprehensive range of automated software tools and user-friendly menus that provide easier data handling, faster calculations, and more accurate results.

Thermal modeling has migrated from system to PWA, component, and environment levels. Modeling was first applied at the system level in applications

Thermal Management

such as computer cabinets, telecommunication racks, and laptop computers. However, as the need for thermal modeling has become more pressing, these same techniques have migrated downward to board- and component-level analysis, which are now commonplace in the electronics design industry. The most advanced thermal modeling solutions allow designers to predict air movement and temperature distribution in the environment around electronic equipment as well as inside it and determine true airflow around the component. A three-dimensional package-level model can include the effects of air gaps, die size, lead frame, heat spreader, encapsulant material, heat sinks, and conduction to the PWA via leads or solder balls. In this way, a single calculation can consider the combined effects of conduction, convection, and radiation.

An example of the complexity of the airflow patterns from the system fan in a desktop PC are illustrated in Figure 1 using a particle tracking technique (see color insert).

To be most effective, thermal analysis tools must also be compatible and work with a wide range of mechanical computer aided design (MCAD) and electronic design automation (EDA) software. This is easier said than done given the wide range of formats and the different levels of data stored in each system. Once modeling is complete and prototype products have been built, thermal imaging is used for analyzing potential problems in circuit board designs (see Fig. 24 of Chapter 3). It can measure complex temperature distributions to give a

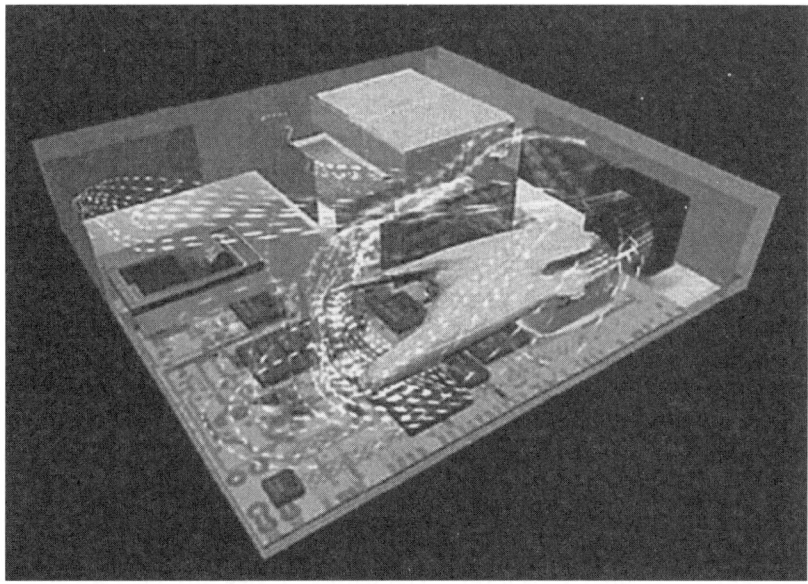

FIGURE 1 Airflow through a PC chassis. (See color insert.)

visual representation of the heat patterns across an application. As a result, designers may find subtle problems at the preproduction stage and avoid drastic changes during manufacturing. The payback is usually experienced with the first thermal problem uncovered and corrected during this stage. Several examples are presented to demonstrate the benefits of various modeling tools.

Low-Profile Personal Computer Chassis Design Evaluation

The thermal design of a PC chassis usually involves compromises and can be viewed as a process of evaluating a finite set of options to find the best balance of thermal performance and other project objectives (e.g., noise, cost, time to market, etc.). The designer need not be concerned about predicting temperatures to a fraction of a degree, but rather can evaluate trends to assess which option provides the lowest temperature or best balance of design objectives (2).

Use of thermal models can give a good indication of the temperature profile within the chassis due to changes in venting, fan placement, etc., even if the absolute temperature prediction for a component isn't highly accurate. Once a design approach is identified with thermal modeling, empirical measurements must follow to validate the predicted trends.

Some common sources of discrepancy between model predictions and measurements that must be resolved include.

1. Modeled and experimental component power do not match.
2. Fan performance is not well known.
3. Measurement error. Common errors include
    a. Incorrectly placed or poorly attached thermocouple
    b. Radiation when measuring air temperature. Nearby hot components can cause the thermocouple junction to heat above the ambient air temperature giving a false high reading.

Figure 2 shows a three-dimensional model of a personal computer desktop chassis. Vents were located in the front bezel, along the left side of the chassis cover, and in the floor of the chassis just in front of the motherboard. Air entering the chassis through the vents flows across the chassis to the power supply (PS), where it is exhausted by the PS fan. A second fan, located at the front left corner of the chassis, was intended to provide additional cooling to the processor and add-in cards. This arrangement is quite common in the PC industry.

Several models were run initially, including one with the front fan on and one with the fan deactivated to determine its effectiveness in cooling the microprocessor. Particle traces are used to show the heat flow across various components and within the chassis enclosure. The particle traces in Figure 3 clearly show that the flow from the fan is deflected by the flow entering from the side

Thermal Management

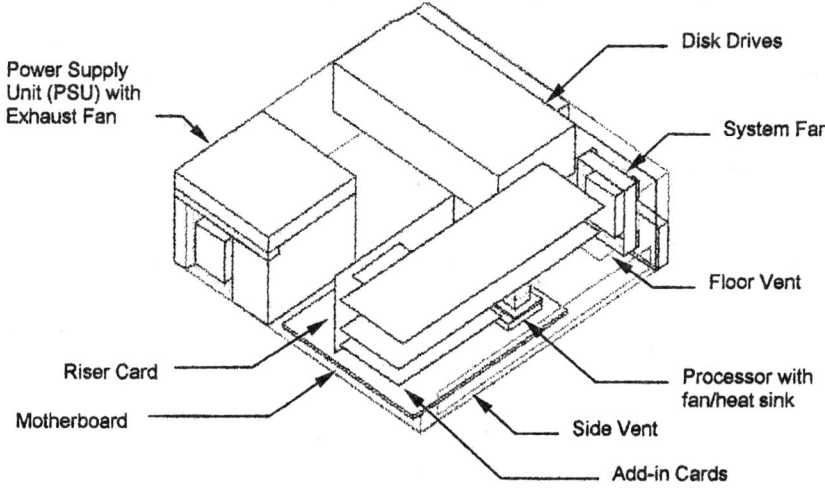

FIGURE 2 Model of personal computer chassis. (From Ref. 2.)

vent, diverting the fan flow away from the processor site and toward the PS. This effect virtually negates any benefit from the second fan as far as the CPU is concerned. The flow at the processor site comes up mainly from the side vents. Therefore, any increase or decrease in the flow through the side vent will have a more significant impact on the processor temperature than a change in flow from the fan.

The ineffectiveness of the system fan was compounded by the fan grill and mount design. Because of the high impedance of the grill and the gap between

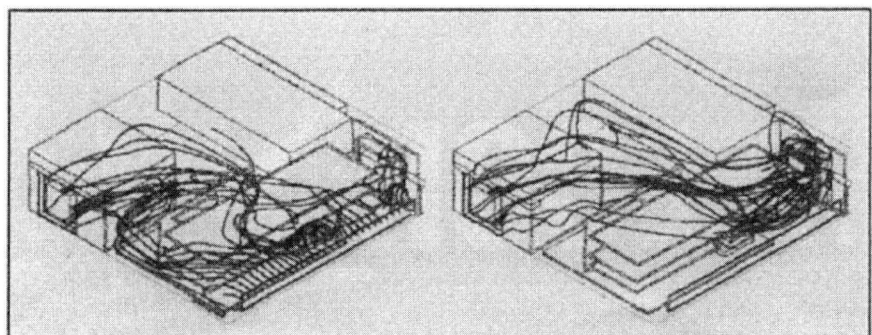

FIGURE 3 Particle traces show that most fan airflow bypasses the microprocessor. (From Ref. 2.)

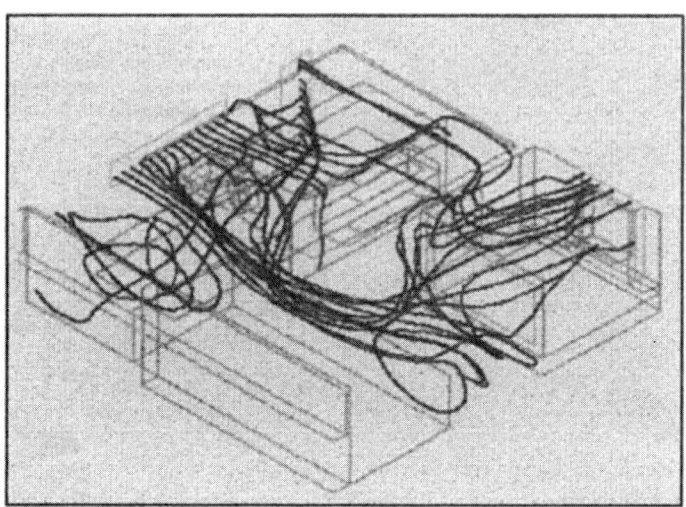

FIGURE 4 Vent modifications to Figure 3 provide strong airflow over microprocessor. (From Ref. 2.)

the fan mount and chassis wall, only 20% of the flow through the fan was fresh outside air. The remaining 80% of the air flow was preheated chassis air recirculated around the fan mount.

The analysis helped explain why the second PS fan reduced the flow through the side vent of the chassis. It also showed that the processor temperature actually declined when the second fan was shut off and demonstrated that the second fan could be eliminated without a thermal performance penalty, resulting in a cost saving. The analysis pointed the way to improving the thermal performance of the chassis. Modifying the chassis vents and eliminating the second PS fan provided the greatest performance improvement. Particle traces of the modified vent configuration demonstrate improved flow at the processor site (Fig. 4).

Use of Computational Fluid Dynamics to Predict Component and Chassis Hot Spots

Computational fluid dynamics using commercially available thermal modeling software is helping many companies shorten their design cycle times and eliminate costly and time-consuming redesign steps by identifying hot spots within an electronic product. An example of such a plot is shown in Figure 5. Figures 6 and 7 show a CFD plot of an Intel Pentium® II processor with a heat sink and a plot of temperature across a PC motherboard for typical operating conditions, respectively (see color insert).

Thermal Management

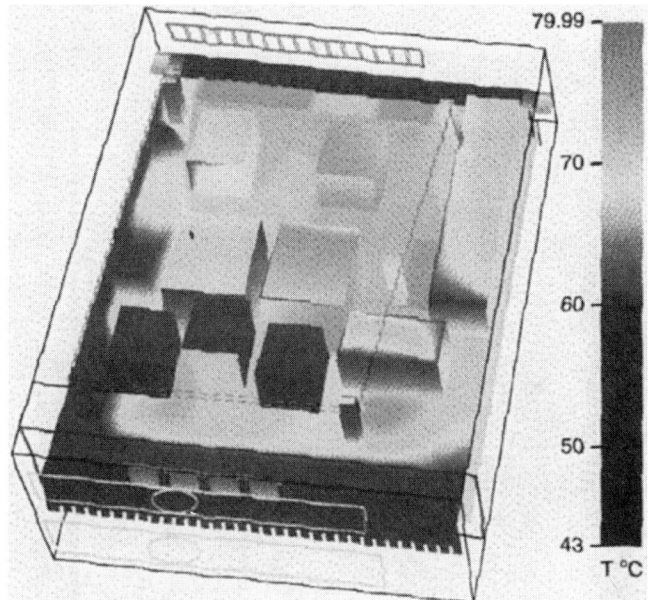

FIGURE 5 Color-coded component surface temperature analysis pinpointing hot spots within a power supply.

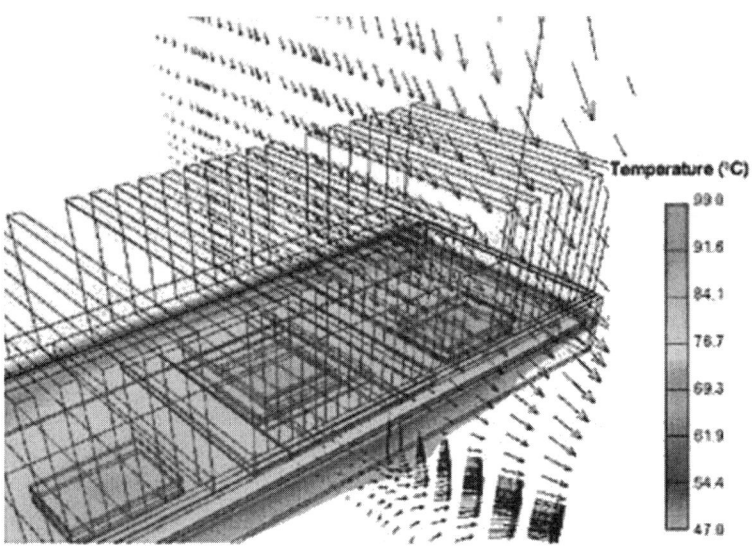

FIGURE 6 CFD plot of Pentium II processor and heat sink. (See color insert.)

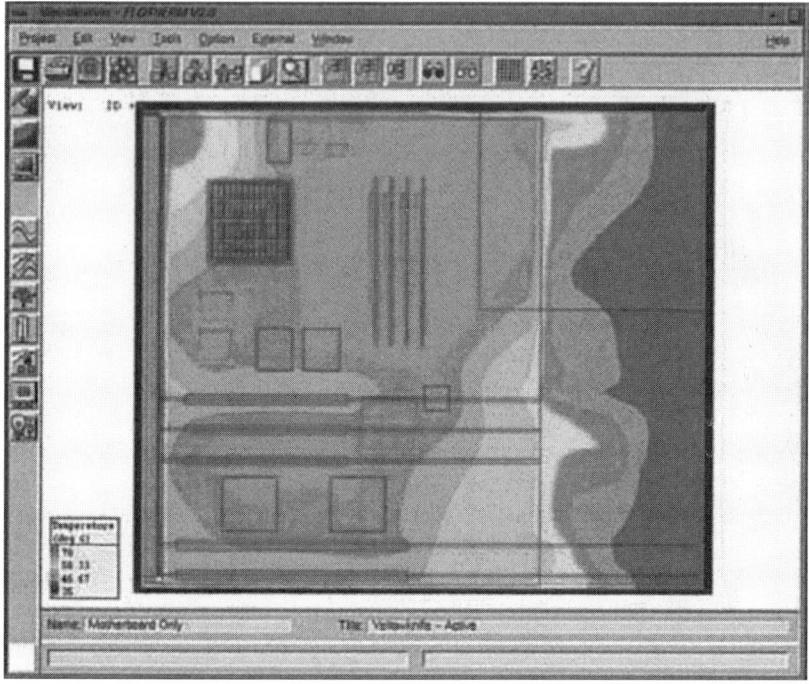

FIGURE 7   Temperatures in a PC motherboard. (See color insert.)

5.3 IMPACT OF HIGH-PERFORMANCE INTEGRATED CIRCUITS

Increased integrated circuit (IC) functional densities, number of I/Os, and operating speed and higher system packing densities result in increased power dissipation, higher ambient operating temperature, and thus higher heat generation. Today's ICs are generating double the power they were only several years ago. Table 1 shows the projected power dissipation trend based on the 1997 Semiconductor Industries Association (SIA) National Technology Roadmap for Semiconductors (NTRS). Notice that in the near future, we will see microprocessors with 100-W ratings (something completely unimaginable in the 1980s and early 1990s). They are here now. This means that today's ICs are operating in an accelerated temperature mode previously reserved for burn-in.

The current attention being given to thermal management at the chip level stems mostly from the quest for higher microprocessor performance gained by shorter clock cycles (that is, higher frequencies) and denser circuits. In CMOS circuits, dissipated power increases proportionally with frequency and capacitance

Thermal Management

TABLE 1 Power Dissipation Trends 1995–2012

	1995-1998	1998-2000	2001-2003	2004-2006	2007-2009	2010-2012
Semiconductor technology (µm)	0.35	0.25	0.18	0.13	0.10	0.07
Power (W)						
Commodity	1	1	1	1	1	1
Handheld PDA	2	2	2	2	2	2
Cost/performance	15	20	23	28	35	55
High-performance	90	110	125	140	160	180
Automotive	3	3	3	3	3	3

Source: SIA NTRS.

as well as with the square of the signal voltage. Capacitance, in turn, climbs with the number of integrated transistors and interconnections. To move heat out, cost-conscious designers are combining innovative engineering with conventional means such as heat sinks, fans, heat pipes, and interface materials. Lower operating voltages are also going a long way toward keeping heat manageable. In addition to pushing voltages down, microprocessor designers are designing in various power-reduction techniques that include limited usage and sleep/quiet modes.

Let's look at several examples of what has been happening in terms of power dissipation as the industry has progressed from one generation of microprocessors to the next. The 486 microprocessor–based personal computers drew 12 to 15 W, primarily concentrated in the processor itself. Typically, the power supply contained an embedded fan that cooled the system while a passive heat sink cooled the processor. However, as PCs moved into the first Pentium generation, which dissipated about 25 W, the traditional passive cooling methods for the processor became insufficient. Instead of needing only a heat sink, the processor now produced enough heat to also require a stream of cool air from a fan.

The Intel Pentium II microprocessor dissipates about 40 W; AMD's K6 microprocessor dissipates about 20 W. The high-performance Compaq Alpha series of microprocessors are both high-speed and high-power dissipation devices, as shown in Table 2. The latest Intel Itanium™ microprocessor (in 0.18-µm technology) dissipates 130 W (3).

Attention to increased heat is not limited to microprocessors. Power (and thus heat) dissipation problems for other components are looming larger than in the past. Product designers must look beyond the processor to memory, system chip sets, graphics controllers, and anything else that has a high clock rate, as well as conventional power components, capacitors, and disk drives in channeling heat away. Even small ICs in plastic packages, once adequately cooled by normal air movement, are getting denser, drawing more power, and getting hotter.

TABLE 2 Alpha Microprocessor Thermal Characteristics

	Alpha generation				
	21064	21164	21264	21264a	23164
Transistors (millions)	1.68	9.3	15.2	15.2	100
Die size (cm^2)	2.35	2.99	3.14	2.25	3.5
Process technology (μm)	0.75	0.50	0.35	0.25	0.18
Power supply (V)	3.3	3.3	2.3	2.1	1.5
Power dissipation (W)	30 at 200 MHz	50 at 300 MHz	72 at 667 MHz	90 at 750 MHz	100 at 1 GHz
Year introduced	1992	1994	1998	1999	2000

In the operation of an IC, electrons flow among tens, if not hundreds, of millions of transistors, consuming power. This produces heat that radiates outward through the chip package from the surface of the die, increasing the IC's junction temperature.

Exceeding the specified maximum junction temperature causes the chip to make errors in its calculations or perhaps fail completely. When IC designers shrink a chip and reduce its operating voltage, they also reduce its power dissipation and thus heat. However, shrinking a chip also means that heat-generating transistors are packed closer together. Thus, while the chip itself might not be as hot, the "power density"—the amount of heat concentrated on particular spots of the chip's surface—may begin to climb.

Although the air immediately surrounding a chip will initially cool the chip's surface, that air eventually warms and rises to the top of the personal computer chassis, where it encounters other warm air. If not ventilated, this volume of air becomes warmer and warmer, offering no avenue of escape for the heat generated by the chips. Efficient cooling methods are required. If not properly removed or managed, this heat will shorten the IC's overall life, even destroying the IC.

Heat buildup from ICs generally begins as the junction temperature rises until the heat finds a path to flow. Eventually, thermal equilibrium is reached during the steady-state operating temperature, which affects the device's mean time between failure (MTBF). As stated previously, a frequently used rule of thumb is that for each 10°C rise in junction temperature, there is a doubling in the failure rate for that component. Thus, lowering temperatures 10 to 15°C can approximately double the lifespan of the device. Accordingly, designers must consider a device's operating temperature as well as its safety margin.

Thermal Management

TABLE 3 Methods of Reducing Internal Package θ_{jc}

Increase the thermal conductivity of the plastic; ceramic; or metal package material, lead frame, and heat spreader.
Improve design of lead frame/heat spreader (area, thermal conductivity, separation from heat source).
Use heat spreaders.
Implement efficient wire bonding methods.
Use cavity-up or cavity-down package.
Ensure that no voids exist between the die and the package. (Voids act as stress concentrators, increasing T_j.)

Junction temperature is determined through the following relationship:

$$T_j = T_a + P_D(\theta_{jc} + \theta_{ca}) = T_j + P_D\theta_{ja} \tag{5.1}$$

Junction temperature (T_j) is a function of ambient temperature (T_a), the power dissipation (P_D), the thermal resistance between the case and junction (θ_{jc}), and the thermal resistance between the case and ambient (θ_{ca}) [junction to ambient thermal resistance (θ_{ja}) = $\theta_{jc} + \theta_{ca}$]. Here the following is assumed: uniform power and temperature distribution at the chip and one-dimensional heat flow.

Maximum junction temperature limits have been decreasing due to higher IC operating speed and thus increased power dissipation. The maximum allowable device operating temperature (T_j) has decreased from a range of 125–150°C to 90°C for reduced instruction set computing (RISC) microprocessors and to less than 70°C at the core for high-speed complex instruction set computing (CISC) microprocessors, for example. This has a significant impact on device and product reliability. As a result, the IC package thermal resistance (θ_{ja}) must decrease. Since θ_{ja} consists of two components: θ_{jc} and θ_{ca}, both θ_{jc} and θ_{ca} must be reduced to reduce θ_{ja}.

Methods of reducing θ_{jc} and θ_{ca} are listed in Tables 3 and 4, respectively. Table 5 lists the thermal resistance of various IC package types as a function of

TABLE 4 Methods of Reducing Package θ_{ca}

Use high-conductivity thermal grease.
Use external cooling (forced air or liquid cooling).
Use high-performance heat sinks to significantly increase volumetric size such that the size benefit of VLSI circuits can be utilized.
Use materials of matching coefficients of thermal expansion.
Use package-to-board (substrate) heat sinks.

TABLE 5 Electrical and Thermal Characteristics of Some Plastic Packages

Package type	No. of pins	Thermal resistance			
		Junction to case (θ_{JC})		Junction to ambient (θ_{JA})	
		A42 L.F.	Cu L.F.	A42 L.F.	Cu L.F.
DIP	8	79	35	184	110
	14	44	30	117	85
	16	47	29	122	80
	20	26	19	88	68
	24	34	20	76	66
	28	34	20	65	56
	40	36	18	60	48
	48	45	—	—	—
	64	46	—	—	—
SOP	8	—	45	236	159
	14	—	29	172	118
	16	—	27	156	110
	16w	—	21	119	97
	20	—	17	109	87
	24	—	15	94	75
	28	—	71	92	71
PLCC	20	—	56	—	20
	28	—	52	—	15
	44	—	45	—	16
	52	—	44	—	16
	68	—	43	—	13
	84	—	42	—	12
PQFP	84	—	14	—	47
	100	—	13	—	44
	132	—	12	—	40
	164	—	12	—	35
	196	—	11	—	30

the lead frame material: Alloy 42 and copper. Notice that copper lead frames offer lower thermal resistance.

Many components and packaging techniques rely on the effects of conduction cooling for a major portion of their thermal management. Components will experience the thermal resistances of the PCB in addition to those of the semiconductor packages. Given a fixed ambient temperature, designers can lower junction temperature by reducing either power consumption or the overall thermal resistance. Board layout can clearly influence the temperatures of components and

Thermal Management

thus a product's reliability. Also, the thermal impact of all components on each other and the PWA layout needs to be considered. How do you separate the heat-generating components on a PWA? If heat is confined to one part of the PWA, what is the overall impact on the PWA's performance? Should heat generating components be distributed across the PWA to even the temperature out?

5.4 MATERIAL CONSIDERATIONS

For designers, a broad selection of materials is available to manage and control heat in a wide range of applications. The success of any particular design with regard to thermal management materials will depend on the thoroughness of the research, the quality of the material, and its proper installation.

Since surface mount circuit boards and components encounter heat stress when the board goes through the soldering process and again after it is operating in the end product, designers must consider the construction and layout of the board. Today's printed wiring assemblies are complex structures consisting of myriad materials and components with a wide range of thermal expansion coefficients.

Tables 6 and 7 list the thermal conductivities of various materials and specifically of the materials in an integrated circuit connected to a printed circuit board, respectively.

The demand for increased and improved thermal management tools has been instrumental in developing and supporting emerging technologies. Some of these include

- High thermal conductivity materials in critical high-volume commercial electronics components
- Heat pipe solutions, broadly applied for high-volume cost-sensitive commercial electronics applications
- Nonextruded high-performance thermal solutions incorporating a variety of design materials and manufacturing methods
- Composite materials for high-performance commercial electronics materials
- Combination EMI shielding/thermal management materials and components (see Chapter 3)
- Adoption of high-performance interface materials to reduce overall thermal resistance
- Adoption of phase-change thermal interface materials as the primary solution for all semiconductor device interfaces
- Direct bonding to high-conductivity thermal substrates

At the first interface level, adhesives and phase-change materials are offering performance advantages over traditional greases and compressible pads.

TABLE 6 Thermal Conductivities of Various Materials

Material	Thermal conductivity (κ), W/cm·°K	Material	Thermal conductivity (κ), W/cm·°K
Metals		Insulators	
Silver	4.3	Diamond	20.0
Copper	4.0	AlN (low O_2 impurity)	2.30
Gold	2.97	Silicon carbide (SiC)	2.2
Copper–tungsten	2.48	Beryllia (BeO) (2.8 g/cc)	2.1
Aluminum	2.3	Beryllia (BeO) (1.8 g/cc)	0.6
Molybdenum	1.4	Alumina (Al_2O_3) (3.8 g/cc)	0.3
Brass	1.1	Alumina (Al_2O_3) (3.5 g/cc)	0.2
Nickel	0.92	Alumina (96%)	0.2
Solder (SnPb)	0.57	Alumina (92%)	0.18
Steel	0.5	Glass ceramic	0.05
Lead	0.4	Thermal greases	0.011
Stainless steel	0.29	Silicon dioxide (SiO_2)	0.01
Kovar	0.16	High-κ molding plastic	0.02
Silver-filled epoxy	0.008	Low-κ molding plastic	0.005
Semiconductors		Polyimide–glass	0.0035
Silicon	1.5	RTV	0.0031
Germanium	0.7	Epoxy glass (PC board)	0.003
Gallium arsenide	0.5	BCB	0.002
Liquids		FR4	0.002
Water	0.006	Polyimide (PI)	0.002
Liquid nitrogen (at 77°K)	0.001	Asbestos	0.001
Liquid helium (at 2°K)	0.0001	Teflon™	0.001
Freon 113	0.0073	Glass wool	0.0001
Gases			
Hydrogen	0.001		
Helium	0.001		
Oxygen	0.0002		
Air	0.0002		

Thermal Management

TABLE 7 Examples of Junction-to-Case Thermal Resistance

Description	Material	Thickness (cm)	κ (W/cm · °K)	ΔT per watt (°C)
Chip	Silicon	0.075	1.5	0.05
Die attach	Silver-filled epoxy	0.0025	0.008	0.313
	Solder	0.005	0.51	0.0098
	Epoxy	0.0025	0.002	1.25
Ceramic package	Alumina	0.08	0.2	0.4
	Copper tungsten	0.08	2.48	0.032
	Aluminum nitride	0.08	2.3	0.035
Interconnect	FR4 board	0.25	0.002	125.0
	Polyimide	0.005	0.002	2.5
Heat spreader	Copper	0.63	4.0	0.158
	Aluminum	0.63	2.3	0.274

Phase-change thermal interface materials have been an important thermal management link. They don't dissipate heat. They provide an efficient thermal conductive path for the heat to flow from the heat-generating source to a heat-dissipating device. These materials, when placed between the surface of the heat-generating component and a heat spreader, provide a path of minimum thermal resistance between these two surfaces. The ultimate goal of an interface material is to produce a minimum temperature differential between the component surface and the heat spreader surface.

5.5 EFFECT OF HEAT ON COMPONENTS, PRINTED CIRCUIT BOARDS, AND SOLDER

Integrated Circuits

Thermal analysis is important. It starts at the system level and works its way down to the individual IC die. System- and PWA-level analyses and measurements define local ambient package conditions. Heat must be reduced (managed) to lower the junction temperatures to acceptable values to relieve internal material stress conditions and manage device reliability. The reliability of an IC is directly affected by its operating junction temperature. The higher the temperature, the lower the reliability due to degradation occurring at interfaces. The objective is to ensure that the junction temperature of the IC is operating below its maximum allowable value. Another objective is to determine if the IC needs a heat sink or some external means of cooling.

Thermal issues are closely linked to electrical performance. For metal–oxide semiconductor (MOS) ICs, switching speed, threshold voltage, and noise

TABLE 8 Characteristics Affected by Elevated Temperature Operation

Materials/junctions	CMOS technology	Bipolar technology
Intrinsic carrier and concentration	Threshold voltage	Leakage current
Carrier mobility	Transconductance	Current gain
Junction breakdown	Time delay	Saturation Voltage
Diffusion length	Leakage current	Latch-up current

immunity degrade as temperature increases. For bipolar ICs, leakage current increases and saturation voltage and latch-up current decrease as temperature increases. When exposed to elevated temperature ICs exhibit parameter shifts, as listed in Table 8. One of the most fundamental limitations to using semiconductors at elevated temperatures is the increasing density of intrinsic, or thermally generated, carriers. This effect reduces the barrier height between n and p regions, causing an 8% per degree K increase in reverse-bias junction-leakage current.

The effects of elevated temperature on field effect devices include a 3- to 6-mV per degree K decrease in the threshold voltage (leading to decreased noise immunity) and increased drain-to-source leakage current (leading to an increased incidence of latch-up). Carrier mobility is also degraded at elevated temperatures by a factor of $T^{-1.5}$, which limits the maximum ambient-use temperature junction-isolated silicon devices to 200°C.

Devices must also be designed to address reliability concerns. Elevated temperatures accelerate the time-dependent dielectric breakdown of the gate oxide in a MOS field-effect transistor (FET), and can cause failure if the device is operated for several hours at 200°C and 8 MV/cm field strength, for example. However, these concerns can be eliminated by choosing an oxide thickness that decreases the electric field sufficiently. Similar tradeoffs must be addressed for electromigration. By designing for high temperatures (which includes increasing the cross-section of the metal lines and using lower current densities), electromigration concerns can be avoided in aluminum metallization at temperatures up to 250°C.

Integrated Circuit Wires and Wire Bonds

The stability of packaging materials and processes at high temperatures is an important concern as well. For example, elevated temperatures can result in excessive amounts of brittle intermetallic phases between gold wires and aluminum bond pads. At the same time, the asymmetric interdiffusion of gold and aluminum at elevated temperatures can cause Kirkendall voiding (or purple plague). These voids initiate cracks, which can quickly propagate through the brittle intermetallics causing the wire bond to fracture. Though not usually observed until 125°C, this phenomenon is greatly accelerated at temperatures above 175°C, particularly in the presence of breakdown products from the flame retardants found in plastic molding compounds.

Voiding can be slowed by using other wire bond systems with slower interdiffusion rates. Copper–gold systems only show void-related failures at temperatures greater than 250°C, while bond strength is retained in aluminum wires bonded to nickel coatings at temperatures up to 300°C. Monometallic systems, which are immune to intermetallic formation and galvanic corrosion concerns (such as Al-Al and Au-Au), have the highest use temperatures limited only by annealing of the wires.

Plastic Integrated Circuit Encapsulants

Plastic-encapsulated ICs are made exclusively with thermoset epoxies. As such, their ultimate use temperature is governed by the temperature at which the molding compound depolymerizes (between 190 and 230°C for most epoxies). There are concerns at temperatures below this as well. At temperatures above the glass transition temperature (T_g) (160 to 180°C for most epoxy encapsulants), the coefficient of thermal expansion (C_{TE}) of the encapsulant increases significantly and the elastic modulus decreases, severely compromising the reliability of plastic-encapsulated ICs.

Capacitors

Of the discrete passive components, capacitors are the most sensitive to elevated temperatures. The lack of compact, thermally stable, and high–energy density capacitors has been one of the most significant barriers to the development of high-temperature systems. For traditional ceramic dielectric materials, there is a fundamental tradeoff between dielectric constant and temperature stability. The capacitance of devices made with low–dielectric constant titanates, such as C0G or NP0, remains practically constant with temperature and shows little change with aging. The capacitance of devices made with high–dielectric constant titanates, such as X7R, is larger but exhibits wide variations with increases in temperature. In addition, the leakage currents become unacceptably high at elevated temperatures, making it difficult for the capacitor to hold a charge.

There are few alternatives. For example, standard polymer film capacitors are made of polyester and cannot be used at temperatures above 150°C because both the mechanical integrity and the insulation resistance begin to break down. Polymer films, such as PTFE, are mechanically and electrically stable at higher operating temperatures—showing minimal changes in dielectric constant and insulation resistance even after 1000 hr of exposure to 250°C; however, these films also have the lowest dielectric constant and are the most difficult to manufacture in very thin layers, which severely reduces the energy density of the capacitors.

The best option is to use stacks of temperature-compensated ceramic capacitors. New ceramic dielectric materials continue to offer improved high-temperature stability via tailoring of the microstructure or the composition of barium titanate–based mixtures. One particularly promising composition is X8R, which exhibits the energy density of X7R, but has a minimal change in capacitance to 150°C.

Printed Circuit Boards and Substrates

Printed circuit boards (PCBs) and substrates provide mechanical support for components, dissipate heat, and electrically interconnect components. Above their glass transition temperature, however, organic boards have trouble performing these functions. They begin to lose mechanical strength due to resin softening and often exhibit large discontinuous changes in their out-of-plane coefficients of thermal expansion. These changes can cause delamination between the resin and glass fibers in the board or, more commonly, between the copper traces and the resin. Furthermore, the insulation resistance of organic boards decreases significantly above T_g.

Optimization from a heat dissipation perspective begins with the PCB. Printed circuit boards using standard FR-4 material are limited to temperatures of less than 135°C, although high-temperature versions (for use to 180°C) are available. Those manufactured using bismaleimide triazine (BT), cyanate ester (CE), or polyimide materials can be used to 200°C or more, with quartz–polyimide boards useful to 260°C. Boards made with PTFE resin have a T_g greater than 300°C, but are not recommended for use above 120°C due to weak adhesion of the copper layer. The use of copper improves the PCB's thermal characteristics since its thermal conductivity is more than 1000 times as good as the base FR-4 material.

Clever design of the PCB along with the thoughtful placement of the power-dissipating packages can result in big improvements at virtually no cost. Using the copper of the PC board to spread the heat away from the package and innovative use of copper mounting pads, plated through-holes, and power planes can significantly reduce thermal resistance.

Since the PCB is, in effect, a conduit for heat from the package to the exterior, it is essential to obtain good thermal contact at both ends: the package attachment and the PCB mounting to the enclosure. At the package end, this can be achieved by soldering the package surface (in the case of a slug package) to the PCB or by pressure contact. At the other end, generous copper pads at the points of contact between the PCB and enclosure, along with secure mechanical attachment, complete this primary path for heat removal.

Not only are ICs getting faster and more powerful, but PC boards are shrinking in size. Today's smaller PC boards (such as those found in cell phones and PDAs) and product enclosures with their higher speeds demand more cooling

Thermal Management

than earlier devices. Their increased performance-to-package size ratios generate more heat, operate at higher temperatures, and thus have greater thermal management requirements. These include

- Increased use of any available metal surface to dissipate heat and move heat to an external surface
- Increased use of heat pipe-based thermal solutions to move heat to more accessible locations for airflow
- Increased demand for highly efficient thermal materials to reduce losses
- More difficult manufacturing requirements for product assembly

Solders

Most engineering materials are used to support mechanical loads only in applications where the use temperature in Kelvin is less than half the melting point. However, since the advent of surface mount technology, solder has been expected to provide not only electrical contact but also mechanical support at temperatures well in excess of this guideline. In fact, at only 100°C, eutectic solder reaches a temperature over 80% of its melting point and is already exhibiting Navier–Stokes flow. Above this temperature, shear strength is decreased to an unacceptable level and excessive relaxation is observed. In addition, copper–tin intermetallics can form between tin–lead solder and copper leads at elevated temperatures, which can weaken the fatigue strength of the joints over time.

There are a number of solders that can be used at temperatures to 200°C. These are listed in Table 9. Thus, the temperature that a PWA can withstand is the lowest maximum temperature of any of the components used in the assembly of the PWA (connectors, plastic ICs, discrete components, modules, etc.), the PCB and its materials, and the solder system used.

TABLE 9 Solidus Levels for High-Temperature Solders

Solder	Solidus (°C)	Solder	Solidus (°C)
Sn63Pb37	183	Pb90Sn10	268
Sn60Pb40	183	Au80Sn20	280
Pb60In40	195	Pn90In10	290
Sn92Ag3Bi5	210	Pb92In3Ag5	300
Sn93Ag5Cu2	217	Pb98Sn1Ag1	304
Sn96Ag4	221	Pb97Ag3	304
Sn95Pb5	223	Pb95Sn5	308
Sn95Sb5	235	Au88Ge12	356
Pb75In25	250	Au97Si3	363
Pb88Sn10Ag2	268		

The change to no-lead or lead-free solder, as a result of the environmental and health impact of lead, presents the electronics industry with reliability, manufacturability, availability, and price challenges. Generally speaking, most of the proposed materials and alloys have mechanical, thermal, electrical, and manufacturing properties that are inferior to lead–tin (Pb-Sn) solder and cost more. To date the electronics industry has not settled on a Pb-Sn replacement. Pure tin is a serious contender to replace Pb-Sn. From a thermal viewpoint, lead-free solders require a higher reflow temperature (increasing from about 245°C for Pb-Sn to >260°C for lead-free solder compounds) and thus present a great potential for component and PWA degradation and damage, impacting reliability. Tables 10 through 12 present the advantages and disadvantages of various lead replacement materials; the melting points of possible alternative alloys; and a comparison of the properties of pure tin with several material classifications of lead-free alloys, respectively.

5.6 COOLING SOLUTIONS

As a result of the previously mentioned advances in integrated circuits, printed circuit boards, and materials—and due to the drive for product miniaturization—it is no longer adequate to simply clamp on a heat sink selected out of a catalog after a PWA or module is designed. It's very important that thermal aspects be

TABLE 10 Lead Alternatives

Material	Advantages	Disadvantages
Bismuth	Lowers melting point	Byproduct of lead mining
	Improves wetting	Forms low-melting eutectic with tin and lead
	Low toxicity	Embrittlement concerns
Indium	Lowers melting point	In short supply
	Good elongation and strength	Expensive
		Corrosion concerns
Zinc	Minimal toxicity	Highly reactive
	Inexpensive	Oxidizes easily
	Lowers melting point	
	Readily available	
Antimony	Strengthen alloy	High toxicity
	Available supply	
Copper	Slightly improves wetting	Forms brittle intermetallic compounds with Sn
	Available supply	
Silver	Sn96.5Ag3.5 experienced alloy	Expensive

TABLE 11 Lead-Free Solders

Alloy composition	Melting point (°C)	Comment
Sn63Pb37	183	Low cost
Sn42Bi58	138	Too low melting point, depending on usage
		Unstable supply of Bi
Sn77.2In20Ag2.8	179–189	Higher cost
		Unstable supply
Sn85Bi10Zn5	168–190	Poor wettability
Sn91Zn9	198	Poor wettability
Sn91.7Ag3.5Bi4.8	205–210	Good wettability
Sn90Bi7.5Ag2Cu0.5	213–218	Poor reliability
		Poor control of composition
Sn96.3Ag3.2Cu0.5	217–218	Good reliability
Sn95Ag3.5In1.5	218	Unstable supply of In
Sn96.4Ag2.5Cu0.6Sb0.5	213–218	Poor control of composition
Sn96.5Ag3.5	221	Much experience

considered early in the design phase. All microprocessors and ASIC manufacturers offer thermal design assistance with their ICs. Two examples of this from Intel are Pentium III Processor Thermal Management and Willamette® Thermal Design Guidelines, which are available on the Internet to aid the circuit designer.

Up-front thermal management material consideration may actually enhance end-product design and lower manufacturing costs. For example, the realization that a desktop PC can be cooled without a system fan could result in a quieter end product. If the design engineer strategizes with the electrical layout designers, a more efficient and compact design will result. Up-front planning results in thermal design optimization from three different design perspectives:

Thermal—concentrating on the performance of the thermal material
Dynamic—designing or blending the material to operate within the actual conditions
Economic—using the most effective material manufacturing technology

Thermal management design has a significant impact on product package volume and shape. Heat generated inside the package must be moved to the surface of the package and/or evacuated from the inside of the package by air movement. The package surface area must be sufficient to maintain a specified temperature, and the package shape must accommodate the airflow requirements.

Early decisions concerning component placement and airflow can help prevent serious heat problems that call for extreme and costly measures. Typically, larger original equipment manufacturers (OEMs) are most likely to make that

TABLE 12 Grouping Pure Tin with Various Lead-Free Solder Alloys

Item	Sn-Pb	Sn-Ag	Sn-Cu	Sn-Bi	Pure tin	Palladium
Melting point (°C)	183	221	227	138	232	1554 diffusion
Solderability	Excellent	Inferior to Sn-Pb	Good	Good	Excellent	Excellent
Joint strength	Good	Good	Good	Good	Good	Good
Characteristics	High reliability	Good heat resistance	Stable liquid	Low melting point	Stable liquid	Full plating (pre-plated application)
Issues	Environmental issue of Pb	Alloy consisting of precious metal layer is formed Complex chemistry makes waste treatment difficult	Discoloration and whiskering Difficult process control	Lift-off Contaminated with Pb; primary source is lead mining	Whiskering, but recent developments yielded deposits that inhibit whisker growth	Availability problems High cost Not applicable for alloy 42
Cost of liquid[a]	1	1.2–1.5	1.0–1.2	1.2–1.5	1	6–7

[a] Normalized cost factor compared with Sn-Pb (1); 1.2–1.5 means that Sn-Ag costs 1.2 to 1.5 times that of Sn-Pb.

investment in design. Those that don't tend to rely more heavily, often after the fact, on the thermal-component supplier's products and expertise. A component-level solution will be less than optimal. If one works solely at the component level, tradeoffs and assumptions between performance and cost are made that often are not true.

In light of the growing need to handle the heat produced by today's high-speed and densely packed microprocessors and other components, five approaches to thermal management have been developed: venting, heat spreaders, heat sinks (both passive and active), package enclosure fans and blowers, and heat pipes.

5.6.1 Venting

Natural air currents flow within any enclosure. Taking advantage of these currents saves on long-term component cost. Using a computer modeling package, a designer can experiment with component placement and the addition of enclosure venting to determine an optimal solution. When these solutions fail to cool the device sufficiently, the addition of a fan is often the next step.

5.6.2 Heat Spreaders

All-plastic packages insulate the top of the device, making heat dissipation through top-mounted heat sinks difficult and more expensive. Heat spreaders, which are typically made of a tungsten–copper alloy and are placed directly over the chip, have the effect of increasing the chip's surface area, allowing more heat to be vented upward. For some lower-power devices, flexible copper spreaders attach with preapplied double-sided tape, offering a quick fix for borderline applications. Figure 8 shows a thermally enhanced flip chip ball grid array (BGA) with an internal heat spreader.

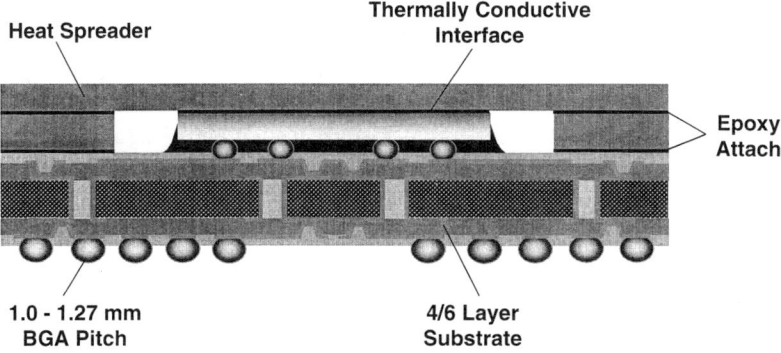

FIGURE 8 Thermally enhanced flip chip BGA.

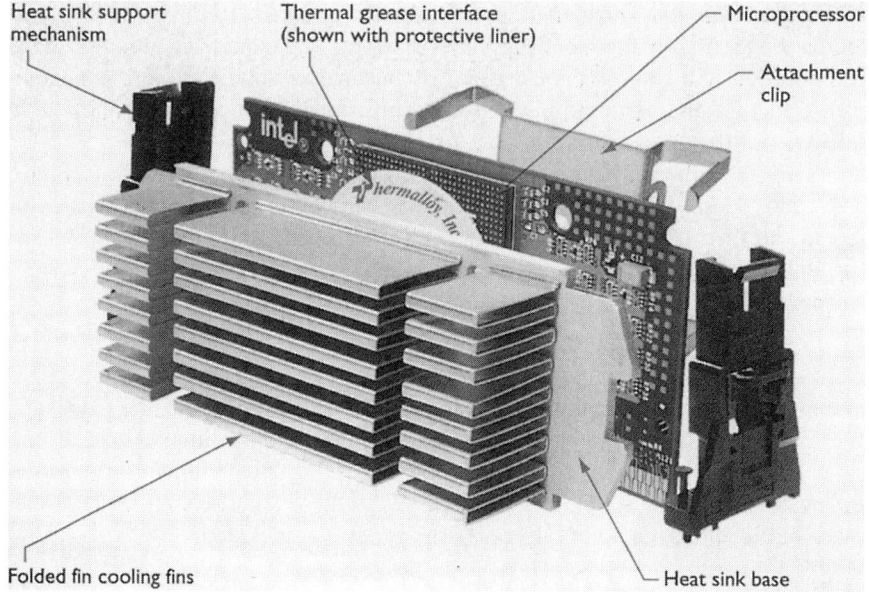

FIGURE 9 Folded fin heat sinks offer triple the amount of finned area for cooling.

Heat spreaders frequently are designed with a specific chip in mind. For example, the LB-B1 from International Electronic Research Corp. (Burbank, CA) measures 1.12 in. × 1.40 in. × 0.5 in. high and dissipates 16.5° C/W.

5.6.3 Passive Heat Sinks

Passive heat sinks use a mass of thermally conductive material (normally aluminum) to move heat away from the device into the airstream, where it can be carried away. Heat sinks spread the heat upward through fins and folds, which are vertical ridges or columns that allow heat to be conducted in three dimensions, as opposed to the two-dimensional length and width of heat spreaders.

Folding (Fig. 9) and segmenting the fins further increases the surface area to get more heat removal in the same physical envelope (size), although often at the expense of a large pressure drop across the heat sink. Pin fin and crosscut fin heat sinks are examples of this solution. Passive heat sinks optimize both cost and long-term reliability.

Heat Sinking New Packages

While BGA-packaged devices transfer more heat to the PWA than leaded devices, this type of package can affect the ability to dissipate sufficient heat to maintain high device reliability. Heat sink attachment is a major problem with

COLOR PLATES

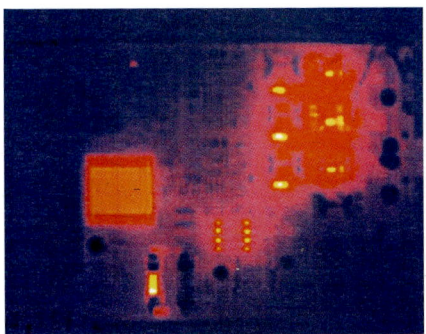 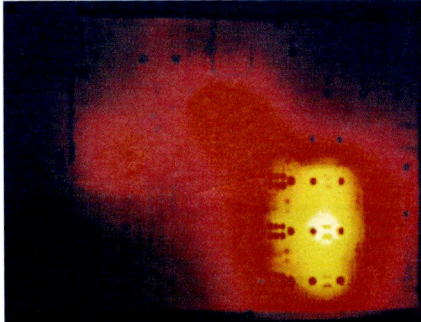

Figure 3.24 Examples of thermal imaging/scan applied to a DC control PWA. Left: temperature scan of component side of PWA showing operating temperature with 250W load. Right: temperature scan of back (solder) side of same PWA showing effect of thermal conduction from "warm" components on the top side. To understand the coordinates of the right scan, imagine the left scan is rolled 180° around its horizontal axis.

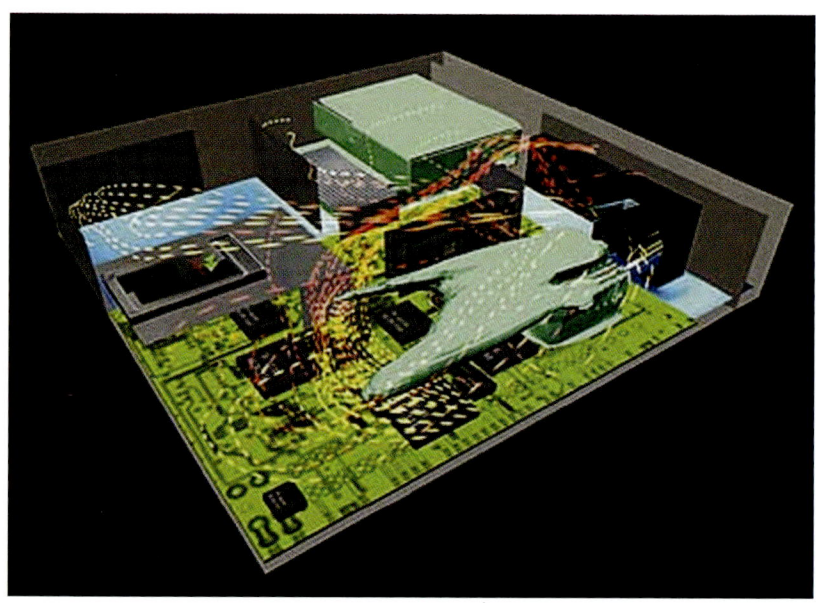

Figure 5.1 Airflow through a PC chassis.

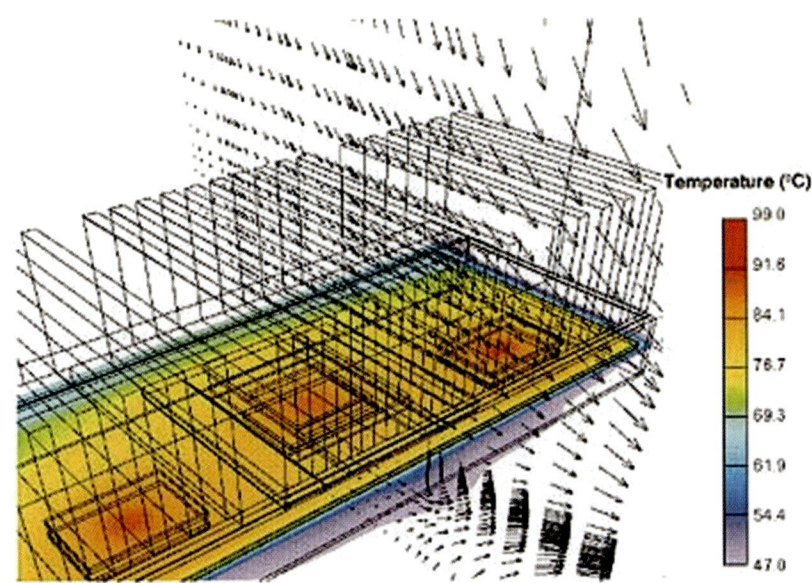

Figure 5.6 CFD plot of Pentium II processor and heat sink.

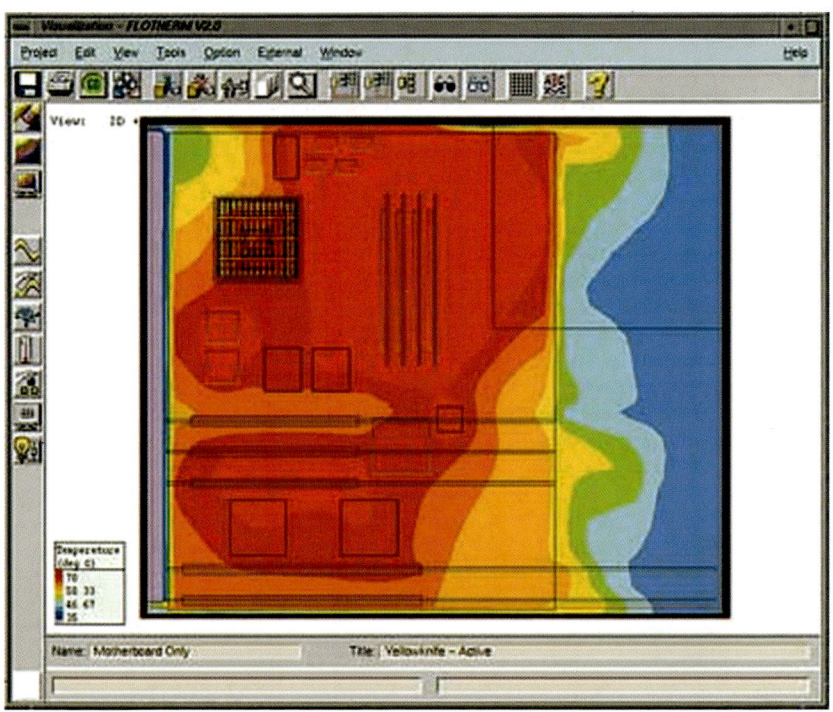

Figure 5.7 Temperatures in a PC motherboard.

Thermal Management

BGAs because of the reduced package size. As the need to dissipate more power increases, the optimal heat sink becomes heavier. Attaching a massive heat sink to a BGA and relying on the chip-to-board solder connection to withstand mechanical stresses can result in damage to both the BGA and PWA, adversely impacting both quality and reliability.

Care is needed to make sure that the heat sink's clamping pressure does not distort the BGA and thereby compromise the solder connections. To prevent premature failure caused by ball shear, well-designed, off-the-shelf heat sinks include spring-loaded pins or clips that allow the weight of the heat sink to be borne by the PC board instead of the BGA (Fig. 10).

Active Heat Sinks

When a passive heat sink cannot remove heat fast enough, a small fan may be added directly to the heat sink itself, making the heat sink an active component. These active heat sinks, often used to cool microprocessors, provide a dedicated

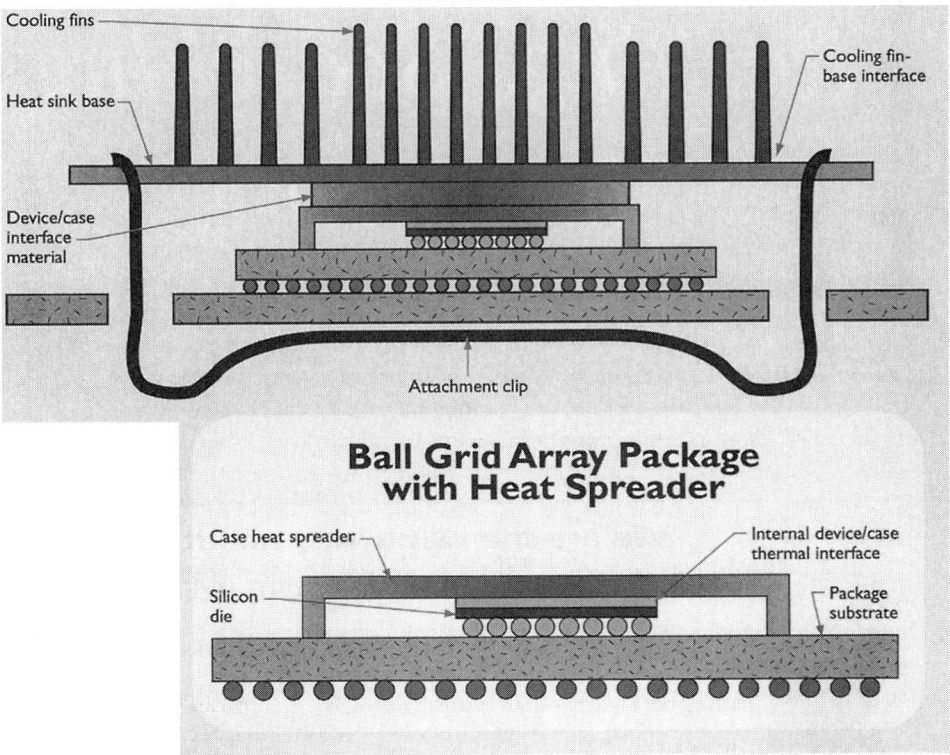

FIGURE 10 Cooling hierarchy conducts heat from the die through the package to the heat sink base and cooling fins and into the ambient.

FIGURE 11 Example of active heat sink used for cooling high-performance microprocessors.

airstream for a critical device (Fig. 11). Active heat sinks are often a good choice when an enclosure fan is impractical. As with enclosure fans, active heat sinks carry the drawbacks of reduced reliability, higher system cost, and higher system operating power.

Fans can be attached to heat sinks in several ways, including clip, thermal adhesive, thermal tape, or gels. A clip is usually designed with a specific chip in mind, including its physical as well as its thermal characteristics. For example, Intel's Celeron™ dissipates a reasonable amount of heat—about 12 W. But because the Celeron is not much more than a printed circuit board mounted vertically in a slot connector, the weight of the horizontally mounted heat sink may cause the board to warp. In that case, secondary support structures are needed. It is believed that aluminum heat sinks and fans have almost reached their performance limitations and will have no place in future electronic products.

Similarly, because extruded heat sinks or heat spreaders may have small irregularities, thermal grease or epoxies may be added to provide a conductive sealant. The sealant is a thermal interface between the component and the heat sink. Because air is a poor thermal conductor, a semiviscous, preformed grease may be used to fill air gaps less than 0.1 in. thick.

But thermal grease doesn't fill the air gaps very well. The problem with thermal grease is due to the application method. Too much grease may "leak" on the other components creating an environment conductive to dendritic growth and contamination, resulting in a reliability problem. This has led designers to use gels. The conformable nature of the gel enables it to fill all gaps between components and heat sinks with minimal pressure, avoiding damage to delicate components.

The Interpack and Semi-Therm conferences and Electronics Cooling Maga-

Thermal Management

zine website (http://www.electronics-cooling.com) deal with thermal grease and heat management issues of electronic systems/products in detail.

5.6.4 Enclosure Fans

The increased cooling provided by adding a fan to a system makes it a popular part of many thermal solutions. Increased airflow significantly lowers the temperature of critical components, while providing additional cooling for all the components in the enclosure. Increased airflow also increases the cooling efficiency of heat sinks, allowing a smaller or less efficient heat sink to perform adequately. Figure 12 is a three-dimensional block diagram of a computer workstation showing the airflow using an external fan.

Industry tests have shown that more heat is dissipated when a fan blows cool outside air into a personal computer rather than when it sucks warm air from inside the chassis and moves it outside. The amount of heat a fan dissipates depends on the volume of air the fan moves, the ambient temperature, and the difference between the chip temperature and the ambient temperature. Ideally, the temperature in a PC should be only 10 to 20°C warmer than the air outside the PC. Figure 13 shows the positive impact a fan (i.e., forced air cooling) provides in

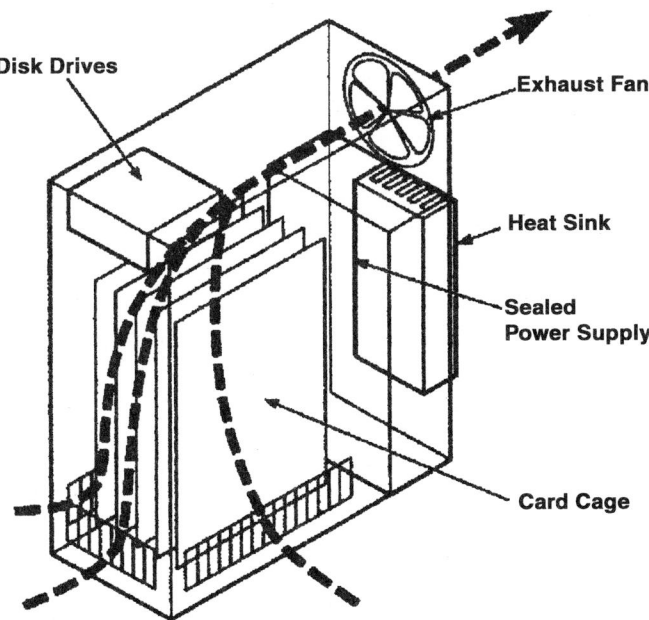

FIGURE 12 Diagram of a typical computer workstation showing airflow paths.

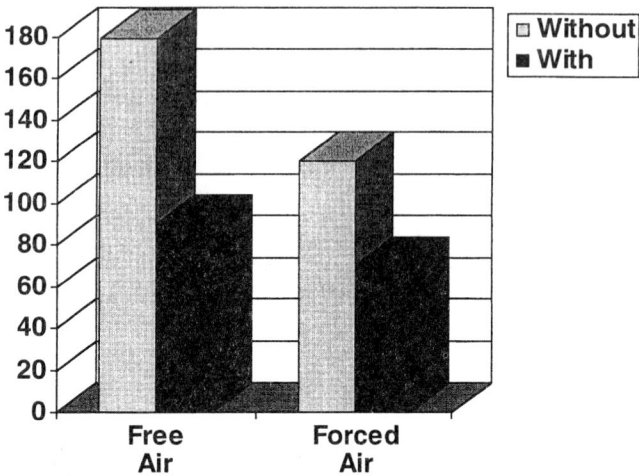

FIGURE 13 Diagram showing thermal resistance improvement with and without both heat spreaders and forced air cooling for a 16-pin plastic DIP. The results are similar for larger package types.

conjunction with a heat spreader, and Table 13 presents various IC package styles and the improvement in θ_{ja} that occurs with the addition of a fan.

The decision to add a fan to a system depends on a number of considerations. Mechanical operation makes fans inherently less reliable than a passive system. In small enclosures, the pressure drop between the inside and the outside of the enclosure can limit the efficiency of the fan. In battery-powered applications, such as a notebook computer, the current drawn by the fan can reduce battery life, thus reducing the perceived quality of the product.

Despite these drawbacks, fans often are able to provide efficient, reliable cooling for many applications. While fans move volumes of air, some PC systems also require blowers to generate air pressure. What happens is that as the air moves through the system its flow is hindered by the ridges of the add-on cards and the like. Even a PC designed with intake and exhaust fans may still require a blower to push out the warm, still air.

Fan design continues to evolve. For example, instead of simply spinning at a fixed rate, fans may include a connector to the power supply as well as an embedded thermal sensor to vary the speed as required by specific operating conditions.

5.6.5 Heat Pipes

While heat spreaders, heat sinks, fans, and blowers are the predominant means of thermal management, heat pipes are becoming more common, especially in

TABLE 13 Junction-to-Ambient Thermal Resistance as a Function of Air Cooling for Various IC Package Styles

Package/lead count	θ_{JA}(°C/W) 0 LFM	θ_{JA}(°C/W) 500 LFM
M-DIP		
16	82	52
18	80	50
20	78	48
24	77	47
40	56	44
PLCC		
20	92	52
44	47	31
68	43	30
84	43	30
124	42	30
PQFP		
132	43	33
TAPEPAK		
40	92	52
64	60	38
132	42	21

notebook PCs. A heat pipe (Fig. 14) is a tube within a tube filled with a low–boiling point liquid. Heat at the end of the pipe on the chip boils the liquid, and the vapor carries that heat to the other end of the tube. Releasing the heat, the liquid cools and returns to the area of the chip via a wick. As this cycle continues, heat is pulled continuously from the chip. A heat pipe absorbs heat generated by components such as CPUs deep inside the enclosed chassis, then transfers the heat to a convenient location for discharge. With no power-consuming mechanical parts, a heat pipe provides a silent, light-weight, space-saving, and maintenance-free thermal solution. A heat pipe usually conducts heat about three times more efficiently than does a copper heat sink. Heat pipe lengths are generally less than 1 ft, have variable widths, and can dissipate up to 50 W of power.

5.6.6 Other Cooling Methods

Spray Cooling

A recent means of accomplishing liquid cooling is called spray cooling. In this technique an inert fluid such as a fluorocarbon (like 3M's PF-5060 or FC-72) is

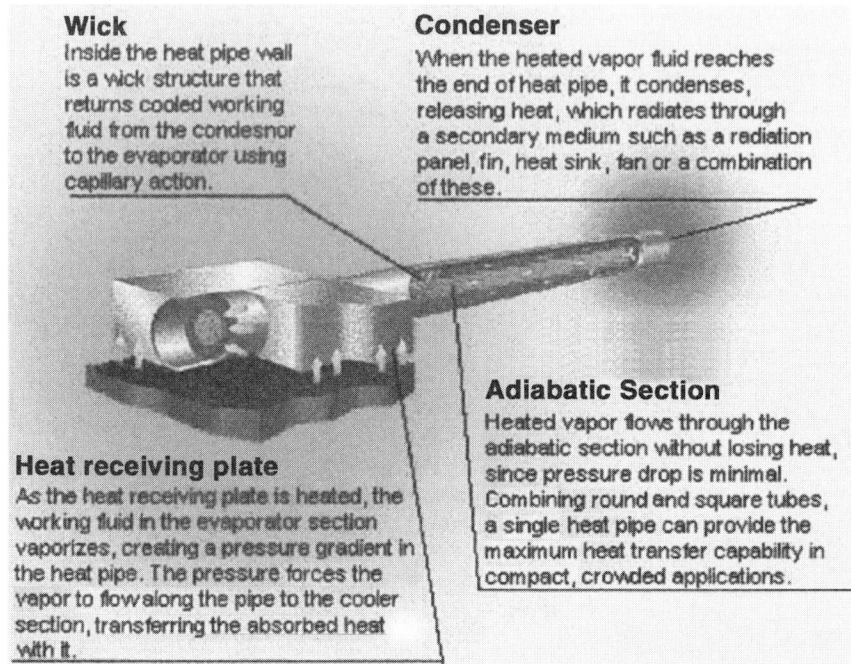

FIGURE 14 A heat pipe is a highly directional heat transport mechanism (not a heat spreading system) to direct heat to a remote location.

applied either directly to the surface of an IC or externally to an individual IC package and heat sink. Spray cooling can also be applied to an entire PWA, but the enclosure required becomes large and costly, takes up valuable space, and adds weight. Figure 15 shows a schematic diagram of the spray cooling technique applied to an IC package, while Figure 16 shows the technique being applied to an enclosed multichip module (MCM). Notice the sealed hermetic package required. In selecting a liquid cooling technique such as spray cooling, one has to compare the complexity of the heat transfer solution and technical issues involved (such as spray velocity, mass flow, rate of spray, droplet size and distribution, fluid/vapor pump to generate pressure, control system with feedback to pump, heat exchanger to ambient fluid reservoir, and hermetic environment) with the increased cost and added weight.

Thermoelectric Coolers

A thermoelectric cooler (TEC) is a small heat pump that is used in various applications where space limitations and reliability are paramount. The TEC operates

Thermal Management

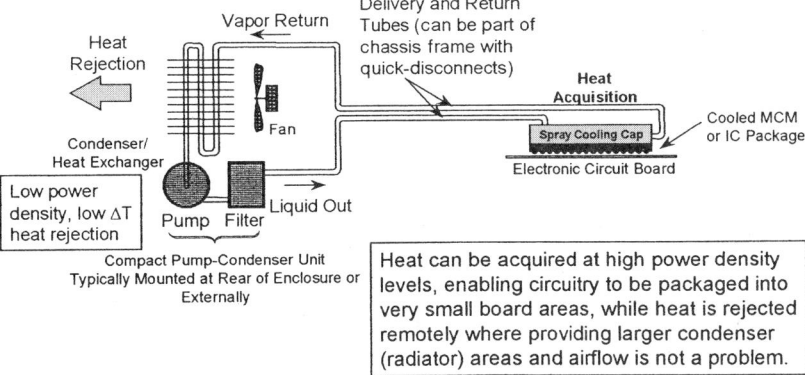

FIGURE 15 Schematic diagram of spray cooling system.

on direct current and may be used for heating or cooling by reversing the direction of current flow. This is achieved by moving heat from one side of the module to the other with current flow and the laws of thermodynamics. A typical single-stage cooler (Fig. 17) consists of two ceramic plates with p- and n-type semiconductor material (bismuth telluride) between the plates. The elements of semiconductor material are connected electrically in series and thermally in parallel.

When a positive DC voltage is applied to the n-type thermoelement, electrons pass from the p- to the n-type thermoelement, and the cold side temperature will decrease as heat is absorbed. The heat absorption (cooling) is proportional to the current and the number of thermoelectric couples. This heat is transferred to the hot side of the cooler, where it is dissipated into the heat sink and surrounding

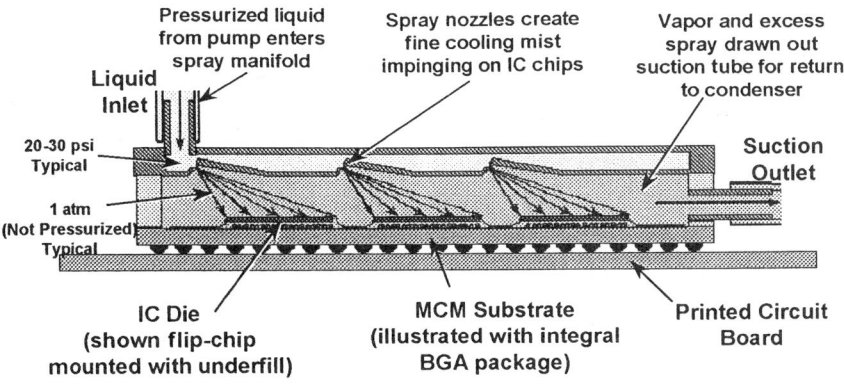

FIGURE 16 Spray cooling a multichip module package.

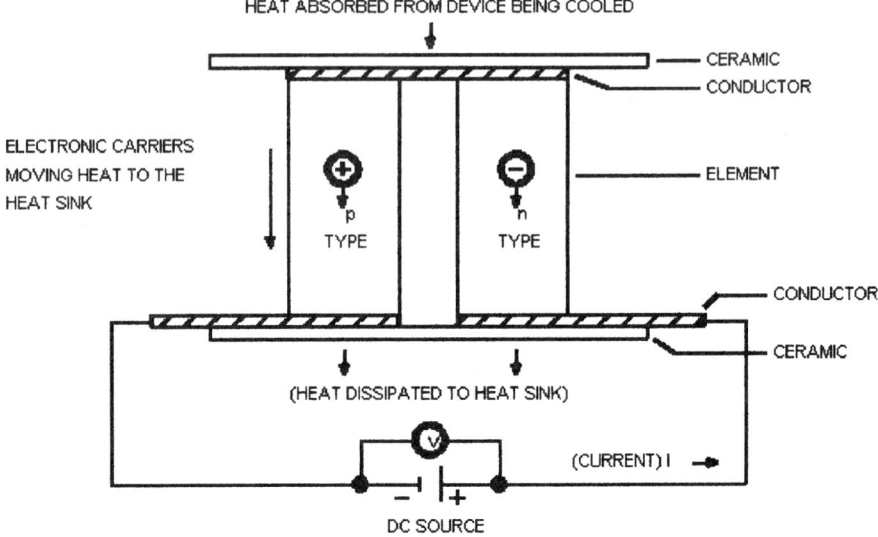

FIGURE 17 Typical single-stage thermoelectric cooler.

environment. Design and selection of the heat sink are crucial to the overall thermoelectric system operation and cooler selection. For proper thermoelectric management, all TECs require a heat sink and will be destroyed if operated without one. One typical single-stage TEC can achieve temperature differences up to 70°C, and transfer heat at a rate of 125 W.

The theories behind the operation of thermoelectric cooling can be traced back to the early 1800s. Jean Peltier discovered the existence of a heating/cooling effect when electric current passes through two conductors. Thomas Seebeck found that two dissimilar conductors at different temperatures would create an electromotive force or voltage. William Thomson (Lord Kelvin) showed that over a temperature gradient, a single conductor with current flow will have reversible heating and cooling. With these principles in mind and the introduction of semiconductor materials in the late 1950s, thermoelectric cooling has become a viable technology for small cooling applications.

Thermoelectric coolers (TECs) are mounted using one of the three methods: adhesive bonding, compression using thermal grease, or solder. Figure 18 shows a TEC with an attached heat sink being mounted with solder.

Metal Backplanes

Metal-core printed circuit boards, stamped plates on the underside of a laptop keyboard, and large copper pads on the surface of a printed circuit board all

Thermal Management

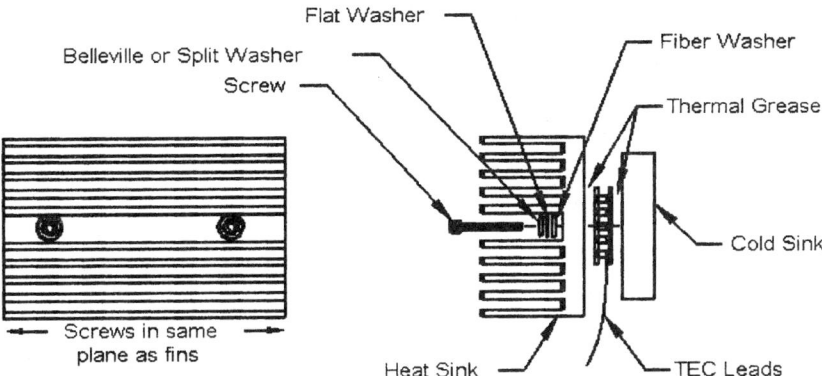

FIGURE 18 Mounting a TEC with solder.

employ large metallic areas to dissipate heat. A metal-core circuit board turns the entire substrate into a heat sink, augmenting heat transfer when space is at a premium. While effective in cooling hot components, the heat spreading of this technique also warms cooler devices, potentially shortening their lifespan. The 75% increase in cost over conventional substrates is another drawback of metal-core circuit boards.

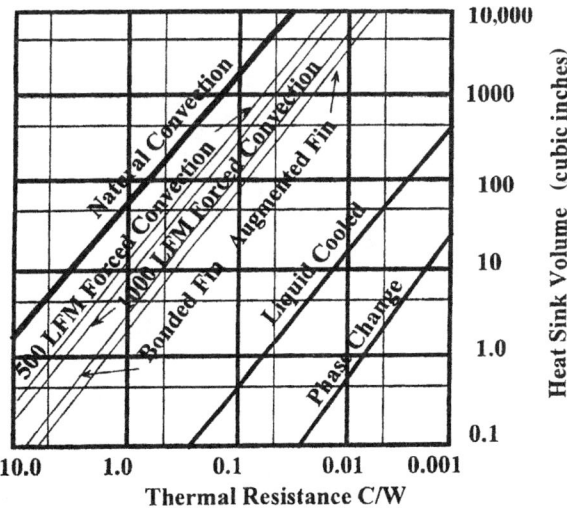

FIGURE 19 Effectiveness of various thermal cooling techniques.

When used with a heat pipe, stamped plates are a cost-effective way to cool laptop computers. Stamped aluminum plates also can cool power supplies and other heat-dissipating devices. Large copper pads incorporated into the printed circuit board design also can dissipate heat. However, copper pads must be large to dissipate even small amounts of heat. Therefore, they are not real-estate efficient.

Thermal Interfaces

The interface between the device and the thermal product used to cool it is an important factor in implementing a thermal solution. For example, a heat sink attached to a plastic package using double-sided tape cannot dissipate the same

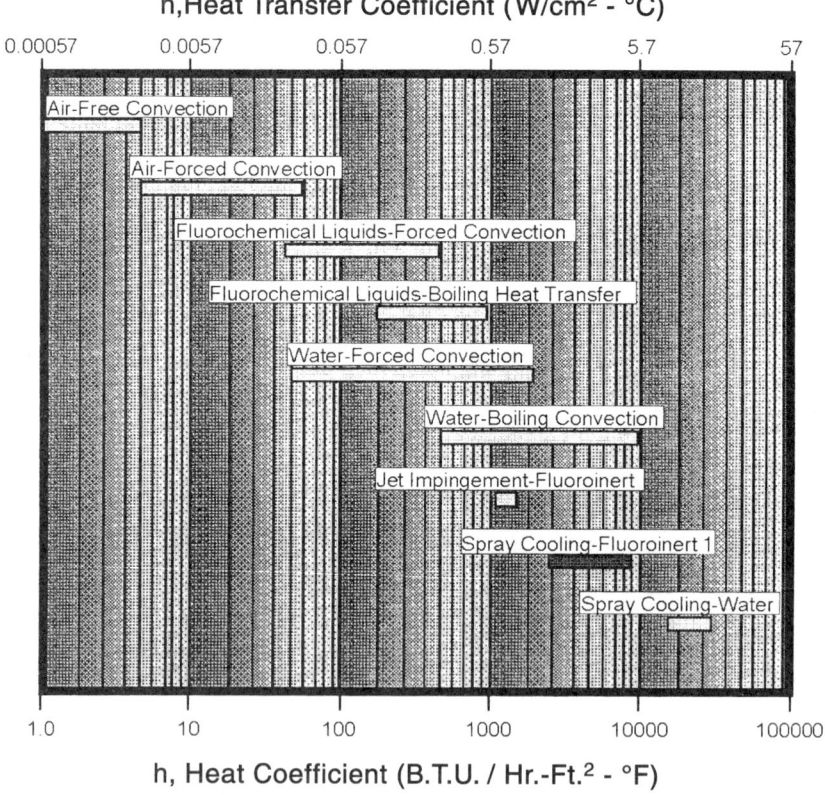

FIGURE 20 Comparison of heat transfer coefficients for various cooling techniques.

amount of heat as the same heat sink directly in contact with a thermal transfer plate on a similar package.

Microscopic air gaps between a semiconductor package and the heat sink caused by surface nonuniformity can degrade thermal performance. This degradation increases at higher operating temperatures. Interface materials appropriate to the package type reduce the variability induced by varying surface roughness. Since the interface thermal resistance is dependent upon applied force, the contact pressure becomes an integral design parameter of the thermal solution. If a package/device can withstand a limited amount of contact pressure, it is important that thermal calculations use the appropriate thermal resistance for that pressure. The chemical compatibility of the interface materials with the package type is another important factor. Plastic packages, especially those made using mold-release agents, may compromise the adherence of tape-applied heat sinks.

In summary, the selected thermal management solution for a specific application will be determined by the cost and performance requirements of that particular application. Manufacturing and assembly requirements also influence selection. Economic justification will always be a key consideration. Figure 19 summarizes the effectiveness of various cooling solutions. Figure 19 is a nomograph showing the progression of cooling solutions from natural convection to liquid cooling and the reduction in both thermal resistance and heat sink volume resulting from this progression. Figure 20 compares the heat transfer coefficient of various cooling techniques from free air convection to spray cooling.

5.7 OTHER CONSIDERATIONS

5.7.1 Thermal Versus Electromagnetic Compatibility Design Tradeoffs

Often traditional thermal and electromagnetic compatibility (EMC) solutions are at odds with one another. For example, for high-frequency processor signals, EMC requirements call for some kind of enclosure, limiting cooling air to hot devices. Since EMC is a regulatory requirement necessary to sell a product and thermal management is not, it is most often the thermal solutions that must be innovative (e.g., use of enclosure surfaces as heat sinks or shielding layers in the PWA as heat spreaders), especially in the shortened design cycles encountered today.

At the package level, a design that may be excellent for electrical performance may be thermally poor or may be far too expensive to manufacture. Tradeoffs are inevitable in this situation and software tools are becoming available that allow package designers to use a common database to assess the effect of design changes on both thermal and electrical performance.

5.7.2 Overall System Compatibility

In the telecommunications and networking industries, a cabinet of high-speed switching equipment may be installed in close proximity to other equipment and so must be shielded to prevent both emission and reception of electromagnetic radiation. Unfortunately, the act of shielding the shelves has an adverse effect on the natural convection cooling, which is becoming a requirement for designers.

These divergent needs can only be achieved with a concurrent design process involving the electrical designer, the EMC engineer, the mechanical designer, and the thermal engineer. Thermal design must therefore be part of the product design process from concept to final product testing.

ACKNOWLEDGMENTS

Portions of Section 5.2 were excerpted from Ref. 1. Used by permission of the author.

I would also like to thank *Future Circuits International* (www.mriresearch.com) for permission to use material in this chapter.

REFERENCES

1. Addison S. Emerging trends in thermal modeling. Electronic Packaging and Production, April 1999.
2. Konstad R. Thermal design of personal computer chassis. Future Circuits Int 2(1).
3. Ascierto J, Clenderin M. Itanium in hot seat as power issues boil over. Electronic Engineering Times, August 13, 2001.

FURTHER READING

1. Case in Point. Airflow modeling helps when customers turn up heat. EPP, April 2002.
2. Electronics Cooling magazine. See also www.electronics-cooling.com.

6
Electrostatic Discharge and Electromagnetic Compatibility

6.1 ELECTROSTATIC DISCHARGE

6.1.1 What Is Electrostatic Discharge and How Does It Occur?

Electrostatic discharge (ESD), a major cause of failure in electronic components, affects the reliability of electronic systems, including the functioning of an electronic component at any stage—during device fabrication, testing, handling, and assembly; printed circuit board (PCB) assembly and test; system integration and test; and field operation and handling of printed wiring assemblies (PWAs) in the field.

Electrostatic discharge is a charge-driven mechanism because the event occurs as a result of a charge imbalance. The current induced by an ESD event balances the charge between two objects. The ESD event has four major stages: (1) charge generation, (2) charge transfer, (3) charge conduction, and (4) charge-induced damage.

Electrostatic discharge occurs from an accumulation of charges on a surface due to reasons such as contact or friction. The process, called triboelectric charging, can occur when one wearing footwear walks onto a carpeted surface, when an integrated circuit slides through a plastic tray or tube, or during handling of components by handlers or robotic machinery.

In triboelectric charging, there is a transfer of electrons from one surface to the other in which one surface gets negatively charged, due to excess of electrons, and the other gets positively charged, due to a deficiency of electrons in equal measure. Materials differ in their capacity to accumulate or give up electrons.

The increasing miniaturization in electronics and the consequent use of small-geometry devices with thin layers has increased susceptibility to ESD damage; ESD is a silent killer of electronic devices that can destroy a device in nanoseconds, even at low voltages. Electrostatic discharge causes damage to an electronic device by causing either an excessive voltage stress or an abnormally high current discharge, resulting in catastrophic failure or performance degradation (i.e., a latent defect in the device that may surface later during system operation and cause device failure).

Taking a few simple precautions during device design, assembly, testing, storage, and handling, and the use of good circuit design and PWA layout techniques can minimize the effects of ESD and prevent damage to sensitive electronic components.

Damage is expensive—in the cost of the part; the processes; detection and repair; and in loss of reputation, as well as lost production time. Walking wounded parts can be extremely expensive and although the exact figures are difficult to establish, real overall costs to industry worldwide are certainly measured in terms of many millions, whatever the currency. Once damage has been done, it cannot normally be undone. Therefore, precautions need to be taken from cradle to grave.

6.1.2 Electrostatic Discharge–Induced Failure Mechanisms

The currents induced by ESD are extremely high. It is the current, directly or indirectly, that causes the physical damage observed in an ESD failure. Direct damage is caused by the power generated during the event. It melts a section of the device causing failure. Indirectly, the current generates a voltage by the ohmic resistance and nonlinear conduction along its path. Small voltages are generated when junctions are operated in a forward bias mode, but large voltages are generated when they are in a reverse bias mode. The reverse bias conduction causes thermal damage at lower current levels because the power dissipation is higher from the higher voltage across the junction. In addition, the voltage generated by this event weakens dielectrics by charge injection. The limiting case for this charge injection is dielectric rupture. Electrostatic discharge can affect an integrated circuit in many ways by causing

> Thermal overstress leading to melting of the metallization and damage to the various transistor junctions in the device.
> Intense electric fields that cause a breakdown of transistor thin gate oxide and the junctions themselves.

ESD and EMC

Latch-up in the internal circuit of complementary metal oxide semiconductor (CMOS) devices due to parasitic p-n-p-n structures and consequent device failure by electrical and thermal overstresses.

Latent defects caused by ESD may cause the device to malfunction or fail under field conditions. Among the commonly used devices, CMOS ICs are especially susceptible to damage due to ESD.

6.1.3 Preventing Electrostatic Discharge Damage

Preventing ESD-induced damage in electronic systems requires a multipronged approach that includes application of better design techniques at board and circuit level and observing appropriate precautions during handling of components, testing, assembly, system integration, shipment, and field operation. Protection from ESD looks first to minimize the charge generation and slow the charge transfer by controlling the environment where parts are handled and stored.

Some of the techniques that can be used to reduce ESD-related failures of electronic devices include

- Use of good circuit design techniques—both by proper choice of components and by using circuit level techniques such as protection networks at critical points in the circuit
- Use of good grounding and layout techniques. Any charge generated should be discharged to ground in a controlled manner
- Careful handling of ESD-sensitive components during assembly, production, and testing operations
- Use of an ESD-controlled production environment
- Use of appropriate antistatic packaging
- Proper shielding of the circuit

All IC manufacturers incorporate various protection circuits on the I/O pins of their ICs. The effectiveness of these protection circuits varies from supplier to supplier and from device type to device type. The fundamental approach in preventing ESD-induced failures is to start by selecting an IC with the appropriate ESD rating. Use a device with a higher ESD immunity that meets the application requirement to reduce the incidence of failures due to ESD. Electromagnetic interference (EMI) and ESD are closely related and can be controlled by using similar methods (ESD can be treated as a subset of EMI). Following are brief descriptions of techniques used to reduce the effects of ESD on electronic systems.

Circuit Design Techniques

High-speed logic transitions cause radiation of high-frequency fields resulting in interference to other devices on the PWA and to sensitive circuits in close proximity. Avoid high-speed devices in the design unless they are needed. But, for today's designs to be competitive requires the use of high speed ICs.

Anticipate problems that could arise in the field and tailor your circuit design appropriately.

Although IC manufacturers provide protective networks consisting of diodes to protect a CMOS device against ESD damage, a higher level of protection using external components is recommended in vulnerable circuit designs.

Use transient suppressor diodes at critical points in the circuit because they respond fast and clamp the voltage to a safe value when an overvoltage transient occurs. Keep the transient suppressor very close to the device terminals. Long leads and long PCB traces have parasitic inductances that cause voltage overshoots and ringing problems (if there is an ESD pulse).

A typical method for suppressing ESD transients, which can be used at the input stage of a circuit, is to slip a ferrite bead on the input lead to the ground.

Use series resistors to limit the rate of discharge.

A good low-impedance ground can divert the energy of the ESD transient efficiently. Maintaining a clean ground holds the key to the proper functioning of many electronic circuits using a mixture of analog and digital circuits.

Proper Printed Circuit Board Design

A properly routed PCB significantly contributes to ESD reduction. Magnetic flux lines exist in all energized cards due to the presence of various components and current flow through the circuit. A large loop area formed by conducting paths will enclose more of the magnetic flux, inducing current in the loop (due to the loop acting as an antenna). This loop current causes interfering fields which affect components in the circuit and the functioning of the circuit (closely routing supply and ground lines reduces loop areas).

Provide a large ground area on a PCB; convert unused area into a ground plane.

Place sensitive electronic components away from potential sources of ESD (such as transformers, coils, and connectors).

Run ground lines between very long stretches of signal lines to reduce loop areas.

Keep sensitive electronic components away from board edges so that human operators cannot accidentally cause ESD damage while handling the boards.

Multilayer boards with separate ground planes are preferred to double-sided boards.

ESD and EMC

Avoid edge-triggered devices. Instead use level-sensing logic with a validation strobe to improve ESD immunity of the circuit.

360° contact with the shield is necessary to prevent antenna effects (i.e., radiated fields).

Packaging guidelines, as applicable to EMI reduction, should be followed to reduce susceptibility to outside fields and to prevent unwanted radiation that can affect nearby equipment.

At the system level, include a marked wrist strap stud where possible.

Proper Materials for Packaging and Handling

Insulating materials have a surface resistivity greater than 10^{14} Ω/square and retain charge. It is advisable to keep insulating materials such as polyethylene, ceramics, and rubber away from electronic components and assembly areas.

Antistatic materials have a surface resistivity of 10^9 and 10^{14} Ω/square and resist the generation of static electricity. These materials have a short life for reuse and are meant for limited reuse applications, such as storing PWAs and electronic components. In view of the high surface resistivity, connecting this material to ground will not be effective in bleeding off any accumulated charge.

Static dissipative materials have a surface resistivity of 10^5 and 10^9 Ω/square. Due to the low surface resistivity, charges on a component can be diverted to ground if the material is used to protect a component against static charge and the static dissipative shield is grounded. Static charges can be generated in such materials by friction, but due to better surface conductivity the charges will spread across the surface. Generally such materials are used to cover floors, tabletops, assembly areas, and aprons.

The surface resistivity of conductive materials is less than 10^5 Ω/square. The charge accumulated on the surface of conductive material can be easily discharged to ground. Materials used for packaging electronic components and PWAs are generally plastics with conductive material impregnated.

Assembly and Production Techniques

No item should be allowed inside the ESD protect work area (EPA) that can generate and hold electrostatic charge. Examples include

- Packaging of polystyrene, untreated cling- and shrink-film, etc., whether for sensitive or nonsensitive parts (this charged packaging can come into contact with sensitive parts)
- Polystyrene or similar cups or other containers
- Photocopiers

- Sensitive parts should be packaged in low-charging, static-shielding containers (such as bags) when taken out of an EPA.
- Sensitive parts should be kept well away from strong electrostatic fields (e.g., CRT monitors) Very strong electrostatic field generators, such as contactors, arc welders and so on, should be kept well away from the EPA.
- Any sensitive part brought into the EPA should be resistively grounded or otherwise discharged before use. This can be through contact with a wrist-strapped person, through a resistively grounded (nonmetallic) bench, through ionization, or by other suitable means.
- Avoid potential differences between device pins during handling.
- CMOS ICs should be stored in antistatic tubes, bins, or conductive foams specially designed for storage. The conductive surfaces in contact with the CMOS devices will bleed off the accumulated charges.
- For soldering CMOS devices in PCBs, use a soldering iron in which the tip has a proper ground connection. Tools used to insert or remove CMOS ICs from boards or sockets should also be properly grounded.
- Do not insert or remove devices when the circuit power is on to prevent damage due to transient voltages.
- Unused communication connectors should be covered with static dissipative material when not in use to prevent charge buildup.
- Everyone in an EPA should be resistively grounded. In many cased, the most effective method is by the use of a wrist strap, but for mobile operators where the floor has a good defined resistance to ground, footstraps or conductive footwear may be used except while seated.
- Many garments can be static generating. Electrostatic discharge–protective garments should be worn to cover any static-generating clothes in the EPA. These garments need a path to ground, which typically will be through the skin and the wrist strap.
- Working surfaces need to be resistively grounded, typically through 1 MΩ, and to have a surface-to-surface resistance of less than 1000 MΩ. In few cases, metal surfaces may be used, but in general a point-to-point resistance of greater than 750 KΩ should be used.
- Where grounding through resistance is inappropriate, air ionizers should be considered as a means of preventing static buildup.
- Slightly higher humidity conditions provide a means to discharge any charge accumulated to ground and provide protection against static electricity buildup.
- Do not use tools with plastic handles (due to the triboelectric effect).
- Printed wire assemblies should be stored in antistatic bags.
- Antistatic precautions should be observed while handling PWAs in the field.

Electrostatic discharge awareness programs should be conducted periodically.

6.1.4 The Electrostatic Discharge Threat at the Printed Wire Assembly Level

Much effort has been put into characterizing the impact of electrostatic discharge on individual integrated circuits and on completed equipment such as computers. However, less time has been spent characterizing the ESD threat to ICs mounted on printed circuit boards. Since completed equipment can only be manufactured by mounting the ICs on subassemblies, the ESD threat to PCB-mounted ICs is an important concern.

Computer simulations have shown that the ESD threat to ICs mounted on PCBs may significantly exceed the threat to unmounted, individual ICs. Factors that impact ESD at the PWA are shown graphically in Figure 1 and are discussed in the following sections.

Sources of Electrostatic Discharge

While ESD threats to PCB-mounted ICs may have many sources, three sources are considered most probable:

1. Charged personnel
2. Charged PCB assemblies
3. The combination of a charged person holding a charged PCB

Charged Personnel. Often, when people think of ESD sources, they think of ESD from personnel. Unless wrist straps or other static-preventive measures

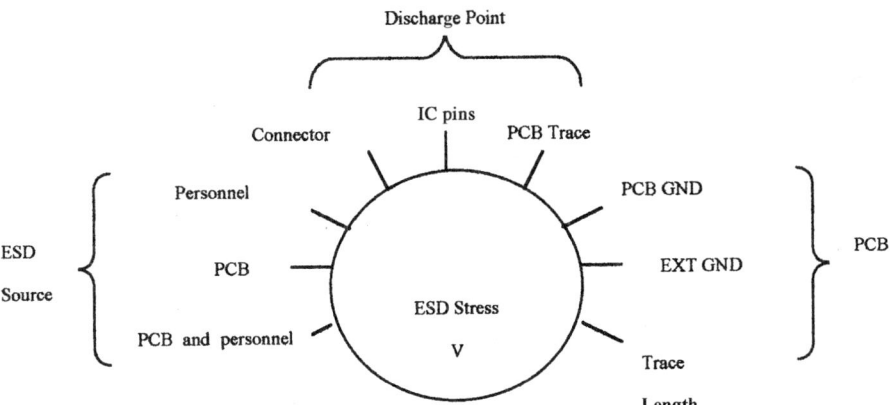

FIGURE 1 Factors that impact the level of electrostatic discharge to ICs on PWAs.

are used, personnel can become charged due to walking or other motion. When charged personnel contact (or nearly touch) metallic portions of a PWA, ESD will occur. The probability of charged personnel causing ESD damage to an IC is especially severe if the discharge from the person is via a metallic object such as a tool, a ring, a watch band, etc. This hand/metal ESD results in very high discharge current peaks.

Charged PWAs. Printed wire assemblies may also be a source of ESD. For example, assemblies can become charged when transported along a conveyer belt, during shipping, or when handled by a charged person. If a charged PWA contacts a conductive surface, or is plugged into a conductive assembly, while not in conductive contact with any other source charge, ESD will occur and discharge the PWA.

Charged PWA and Charged Personnel. If a charged person and a PWA are in conductive contact during the ESD event, the ESD will discharge both the person and the PWA. If a person walks across a carpeted floor while in conductive contact with a PWA, the person and the PWA may become charged. If the PWA then contacts a conductive surface, or is plugged into an equipment assembly, while still in conductive contact with the person, charged-PWA-and-person ESD occurs.

Places of Electrostatic Discharge

The places where ESD impinges on a PWA are called discharge points. Discharge points to PWAs can be grouped into three categories:

1. Directly to IC pins
2. Printed circuit board traces between ICs
3. Printed circuit board connector pins

With the possible exception of connector pins, discharge points could be expected to be physically located almost anywhere on the surface of the PCB assembly (PWA).

Integrated Circuit Pins. The pins of a PCB-mounted IC extend above the surface of the board itself. Because of this, an ESD arc can actually terminate on the pins of an IC. In this case, the ESD current will not travel to the device via a PCB trace. However, any trace connected to the IC pin may alter the character of the ESD threat.

PCB Traces. Since ICs do not cover the entire surface of a PCB assembly, a PCB trace may be the nearest metallic point to which an ESD threat may occur. In this case, the ESD arc will terminate not on an IC pin, but on a PCB trace between IC pins. This is especially true if an ESD-charged electrode (such as a probe tip) is located very close to a PCB trace, at a point equidistant between

ESD and EMC

two ICs. In this case, the ESD current flows to the IC via the PCB trace, modifying the ESD current waveform.

PCB Connector Pins. The connector pins of a PCB assembly are extremely likely to be subjected to ESD when the assembly is being installed in equipment or in a higher-level assembly. Thus, ESD to or from connector pins is often associated with ESD from a charged PCB or a charged PCB-and-person. Like ESD to traces, ESD to connector pins must flow via a PCB trace to the IC.

PCB Structures That Influence Electrostatic Discharge

In any ESD event, the character of the ESD threat is determined not only by the source of the ESD, but by the ESD receptor, which in this case is the PWA. When the receptor is a PWA, the path from the discharge point to the IC and the path from the IC to the ground reference are important. Also important is the structure of the ground reference. This structure includes the local ground and external ground reference common to both the IC and the ESD source.

Local Ground Structure. The local ground structure of a PCB is the section of the PCB's ground reference that is part of the PCB assembly itself. Multilayer PCBs with ground plane layers have the most extensive local ground structures. At the other extreme are PCB assemblies where the only local ground reference is provided by a single ground trace to the IC.

External Ground Structure. The local ground of a PCB may be connected to an external ground. This connection may be intentional (a direct connection to the AC power "green" ground) or unintentional (placing a PWA on a grounded metallic surface). Often PWAs have no metallic connection to any external ground (e.g., during transportation). In this case the only reference to an external ground is via stray capacitance from the PWA to the external ground.

Length of PCB Traces. In addition to the PCB ground structure, lengths of the PCB traces are also important. The trace lengths help determine the impedance and resonant frequency of each of the possible ESD current paths to ICs in a PCB. Studies have shown that ESD conducted to a device through short traces was worse than ESD conducted through long traces. Also the ESD threat was worse for PCBs with no ground plane than for multilayer PCBs using ground and power planes. Finally, the ESD energy threat was consistently worse for PWAs connected to external ground, rather than for floating PWAs. These results show the structure and layout of a PCB have a significant impact on the ESD threat level experienced by devices mounted on the PCB. The ESD threat to devices mounted on a PCB significantly exceeds the threat for unmounted devices. For a given PCB the sensitivity of the entire assembly is much worse than that of the individual device.

6.2 ELECTROMAGNETIC COMPATIBILITY

6.2.1 Introduction to Electromagnetic Compatibility

Electromagnetic compatibility (EMC) deals with the proper functioning of electronic equipment in an electromagnetic interface environment and compliance with various regulatory agency equipment interference generating and control requirements ("don't interfere with others; don't be interfered with; don't interfere with yourself"). Electromagnetic compatibility has been called the good housekeeping of electrical circuit design. It is applicable in all industries where electrical design occurs.

Electromagnetic compatibility is all about designing products that function within their intended environments in electromagnetic symbiosis with other products that must share the same environment. Products must not cause other products to malfunction or, in the most basic sense, disrupt communications. Chaos would result.

Today, products must be capable of unimpaired operation in an ever-changing electromagnetic world. Immunity to external electromagnetic events is no longer a design advantage; it is a product requirement. Many subassemblies and components that are used in the design of new products create their own microelectromagnetic environments which can disturb normal operation of nearby circuits. Learning to identify, avoid, and/or troubleshoot these electromagnetic problems is essential.

National regulatory agencies first promoted EMC because of concern over communications interference. Satisfying these agencies' requirements is still very much an essential task. The list of countries that have legislated mandatory compliance to EMC limits grows each year. Additionally, many country regulators are requiring EMC compliance reports be submitted for review and approval. Laboratory accreditation by country authorities is another requirement that must be addressed.

Regardless of the mandatory requirements for compliance to EMC standards, competitive pressures and customer demands for robust products are good justification for improving EMC designs.

Electromagnetic Interference

Electromagnetic interference (EMI) occurs when an electrical or electronic device's normal intended operation is compromised by another electrical or electronic device's usage.

Conducted EMI is caused when an interfering signal is propagated along power or signal lines. An example of conducted interference would be when your television goes fuzzy at the time your refrigerator compressor turns on.

Radiated electrical field (*E-field*) *interference* is an electric field that is propagated through the air. This field then couples into a receptor. An example

ESD and EMC

of E-field interference would be the noise you hear on your portable telephone when you bring it near your computer.

Radiated magnetic field (H-field) interference is a magnetic field that is propagated through the air. An example would be two computer monitors placed sufficiently close so that their displays appear to bounce and distort.

Recapping the previous section, electrostatic discharge occurs in the following manner: An object (usually conductive) builds up an electrostatic charge through some form of motion or exposure to high velocity air flows. This charge can be many kilovolts. When the charged object comes in contact with another (usually conductive) object at a different charge potential, an almost instantaneous electron charge transfer (discharge) occurs to normalize the potential between the two objects, often emitting a spark in the process. Depending upon the conductivity of the two objects, and their sizes, many amperes of current can flow during this transfer. Electrostatic discharge is most often demonstrated by shuffling across a carpet and touching someone, causing the signature electrical zap to occur. Confusion occurs with plastic insulators that can become charged to many kilovolts, but because of their high resistance dissipate low current levels very slowly.

Example of EMI-Caused Problems. Many switching power supplies are operating with a fundamental switching frequency between 100 kHz and 1 MHz. These designs emit very high electrical and magnetic fields and may disturb sensitive circuits by coupling noise. Digital designs using high-speed ICs (such as microprocessors and memories) with their fast rising and falling edges cause the same problems but at a more severe level. The microprocessor is by far the largest source of electromagnetic energy. Containing radiated and conducted emissions at the source (the CPU package) would make the system design easier for computer original equipment manufacturers (OEMs). To comply with FCC regulations, the system must be tested for emissions at up to five times the CPU operating frequency or 40 GHz, whichever is lower. This is necessary because of the harmonics that are generated from the base CPU frequency. The main component of EMI is a radiated electromagnetic wave with a wavelength that gets smaller as the frequency increases. Disk drive read/write heads are occasional victims. The trouble often manifests as high error rates or retries slowing data transfer rather than exhibiting a complete failure.

Electrostatic discharge sensitivity is often an elusive problem to solve, but most times it is the result of a compromised ground system. An example would be a product with high sensitivity to physical contact by operators such as a semiconductor wafer handler that randomly stops.

Cable television reception can easily be degraded by a pigtail connection of the shield to the input signal connector and further degraded by a nearby computer, monitor, or microwave oven.

Automotive electronic ignition can be both a nuisance to radio and a victim of radio signals. Some time back Porsches were known to be sensitive to certain ham radio frequencies, and the knowledgeable fun-loving EMC types would key their microphones as they neared such vehicles on the highways.

Cellular phones don't work so well in tunnels or in mountainous terrain.

Electromagnetic Compatibility Basics

Recognizing EMC problems in a design can be straightforward or extremely subtle. Focusing on problem avoidance by practicing some simple yet effective design rules will minimize or eliminate the need for corrective efforts late in the development cycle. These EMC design practices are applicable throughout all aspects of electrical/electronic product design.

Fourier analysis shows that the signals most focused on in EMC design are the clock and clock harmonic signals. This is because the emitted signals detected are continuous and periodic, typically square waves. The Fourier series model for a square or trapezoidal wave is a sine wave expansion of the fundamental frequency, explaining why digital signals are really complex analog signals. This is an important point to remember in EMC design. It is clock and clocklike signals that cause most common-mode noise. Random signals, such as data, do not cause the emissions that are measured. However, we must still be concerned with these signals from a signal integrity perspective.

Common Mode Noise. *Common mode noise* is undesirable signals or changes in signals that appear simultaneously and in phase at both input terminals and/or at all output terminals. These signals will be equal in amplitude and phase when measured at the input or outputs terminals, but the input amplitude may not equal the output amplitude because of the different characteristics of the common mode filters located at the input and output terminals. Most emissions-related problems are common mode.

Common mode filters must act on all conductors in an input or output circuit to be effective. A common toroidal core is usually included in the design.

Inside the product, both the power supply and the printed circuit board(s) will generate their own common mode noise profiles. Note: if the power supply output return (ground) and printed circuit board ground are conductively tied to chassis through a very low impedance bond, the internal common mode noise will be greatly reduced by the absence of a dV/dt between return current paths and chassis.

Differential Mode Noise. *Differential mode noise* is an undesirable signal appearing on the input or output terminals external to the enclosure versus one another and is measured with respect to chassis (ground). Any one of these signal measurements (noise) is not equal in amplitude or phase to any other one. Differential mode noise may also occur internally within the product.

Differential filtering is usually easy to implement, typically a simple inductive (L) or capacitive (C) component, assuming the ground system is working properly.

Ferrites. A *ferrite* is an EMI suppression component used to provide broadband, low-Q, and high-frequency impedance. Ferrites have both reactive (inductive) and lossy (resistive) properties. They have high resistivity and are nonconductive at low voltage.

Ferrites consist of sintered fine manganese–zinc or nickel–zinc ferrite powders that exhibit magnetic properties related to the frequency at which they are excited. By varying the permeability of these powders, it is possible to change the frequency response curve of ferrites. When a high-frequency current passes through the ferrite, these powders will attempt to reorient themselves much like what happens when a magnet is placed under iron particles on a piece of paper. Since the ferrite powder particles cannot move in place, the high frequency energy they absorb is converted into heat. Ferrites are very low maintenance and resistant to corrosion, rust, chemicals, and heat in the range of electronic circuits.

Ferrites are frequently used on cables for common mode noise reduction. They may also be designed into printed circuit boards as common mode I/O filters for unshielded I/O cables or in power or signal lines as differential or common mode filters. When used on cables, concerns for permanency must be addressed. A ferrite can become dislodged, misplaced, or forgotten during maintenance. Use design practices to avoid the need for ferrites on cables whenever possible.

Shielding. *Electromagnetic shielding* is the use of conductive barriers, partitions, or enclosures to improve a product's EMC performance. Shields are a design tool that completes an enclosure to meet EMC design goals and should not be considered to be a band-aid solution.

Shielding solutions may be applied internally to the product, on cables, or as the external enclosure. Parts of the external enclosure that form a shield require continuous conductive connections along their intersecting surfaces. Additionally, they must bond to chassis (ground) using similar low-impedance techniques. These connections are important to reduce radiated emissions, provide protection from externally radiated influences, and increase immunity to electrostatic discharge events.

Local shields may be valuable to reduce the EMI effects from noise sources that may otherwise disrupt normal use of the product or benefit the overall EMC performance.

Shields provide a conductive path for high-frequency currents, minimize the number of apertures, and function as design seams for continuous conductive interfaces. Conductive gaskets should be used on doors and access panels that are part of shielding. Noncontinuous conductive seams along enclosure shields have the tendency to form a dV/dt at the gap and radiate from the slot antenna formed.

Note: it is advisable to derate vendor claims by at least 10 dB since the tests are typically conducted in ideal situations.

Bonding. Bonding is the low-impedance interconnection to conductive ground potential subassemblies (including printed circuit board grounds), enclosure components to each other, and to chassis. Low impedance is much more than an ohmmeter reading 0.0 Ω. What is implied here is continuous conductive assembly to provide high-frequency currents with continuous conductive surfaces.

Some practical design hints to ensure proper bonding include the following:

Painted, dirty, and anodized surfaces are some of the biggest problems to avoid up front. Any painted surfaces should be masked to prevent overspray in the area where bonding is to occur. (The exception is where conductive paints are used on plastic enclosures; then paint is needed along the bond interface.)

Screw threads are inductive in nature and cannot be used alone to assure a suitable bond. They are also likely to oxidize over time.

Care should be used to avoid the use of dissimilar metals. Galvanic corrosion may result in very high–impedance bonds.

Conductive wire mesh or elastomer gaskets are very well suited to assure continuous bonds over long surface interfaces.

Choose conductive platings for all metal surfaces.

Designs that take advantage of continuous conductive connections along surface interfaces that are to be bonded are the most effective to support high-frequency currents. Sharp angle turns along a bond should be avoided since they will cause high-frequency current densities to increase at such locations, resulting in increased field strengths.

Loop Control. Designers need to consider that current flows in a loop and that the return current tries to follow as close to the intentional current as possible, but it follows the lowest impedance path to get there. Since all these return currents are trying to find their different ways back to their source, causing all kinds of noise on the power plane, it is best to take a proactive multipronged design approach. This includes designing an intentional return path (i.e., keeping to one reference plane), preventing nets from crossing splits in adjacent planes, and inserting bypass capacitors between planes to give the return current a low-impedance path back to the source pin.

Often in CPU server designs (see Fig. 2) the overall loop and loop currents are of the same size, but individual loop areas are dramatically different. It is the loop area that is the cause for concern with respect to ground loops, not the

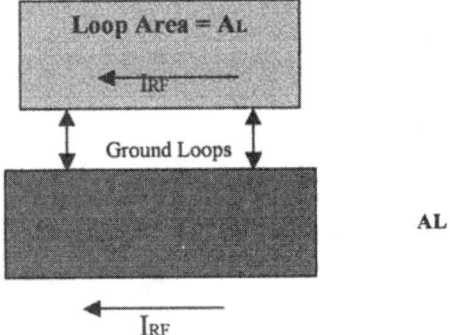

FIGURE 2   Loops and loops currents.

loop. It is also important to note that a ground plane has an infinite number of ground loops.

Each small ground loop in the lower block is carrying a loop current I_{RF}. What is happening is that the loop currents cancel each other everywhere except along the perimeter of the loop. The larger ground loop also carries loop current I_{RF} but has a greater loop area. Larger loops tend to emit and cause interference and are also more susceptible to disruption from external electromagnetic fields. This is why EMC design of two-sided printed circuit boards is improved by forming grids with both power and ground. The loop area can also be reduced by routing cables along ground potential chassis and enclosure walls.

Skin Effect.   *Skin effect* is the depth of penetration of a wave in a conducting medium. As frequency increases the "skin depth" decreases in inverse proportion to the square root of the frequency. Thus, as frequency increases, current flows closer to the surface of a conductor. This is one reason why bonding of conductive surfaces requires careful attention. It is desirable to provide as much continuous surface contact as possible for high-frequency currents to flow unimpeded.

Current Density.   The current density J in a wire of cross-sectional area A and carrying current I is

$$J = \frac{I}{A}$$

Thus, as the area increases, the current density will decrease, and the current in any one section of this area will be low.

Bonding and shielding attempt to distribute high-frequency return currents throughout the surfaces of the ground system. When this is completed, the high-

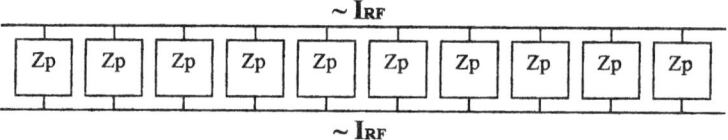

FIGURE 3 Distributing the current across a PWA.

frequency current density (dI/dt) is reduced to very low levels and results in a near equipotential surface, causing dV/dt to approach zero. By minimizing the current density, the generated field strengths can be reduced to a minimum. Using a distributed, low-impedance conductive current path to accomplish this is a very effective, economical, and reliable design approach.

Distributed Current Path. As I_{RF} crosses any boundary we know that it is flowing on the surface of the conductors. If the path impedance Z_P at any contact is low, and we have many contacts, then we have a distributed current path (Fig. 3). The best way to keep high-frequency currents from radiating is to maintain a distributed low-impedance current path where currents can freely flow. This is one of the design variables that are controlled and is effective, cost efficient, and very reliable.

6.2.2 Electromagnetic Compatibility Design Guidelines

Grounding

The grounding system is the backbone to EMC. If its integrity is preserved, a given product will achieve best EMC/ESD performance, lowest common mode noise levels, best EMC cost efficiency, and competitive advantage through customer satisfaction. Robust grounding is the design practice of conductively connecting all ground referenced circuitry and the chassis together through very high–conductivity, low-impedance (broadband) bonds. This approach attempts to establish a near equipotential electrical reference (system) ground by recognizing and leveraging from the fact that all grounds are analog.

In a grounding system, the "green wire" has no essential role in EMC design. Its sole role is to provide a current path to ground in the event of a fault condition that would result in a shock hazard. This conductor is of relatively high impedance at high frequency and looks much like an inductor. For EMC, a given product's radiated emissions profile should be unaffected by the presence of the green wire.

The conducted emissions profile may be impacted by the presence of the green wire if the design depends upon Y (line to ground) capacitors to reduce conducted emissions because Y capacitors increase leakage current in the green wire. A common mode choke used in the supply input can reduce conducted emissions without increasing leakage current. Since one generally designs with leakage current limits in mind, most input filter designs will use both Y capacitors and common mode chokes.

X (line-to-line or line-to-neutral) capacitors are used in the input filter circuit to reduce differential mode conducted emissions. X capacitors do not affect leakage current.

An improperly attached green wire can cause the power cable to radiate emissions. If the green wire loop is substantially long, it can pick up internal noise and act as a radiating antenna outside the enclosure. Design hint: keep the green wire within the product as short as possible (one inch long is a good rule of thumb) and routed next to chassis to reduce loop area.

So what is this equipotential system reference? Ground continuity is provided along as many interfaces as is reasonably possible, so that multiple low-impedance current paths are joined together (Fig. 4). The result is a distributed low-impedance ground system. High-frequency currents flow on the surface of these paths (skin effect) and have very low current densities. If the dI/dt approaches zero, then the dV/dt will also approach zero.

This begs the question at what potential is ground. If it is equipotential, who cares? The less dependent our system is on the potential of the ground system, the better it will perform. A satellite has an excellent ground system yet it has no conductive connection to earth. The same applies to a handheld, battery-powered electronic device or a large, floor-standing, earthquake-anchored, AC-powered electronic system.

Electrostatic discharge events rarely disturb a product with a robust ground system. This is because the entire product responds using its very near equipoten-

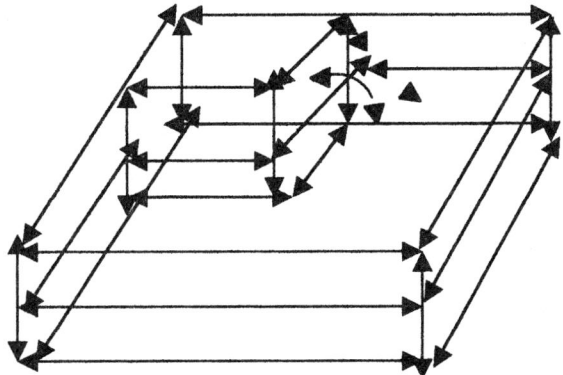

FIGURE 4   Equipotential system provides ground continuity across many interfaces.

tial reference to ground. Upon discharge, the high-frequency, high-current event uniformly distributes its charge throughout the ground system. If the product has a green wire, the charge will drain off to earth, otherwise the ground system of the still operating product will remain charged until it encounters a discharge path.

Chassis/Enclosures

The chassis or enclosure forms the outer protective and/or cosmetic packaging of the product and should be part of the EMC design. Metal enclosures should use plated protective coatings, but they must be conductive. Do not use anodized metals as they are nonconductive.

Design hints for metal enclosures:

1. Design assemblies such that connecting surfaces make a continuous conductive connection (bond) over the full length of interconnecting intersections. Seams of external covers should overlap. Hinges do not make suitable bonds. Conductive gaskets are often needed.
2. Select materials to minimize galvanic corrosion potential and assure that bonds remain intact over the product life.
3. Do not rely on mu metal; it is minimally effective at best.

Plastic enclosures may need to have a conductive coating if relied upon to reduce radiated emissions. Design hints for plastic enclosures:

1. Conductive coatings require clean plastic surfaces, well-radiused intersecting planes, and supports to allow coating to adhere. Continuous conductive mating assembly is required.
2. Coatings work well, but can be sensitive to abrasion and wear due to multiple assembly operations.

Power entry enclosure guidelines:

1. The green wire should terminate as immediately as possible once inside the enclosure. Typically within one inch. Minimize any loop area by routing next to chassis.
2. Power line filter modules should bond to the enclosure at the entry or as close as possible. A filter's output leads must be physically routed away from the input leads. If using a power filter with an integral IEC connector for a power cable, the filter must bond to the chassis 360° around their common interface. If an AC or DC input power cable is radiating with the filter in place, you may just need to add a common mode (common core toroid) ferrite on the leads exiting the filter (not to include the green wire).

3. At peripheral interfaces connector shells for shielded I/O cables must make a continuous 360° conductive bond with the enclosure where they penetrate the enclosure.
4. Unshielded interface cables will require some form of I/O filtering to prevent the I/O cable from becoming a radiating antenna. I/O filters, when used must be very near the I/O connector.

Power Supplies

Because of the need for high-efficiency power supply designs, the switching power supply is probably the most common technology encountered today. Switching supplies emit very strong electrical and magnetic fields. These can be minimized by using a ferrite common mode core on the AC input leads (not including the green wire) and by bonding the output return(s) to chassis.

Linear and ferroresonant supplies do not switch at high frequencies and are therefore not a radiated electric field concern. Their magnetic fields are, however, very strong and can be disruptive. A metal enclosure or localized barrier usually contains this problem. Generally these devices only require input filtering if they have high-frequency loads.

When using off-the-shelf power input filter modules, place them as near to the power entry as possible, and provide them with a good bond to chassis. If the AC power cable is radiating, try placing the AC input leads through a ferrite core (this is a high-frequency common mode filter). The green wire is never filtered, but instead routed next to chassis and attached to the ground lug using the shortest wire length possible (maximum 1 in. long).

The small modular DC/DC converters available today are switching at speeds from 100 kHz to over 1 MHz and because of their very fast rise times can contribute to both conducted and radiated emissions. The output returns should always be bonded to ground. If the input is safety extra low voltage (SELV), the input return should also be bonded to ground. A high-frequency filter capacitor should be placed at the input and output terminals. Also an input choke is good practice (differential if grounded return, otherwise common mode).

Printed Circuit Boards

Printed circuit boards can be safely designed for EMC by adhering to some basic rules. For multilayer PCBs,

1. Provide a perimeter conductive band around all layers within the printed circuit board except the ground (return) layers. These ground bands are of arbitrary width, but are normally 0.25 in. wide. Where a conflict exists with trace routing or components, the ground band may be necked down to a narrower dimension or selectively omitted to avoid component conflicts. The external perimeter ground bands

are to be free from solder mask, if possible, or masked with a pattern that exposes conductive access. This provides a means for persons handling the completed printed circuit assembly to bring the board to their electrostatic potential by accessing ground first, thus minimizing any ESD damage possibilities.
2. Conductively connect all perimeter ground bands in every layer to each other and to the ground plane(s) using vias stitched at an approximate 0.250-in. pitch, assuming 0.020-in. diameter vias. Note: vias can be solder pasted to accomplish the desired conductive access on each external surface.
3. The mounting means for the printed circuit board shall provide a low-impedance bond between the printed circuit board's ground and chassis. Standoff's are higher impedance than L brackets. Keep the skin effect in mind.
4. There shall be *no* isolations in ground planes, *no* traces, *no* moats, and *no* slots or splits. Voids are only allowed in conjunction with the use of a filter at an I/O connector. The ground plane is otherwise a complete copper plane with via holes or connects.
5. Where a printed circuit board has more than one ground plane, all connections to ground shall connect to every ground plane.
6. Some devices (components) are inherently noisy emitters, and when they are used it may be necessary to construct a Faraday enclosure (a conductive cage bonded to ground) around them. The printed circuit board's ground may be used to provide one of the six sides of the enclosure by designing a continuous, exposed, conductive outline (via stitched) ground band around the component's perimeter on the printed circuit board, much like the perimeter ground. The remainder of the enclosure can then be attached to this exposed conductive outline on the printed circuit board to form the remainder of a Faraday enclosure around the offending component.
7. For clock oscillators, provide a ground pad in the shape of the footprint of the clock oscillator that is located directly beneath the oscillator. This ground pad should then be tied to the ground pin. This pad provides a "keep-out" that cannot be accidentally violated when routing traces.
8. Clock and high-speed signal traces shall be routed on internal layers whenever possible (sandwiched between power and ground planes). For example, with a four-layer printed circuit board it is preferred to route clock traces nearest the ground planes as much as possible, although it is also acceptable to route next to the voltage plane.
9. If an I/O port is intended for use with a shielded cable, then the I/O connector shell must be conductively connected to the printed circuit board's ground plane(s).

ESD and EMC 303

10. When an I/O port is intended for use with unshielded cable, an I/O filter is required, and it must include a common mode stage that incorporates a common core shared by all I/O lines. The filter must be physically located as close to the I/O connector as possible. Traces to the input side of the filter cannot cross traces to the output side of the filter on any layer. Traces from the filter output to the I/O connector should be as direct as possible. A void is then designed in all layers (including the power and ground planes) of the printed circuit board between the filter input connections and the I/O connector pins. No traces may be routed in the void area except for those going from the filter to the I/O. A minimum separation of five trace widths of the I/O signal lines must be maintained between the void edges and the I/O traces. The void represents a transition zone between the filter on the printed circuit board and free space. The intent is to avoid coupling noise onto the filtered I/O lines.

Design Rules for Single- and Double-Sided PCBs. Single- and double-sided printed circuit boards do not enjoy the luxury of a continuous power and ground plane, so the preferred practice is to create two grids, thus simulating these planes using interconnected small loops. The method of doing this is to run traces in one direction on the top side of the board and perpendicular traces on the bottom side of the board. Then vias are inserted to conductively tie the traces of each, forming a power grid and a ground grid at the intersection of these respective traces if viewed perpendicular to the board plane. Power and ground grid linewidths should be at least five times the largest signal trace widths. The ground must still bond to the chassis at each mounting, and the perimeter band is still preferred.

With single-sided boards, the grid is formed using one or both sides of the board using wire jumpers soldered in place versus traces on the opposite side and via bonds.

Clock lines are usually routed with a ground trace (guard band) on one or both sides following the entire clock trace lengths and maintaining very close proximity to the clock trace(s). Guard bands should be symmetrical with the clock trace and should have no more than one trace width between the clock and guard traces.

All other rules previously described are essential and should be observed.

Grounds, Voids Moats, and Splits

It is important that robust grounding be used. The reason for repetition is that there are EMC design approaches that conflict with one another. *Always design the ground system first.* Selectively emphasize the use of voids, and never moats, splits, nor slots in the ground plane.

Voids are a means of providing a transition for unshielded cable I/O's.

Voids permit one to substantially reduce capacitive coupling of internal common mode noise to unshielded wire that should otherwise behave as transmitting antennas. *There is no other EMC use for voids in printed circuit boards.*

Moats are a recent concept that has been overused. If the common mode noise has been reduced to the lowest levels by tying the ground return and chassis together with low-impedance bonds, one must question why common mode noise should be increased by allowing a dV/dt in the ground. The use of moats may have some benefits, but only if done in the voltage plane and only if the signal traces to and from the moated components are routed over or under the bridge joining the moated voltage section. *Never moat a ground (return) plane.*

Splits or slots should never be considered in a ground plane. Splits or slots cause an impedance discontinuity seen by a trace with respect to ground, thereby distorting the desired signal quality. Additionally, they encourage a dV/dt between the ground reference on either side of the gap, affecting radiated emissions and assuring increased sensitivity to an external event such as ESD. *Splits or slots do not belong anywhere in ground planes.*

Component Placement

Component placement requires that careful attention be given to the performance and sensitivities of components to be used in a design. From an EMC perspective it is desirable to arrange components in zones so that related components are placed near the components they interact with. It is important to segregate sensitive analog components that might disrupt one another's performance. In the same way, power components can also be grouped so that their higher fields will not interact with other components or subassemblies. However, from other perspectives, such as thermal and mechanical design, it is necessary to distribute both the heat-generating components and the massive (physical size and weight) components across the PWA to prevent hot spots (large thermal gradients across the PWA) and mechanical stress (leading to physical damage), respectively.

It is important to ensure that traces are routed such that sensitive signal traces are routed adjacent to the ground plane and physically separated from all other traces (especially high-speed digital signals) by at least five trace widths. Remember that the magnetic properties of high-speed current flow will cause return currents to follow the mirror image of a trace in the ground plane.

High-frequency capacitors need to be placed at every oscillator and IC, and often several may be needed around larger arrays or application-specific integrated circuits (ASICs). The choice of capacitor value will depend upon the operating frequencies and the rise times of switching signals. Additionally, bulk electrolytic capacitors should be distributed throughout the board to supply large local demands for current and mitigate voltage sag.

Let's consider the issue of resistor–capacitor (RC) time constant and using capacitors to slow rise time. While it is true that in order for signals to propagate

ESD and EMC

faster, the values of the capacitor or series termination resistor need to be lower so that the rise time is shorter, the circuit designer needs to remember the noise comes from uncontrolled current. The shorter the rise time, the higher the dI/dt of the signal, allowing higher frequency content noise. Timing budgets might be met, but EMI can be running rampant. Tuning the rise time so that it's the minimum required to make timing requirements minimizes the noise current created. Take the example of a trapezoidal signal. At frequencies beginning at $1/\pi$ (τ − r) where τ is the pulse width, the spectrum of the waveform falls off at −20 dB/decade. For frequencies above $1/\pi$ (τ − r), where (τ − r) is the rise time, the fall-off is −40 dB/decade, a much greater level of attenuation. In order to reduce the spectral content quickly, the frequency of the −40-dB/decade region needs to be as low as possible, and pulse width and rise times need to be as large as possible, again trading this off with the timing budgets.

Clock terminations are essential at the origin and at every location where a clock signal is replicated. Whenever possible, a series of damping resistors should be placed directly after a clock signal exits a device. While there are many termination methods—pull-up/pull-down resistors, resistor–capacitor networks, and end series resistor terminations—the use of series resistors at the source terminations are preferred. The series clock termination resistor values are usually in the range of 10 to 100 Ω. It is not advisable to drive multiple gates from a single output since this may exceed the capability to supply the required load current.

Similarly, series terminations are effective in high-speed data or address lines where signal integrity is an issue. These resistors flatten the square wave by diminishing the overshoot and undershoot (ringing). Occasionally a component or subassembly will interfere with other parts of a product or will require supplemental shielding to achieve design goals. Local shields are effective in situations where there are high levels of magnetic or electric field radiation. These shields should be bonded to ground. Local shields may also be used to supplement the shielding effectiveness of the enclosure, thereby reducing overall radiated emissions.

Conductive shells of I/O connectors intended to mate with shielded cables must be conductively bonded to the printed circuit bond ground plane, and the conductive connector shell must bond to the enclosure 360° around as it exits the enclosure.

Cables

Cables can affect both the emissions and desired signal integrity characteristics. The following are some EMC design hints for cables:

1. Shielded cables rely upon 360° shield terminations at each end of the cable and at the devices to which they attach.
2. Ribbon cables may increase or alternate the number of ground conduc-

tors to reduce crosstalk. Shields on ribbon cables are difficult to terminate to ground to be beneficial at high frequencies.
3. Whenever possible route cables next to chassis or ground potential surfaces to minimize noise coupling or emissions. Also route cables away from other cables or circuits that may be noisy or which carry large currents. It is a good practice to route signal cables away from AC power cables.
4. Ferrites can be added to cables to reduce common mode noise. It is important to first attempt to correct the underlying cause for wanting to use a ferrite. Is it because a shielded cable has a rubber grommet between the cable shield and the connector housing? Do you have pigtail shield termination? Make sure that the ferrite solution is maintainable and secure.
5. At lower frequencies, twisting cable leads reduces lower frequency common mode noise.
6. Fiber optic cables are not susceptible to EMI, nor do they radiate EMI.

Bonding

I/O connectors intended for use with shielded cables must make a 360° bond with the enclosure at the place where they exit the enclosure. Where possible, choose connectors that can be easily bonded to the enclosure without leaving any unbonded gaps. The concept of 360° bonds must be carried through to include the mating connector shell, its housing, and its subsequent bond with the cable shield exactly at the cable's exit from the housing and similarly at the other end of the cable. Shields must be terminated in this manner at both ends of a cable to be effective. Otherwise the cable will radiate.

Some of the other connector styles commonly in use do not readily lend themselves to this type of bond because of their physical designs. When this occurs a conductive bond may have to be provided by a thin spring finger–like gasket or some other gasket cut to tightly fit the connector ground shell and reach the enclosure.

Printed circuit board grounds must achieve a good low-impedance bond with the enclosure along the I/O boundary exiting the enclosure.

ACKNOWLEDGMENT

The material in Section 6.2 courtesy of EMC Engineering Department, Tandem Computer Division, Compaq Computer Corporation.

FURTHER READING

1. Archambeault B. Eliminating the myths about printed circuit board power/ground plane decoupling. Int J EMC, ITEM Publications, 2001.

2. Brewer R. Slow down—you're going too fast. Int J EMC, ITEM Publications, 2001.
3. Brewer RW. The need for a universal EMC test standard. Evaluation Engineering, September 2002.
4. Dham V. ESD control in electronic equipment—a case study. Int J EMC, ITEM Publications, 2001.
5. Gerfer A. SMD ferrites and filter connectors—EMI problem solvers. Int J EMC, ITEM Publications, 2001.
6. ITEM and the EMC Journal, ITEM Publications.
7. Mayer JH. Link EMI to ESD events. Test & Measurement World, March 2002.
8. Weston DA. Electromagnetic Compatibility Principles and Applications. 2nd Ed. New York: Marcel Dekker, 2001.

7

Manufacturing/Production Practices

Figure 1 of Chapter 1 presented a block diagram of the major steps involved in delivering a product with end-to-end reliability (i.e., from customer requirements and expectations to customer satisfaction). This chapter addresses the details of the block labeled manufacturing/production. Figure 1 of this chapter expands this block to present some of the key elements that constitute the manufacturing/production of electronic products prior to delivery to a customer.

7.1 PRINTED WIRING ASSEMBLY MANUFACTURING

7.1.1 Introduction

Manufacturing/production of electronic products by its nature involves the assembly of components on a printed circuit board (PCB). The entire assembly is called a printed wiring assembly (PWA). In the electronics industry *printed wiring assembly*, *Printed Card Assembly* (PCA), and *circuit card assembly* (CCA) are used interchangeably to describe a fully populated and soldered PCB.

In the fast-paced and cost-sensitive world of electronic manufacturing major changes continually take place. Original equipment manufacturers (OEMs) no longer view manufacturing as a core competency or a competitive advantage. They want to focus on technology, product innovation, brand marketing, and sales. So they divested themselves of their manufacturing facilities, selling them

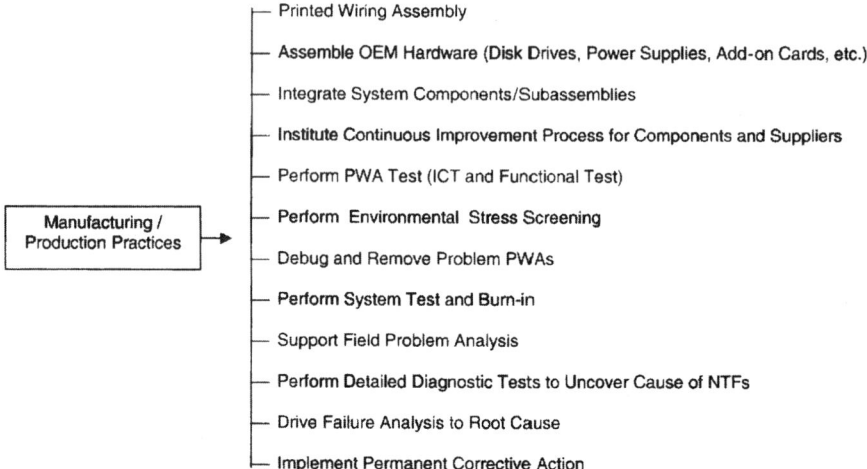

FIGURE 1 Electronic equipment manufacturing tasks.

to contract manufacturers (CMs) and established long-term relationships with them. [Contract manufacturers are also referred to as electronic manufacturing service (EMS) providers and contract electronic manufacturers (CEMs). I use these terms interchangeably.] As a result, OEMs have become increasingly dependent on contract manufacturers for some or all of their manufacturing requirements, especially PWA services.

Computer makers were the first to adopt an outsourcing manufacturing model, followed by both telecommunication equipment and wireless (cell phone) manufacturers, nimbly reacting to instantaneous changes in the marketplace. It is no longer unusual for a half dozen different brand-name computers to come off a single assembly line. (However, such products are not identical. Each OEM designs products with unique characteristics and specifications, while the EMS provider keeps customer jobs strictly separate to avoid conflicts of interest and other difficulties.)

7.1.2 Outsource Manufacturing

When considering outsourcing as a manufacturing strategy, it is important for a company to understand what is driving its need for outsourcing. Is it simply cost? Is it that management feels distracted by having to deal with many functions and activities it feels are ancillary to its true mission? Is it a strategic decision to not maintain a rapidly changing technology that, when closely examined, is associated with a process that can be deemed as a support or management process?

Manufacturing/Production Practices 311

Outsourcing originally began with the OEMs' need to manage the manufacturing peaks and valleys resulting from volatile, often unpredictable, sales volumes. In order to perform their own manufacturing, OEMs had to face three difficult choices:

1. Maintain sufficient staff to deliver product at sales peaks, knowing that workload would later drop along with volume.
2. Staff at some compromise level, carefully scheduling and distributing tasks to accommodate peak loads.
3. Hire staff at peak times and lay them off when sales drop.

Obviously, each of these options presented drawbacks. Contract manufacturers provided a solution: they offered to handle these peaks with their production capacity, and as OEMs gained more experience in dealing with them, routine manufacturing also shifted to EMS companies.

Since manufacturing operations are notoriously expensive to build and maintain, OEMs initially got out of manufacturing to reduce costs (reduce fixed assets and people). The inherent unpredictable market demand, sales volumes, product profitability, and resulting financial returns made an OEM's manufacturing operations exceedingly difficult to manage and maintain and thus led many OEMs to outsource that function. They needed to stop a hemorrhaging bottom line because they couldn't control their own processes; they refused or were unable to address deficiencies; they wanted to get rid of a headache; or they just jumped to outsourcing because everyone else was doing so. What initially weren't well thought out were the ramifications of and support required to manage their outsourcing decision/strategy. They just did it. Many companies eliminated assembly line workers and purchasing personnel in a carte blanche manner when they implemented outsource manufacturing, thinking they no longer needed them. The ramifications included lost critical skills, negative impact on morale of both affected and remaining employees, learning curve loss, inability to both innovate and provide flexibility and fast response to customers, and loss of control of manufacturing quality to a third party. However, the OEMs painfully learned that more (not less) skilled and technically competent workers were required to support the outsource strategy.

The CMs' focus on manufacturing allows them to achieve the lowest product costs and high product quality. The OEMs take advantage of the CMs' strengths to stay competitive in the electronics industry without most of the manufacturing overhead costs. The use of EMS providers adds enormous flexibility to an OEM's arsenal of available tools. Because they make a broad range of products for an equally varied group of customers, EMS providers have accumulated a wider array of knowledge, experience, and expertise than their OEM customers. As a result, EMS providers often suggest other manufacturing (including test) strategies and tactics to improve product performance, manufacturability, quality, reliability, and cost.

Moreover, variations in throughput requirements, product changes, and product mix generally present less of a challenge to an EMS provider than to an OEM. Many products go through regular manufacturing cycles. In the automobile industry, for example, test development and product requirements peak at the beginning of each model year. Others, including computers and cell phones, experience a volume bump during the holiday season. The EMS providers smooth out their own production schedules and optimize their equipment utilization by carefully managing the needs of individual customers and market segments so that their peaks seldom coincide, leveraging the advantage of year-round full production. This is because the demands of both the aggregate customer base and different market segment needs tend to provide an averaging function over a large number of manufacturing opportunities.

Table 1 summarizes the benefits of OEMs' outsourcing their manufacturing requirements. It is important to understand that once established, OEMs hardly ever dismantle these relationships.

Despite its growing popularity, outsourcing presents its own challenges, as Table 2 illustrates. Working with EMS providers requires tighter management practices over the entire manufacturing operation as well as better process documentation. Both are imperative to ensure that what the OEM wants and what the EMS provider builds coincide. Although many engineers regard these requirements with disdain, tight management can minimize the likelihood of design

TABLE 1 Benefits of Outsourcing Manufacturing

Increased focus on core competencies
Improved manufacturability and operating efficiencies resulting in reduced manufacturing costs
Improved quality
Faster time to market and time to technology
Increased flexibility and ability to respond more quickly to market changes
Improved competitiveness
Reduced capital investment
Increased revenue per employee
Latest technology availability/implementation (manufacturing and test tools)
More choices of manufacturing and test strategies
Design for manufacture strategies based on large volume of many different designs from different OEMs
Reduced and/or spread of manufacturing risk
Better assurance of supply in dynamic and changing market conditions

Manufacturing/Production Practices

TABLE 2   Challenges of Outsourcing

Greater exposure and risk when problems occur.
Lack of control of manufacturing process and activities.
Loss of manufacturing expertise and technology knowledge.
Little or no headcount reduction. One of the proposed early advantages of outsourcing PWAs was to reduce personnel headcount since it was thought that the manufacturing support functions were no longer required with an outsourcing strategy. This was found to be false. Various OEM manufacturing, components, and business experts are needed to support, backfill, and validate CM issues. Just as much, if not more, support is required to conduct and manage an effective PWA outsourcing strategy.
Greater coordination required with CM.
Inaccurate and changing build forecasts require much handholding.
A dedicated champion and central interface (point of contact) at the OEM is required to manage the outsourcing activity to prevent mixed messages to the CM.
Long distances for travel and communication.
Additional time required for problem solving.
Ability of CM to perform failure analysis and drive corrective action.
Dealing with proprietary information.
Need for appropriate support (normally not considered when outsourcing):
    Manufacturing engineering
    PWA test engineering
    Component engineering
    Failure analysis and problem resolution

errors and ensure that production processes remain in control, thereby maximizing product quality and functionality and increasing chances for market success.

Outsourcing creates a new and complex relationship system in the supply management chain. In the case of printed wiring assemblies, the quadrangle shown in Figure 2 best describes the relationship between OEMs and CMs. The insertion of the CM and the component supplier's authorized distributor between the OEM and component supplier serves to both separate the supplier and OEM and complicate the supply chain. This figure lists some of the key functions of each link in the supply chain and indicates the changing nature of these functions.

The Changing Role of the Contract Manufacturer

Early on (late 1980s and early 1990s) the sole service provided by the CM was consigned assembly services; the OEM provided the components and the bare PCB, while the EMS provider performed the assembly operation. As time passed, more and more value-added services have been added to the EMS provider's portfolio. These include component engineering [component and supplier selec-

OEM

- Circuit and system design
- Supplier/part selection and qualification for critical commodities
- System build
- System test
- Resident supplier
- System design, sales, marketing

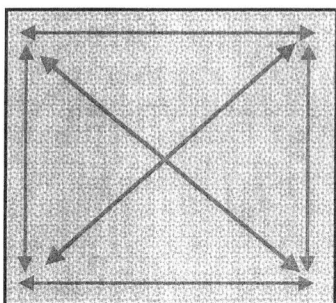

CONTRACT MANUFACTURER

- PCB assembly
- Supplier/part selection and qualification
- System build
- PWA test, troubleshooting, rework services
- Buys parts
- Product design
- Logistics

DISTRIBUTOR

- Buys parts
- Provides materials stocking and hub programs

PARTS SUPPLIER

- Supplies parts and application information and support

FIGURE 2 Electronic equipment manufacturing relational structure and dependencies.

tion and qualification, managing the approved vendors list (AVL), alternative sourcing issue resolution, etc.]; test development and fixturing; material procurement; and most recently full product design, integration, and assembly services as well as shipment and distribution of the product to the end customer. Component costs are reduced because OEMs benefit from the purchasing power, advanced processes, and manufacturing technology of the CM.

Today, many OEMs perform little or no manufacturing of their own. Organizations are using outsourcing to fundamentally change every part of their business because no company can hope to out-innovate all the competitors, potential competitors, suppliers, and external knowledge sources in its marketplace worldwide. Companies are focusing on their core competencies—what they are best in the world at—and then acquiring everything else through a strategic relationship with a CM, in which the "everything else" is the outsource service provider's core competencies. Some have referred to the CM as "infrastructure for rent." With this shift in functions, the OEM looks like a virtual corporation focusing on overall system design, marketing and sales, and managing the outsource service providers to ensure delivery of the product to the customer when requested. The EMS provider becomes a virtual extension of the OEM's business and looks more and more like the vertically integrated OEM of old. The trend toward the "virtual corporation"—in which different parts of the process are handled by different legal entities—will continue to accelerate.

The role of the CM is becoming more critical in the global electronics manufacturing sector. Currently (2002), the majority of outsourcing is given to EMS companies that provide more value-added services rather than single-function assembly houses. High-tech product development requires many different technical disciplines of expertise (industrial design, mechanical and thermal design, electronic circuit design, specification and documentation, test development, and turnkey manufacturing). For electronic equipment manufacturers, the many new technologies becoming available have caused them to consider outsourcing new product development and new product introduction (NPI), requiring a major adjustment in the strategic thinking and structure of modern companies.

One of the most important ways for large, complex-system OEMs to control product cost is to outsource systems build. The trend for OEMs to use CMs to build complex, multitechnology systems is accelerating. At the system level, integrating these subsystems into a complete functional system or larger subsystem presents a significant challenge. Until recently, complex systems have required engineering skills beyond the scope of most CMs, which is why these systems have remained integral to an OEM's engineering and manufacturing staffs. But this has changed as CMs have acquired these skills. A representative from a major aerospace corporation stated that it is now more cost effective to purchase completed systems than it is to purchase components and assemble them in-house.

The contract manufacturing services industry is a high-growth dynamic one. The services provided by contract manufacturers can be broadly segmented into PCB assembly, box or system build, and virtual corporation. The traditional role of the CM (PCB assembly) is decreasing as the scope of responsibilities and services offered increases, with more emphasis being placed on box build and system design. A more detailed description of each of these contract manufacturing services is now presented:

> *PCB assembly.* The CM assembles the components on the PCB to create a PWA. The PWA is tested for opens/shorts and to verify proper component placement. Functional testing of the PWA may be included. The OEM retains PWA design responsibility.
>
> *Box or system build.* This includes the PCB assembly as well as system integration to create the final product. The CM will functionally test the final product or "box" and ship it to the OEM or the OEM's customer. The OEM retains product design responsibility.
>
> *Virtual corporation.* The CM designs the product, assembles the PCB and builds the box. This is a turnkey relationship in which the CM assumes design responsibility for the product.

The trend depicting the changing nature of CM services provided is shown in Figure 3.

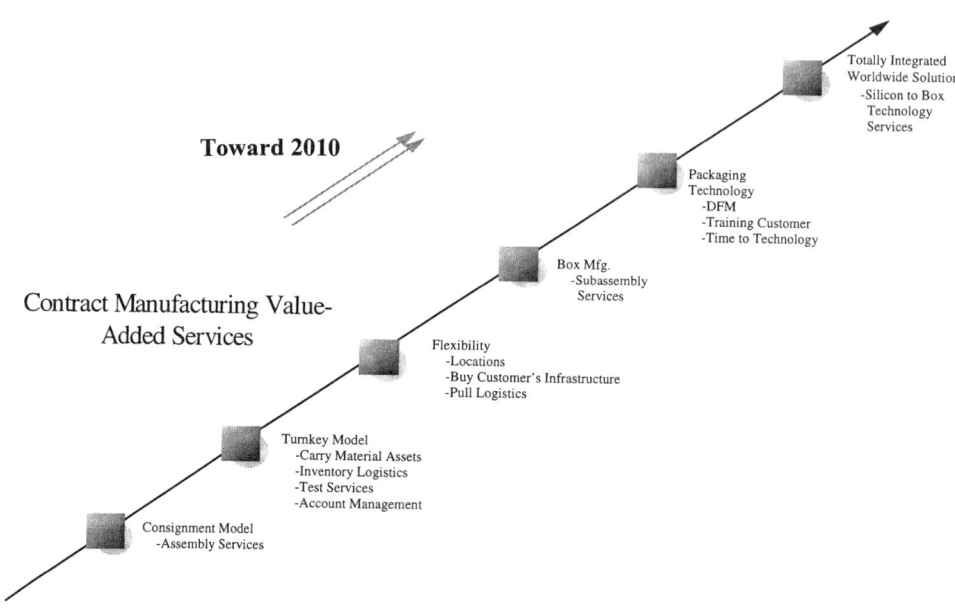

FIGURE 3 Evolution of contract manufacturing services.

Practical Aspects of the OEM–EMS Relationship

With ever-increasing competitive pressure on time to market, functionality, cost, and quality, it is essential that the OEM relationship with its CM is built on a solid foundation that includes organization, strategy, systems, and supplier management as its cornerstones. For an OEM–EMS provider relationship to be effective, it should be a win–win relationship that supports key corporate objectives of both partners. At a minimum, these include the improvement of customer loyalty; optimization of return on assets; and continuing growth of revenues, profits, and corporate value. Let's look at some of the practical issues involved in an OEM–EMS provider relationship that expands on the summaries of Tables 1 and 2.

Loss of Control. For the OEM, the primary drawback to outsourcing is loss of control over the manufacturing process. Geographical distances affect delivery schedules for both inventory and final product, resulting in higher transportation costs than with in-house operations. Calling in a product designer or manufacturing engineer to solve a production problem generally is slower and more expensive when the process involves an EMS provider. Situations in which the EMS provider's manufacturing facilities are far from the OEM's design team underscores the need for local control by the provider. Manufacturing and test strategy engineering and modification also must occur locally because manufacturing departments cannot afford to wait 2 days or longer for answers to engineering problems. Time zone distances aggravate product shipment delays.

OEM Proprietary Information. Out of necessity, an EMS provider will learn intimate details of product design and construction. Revealing such proprietary information makes many OEMs uncomfortable, especially when the same provider is typically working for competitors as well (no OEM puts all of its eggs in one CM basket). This situation has become increasingly common and offers no easy solution. The OEMs must select EMS providers that have a reputation for honesty and integrity. This requires a leap of faith that the provider's reputation, potential loss of business, or (as a last resort) threat of legal action will sufficiently protect the OEM's interests.

Quality. Probably the largest concern with outsourcing is that of quality. The continuing exodus to outsource manufacturing can be a source of quality issues. When making a transition from in-house manufacturing to outsourcing, some things may be omitted. In an environment of rapid technological change, the complex OEM–CM relationship raises a number of valid questions that must be addressed: Will the quality level of the OEM's product prior to the outsource decision be maintained or improved by the outsource provider? Are corners being cut on quality? How much support will the OEM need to provide? What happens and who is responsible for resolving a component issue identified by the OEM or end customer in a timely manner? Who is responsible for conducting a risk assessment to determine the exposure in the field and if a given product should be shipped, reworked, retrofitted, or redesigned? When a low price is obtained for a computer motherboard from an offshore supplier, what is the quality level? When the customer initially turns a product on, does it fail or will it really run for 5 years defect-free as intended? If a CM does a system build, who is responsible for resolving all customer issues/problems? Obviously, it is imperative for OEMs to work closely with contract manufacturers to ensure that relevant items are not omitted and that corners are not cut concerning quality.

The key to managing supplier quality when an OEM outsources manufacturing is to integrate the outsource provider into the OEM's design team. Even before a product is designed, the supplier has to be involved in the design and specifications phase. Quality can best be enhanced by connecting the EMS provider to the OEM's design centers and to its customers' design centers before the supplier starts producing those parts. One can't decree that a product should have a certain number of hours of reliability. Quality and reliability can't be screened in; they have to be designed in.

Today's OEMs are emphasizing the need for robust quality systems at their CMs. They must maintain a constant high level of quality focus, whether the work is done internally or externally, and be actively involved with the quality of all CM processes. The reason for this is that the OEM's quality reputation is often in the hands of the CM since the product may never come to the OEM's factory floor but be shipped directly from the CM to the customer. This requires that a given CM has a process in place to manage, among other things, change in a multitude of production locations. Outsourcing requires extra care to make sure the quality process runs well. Technical requirements, specifications, performance levels, and service levels must be carefully defined, and continuous communication with the CM must be the order of the day. Outside suppliers must understand what is acceptable and what is not. Quality problems are often experienced because an outsource supplier works with one set of rules while the OEM works with another. Both have to be on the same page. The big mistake some OEMs make is to assume that by giving a contract manufacturer a specification they can wash their hands of the responsibility for quality. To the contrary, the OEM must continuously monitor the quality performance of the CM.

Communication. An OEM–CM outsource manufacturing and service structure requires a delicate balancing act with focus on open, accurate, and continuous communication and clearly defined lines of responsibility between parties. (See preceding section regarding quality.)

OEM Support. Effective outsourcing of any task or function typically requires much more support than originally envisioned by the product manufacturer. In many companies the required level of support is unknown before the fact. Many companies jumped into outsourcing PWA manufacturing with both feet by selling off their internal PWA manufacturing facilities, including employees, CMs without fully understanding the internal level of support that is required for a cost-effective and efficient partnership. The product manufacturers felt that they could simply reduce their purchasing, manufacturing engineering, and manufacturing production staffs and turn over all those activities and employees associated with printed wire assembly to the outsource provider. Wrong!

In reality it has been found that outsourcing does not eliminate all manufacturing costs. In fact the OEM has to maintain a competent and extensive technical and business staff (equal to or greater than that prior to the outsourcing decision) to support the outsource contract manufacturer and deal with any issues that arise. More resources with various specialized and necessary skill sets (commodity purchasing specialists, component engineering, failure analysis, manufacturing engineering, etc.) are required at the OEM site than were often considered. Originally, one reason for outsourcing was to eliminate the internal infrastructure. Now a more focused internal support infrastructure is required. Unfortunately, without a significant manufacturing operation of their own, OEMs may find a dwindling supply of people with the necessary skills to perform this function, and they often compete for them with the EMS providers.

Manufacturing/Production Practices

Most large OEMs place no more than 20% of their outsource manufacturing needs with a single EMS provider. This means that OEMs need to have the personnel and support structure in place to deal with at least five EMS providers, all with different infrastructures and processes.

Selecting a CM. Selecting a CM requires as much diligence as any other large investment because all CMs are not alike. Just as there are tiers or categories of competent suppliers, so are there similar distinct tiers of CMs, each with different base characteristics and core competencies. A given CM's manufacturing processes, technologies, and infrastructure must be synergistic with the needs of the OEM's products and its customers. The right EMS partner can provide a smooth, efficient, and problem-free manufacturing process that produces a high-quality, profitable product. The wrong choice can hurt even a good product's quality and financial performance.

Choosing the right EMS partner depends a great deal on the nature of the OEM's product. Some EMS providers concentrate on reducing costs. They specialize in high-volume, low-mix, low-margin products such as cell phones and personal computers. Their strength lies in reducing cost and product-to-product variation. These CMs expend their resources developing efficient manufacturing and test strategies (creating, debugging, and optimizing test programs), preferring to concentrate their efforts on programs and strategies rather than on the products themselves. The automobile industry, for example, prefers this approach. Most automakers would rather spend extra time and money optimizing fixtures, programs, and developing self-tests than incur an increase in component cost.

Other EMS providers emphasize achieving high quality and reliability. For example, some complex, low-volume products offer higher margins and require higher quality, making them less sensitive to manufacturing costs. Typically, product failure consequences demand that defective products never reach customers. Manufacturing pacemakers, for example, is not inherently more difficult than making cell phones or PCs. But test program development costs, as well as product upgrades and other changes, must be amortized over a much smaller product volume. While PC and cell phone failure may cause user dissatisfaction and annoyance, even incur warranty costs, pacemaker failure can cause user death. The EMS providers address these customer requirements as an extension of the OEM's own resources, often serving as partners during the design stages and early stages of manufacture. The link between the OEM and EMS provider is tighter here than in high-volume/low-mix cases.

So there are several perspectives in selecting a CM: high-volume, low-mix production versus low-volume high-mix and large- versus small-sized EMS providers. At one extreme, an EMS provider may populate boards from OEM-supplied bare PCBs and parts. The provider may perform some kind of manufacturing defects test (such as in-circuit test or x-ray inspection). The OEM receives finished boards, subjects them to environmental stress screening and functional testing as appropriate, then assembles, tests, and packages the systems for shipment. Sometimes several EMS providers manufacture different boards for large systems. Here, the OEM retains control over the manufacturing and test process, including the methods employed at each step.

At the other extreme, an OEM may provide the CM with only design and functional specifications, leaving every aspect of the production process, deciding strategies, and tactics to the EMS provider. Many EMS providers offer extensive engineering services extending far beyond buying bare boards, parts, chassis, and performing assembly operations. These providers can essentially reengineer boards and systems to enhance their manufacturability, testability, quality, and reliability. For example, a board measuring approximately 24 in. on a side that boasts 26% node access for an in-circuit test (ICT) may defy conventional attempts to achieve acceptable yields. Sometimes the solution requires changing strategy, e.g., supplementing ICT with x-ray inspection. Then too EMS provider's engineers may be able to redesign the board to provide additional access, add self-test coverage to supplement conventional test, and thus make the board more "manufacturable" and testable. Where reengineering is appropriate, an EMS provider that specializes in such activities generally will produce higher-quality, higher-reliability products at lower cost. The EMS provider may even box systems up and ship them directly to distributors or end users. This arrangement requires that the provider share engineering responsibility with the OEM, pooling expertise to best benefit both parties. The OEM gets a better-engineered product for less, thereby increasing profit margins.

A large EMS company generally encounters a wider variety of manufacturing and test strategies and philosophies than a smaller company. Bringing an unusual situation to a small EMS provider may not lead to the best solution simply because of a lack of sufficient experience with similar problems. Yet the wider experience cuts both ways. Rather than evaluate each situation on its own merits, the large provider may attempt to fit every problem presented into one of the alternatives in a menu of suggested strategies. The OEM must always ensure that the EMS provider's strategic decisions address the constraints and parameters in the best possible way.

Conversely, a small OEM generally does not want to engage a CM so large that the amount of business generated gets lost in a sea of much larger projects. A small company often can get much more personal attention from a small EMS provider (subject to the same size caveat as before). Rather than impose a "best" solution, the small contractor may be more willing to design a manufacturing and test strategy in cooperation with the OEM. Smaller CMs are generally more flexible than their larger counterparts. Although a small CM may have fewer resources and less experience, if the CM's skills match the OEM's needs, the result should be a strong and mutually beneficial relationship.

Manufacturing/Production Practices

Between these points lies a continuum of alternatives. Deciding where along that continuum an OEM should "drop anchor" represents the crux of the CM selection and planning process.

EMS Resources Required. It is important that EMS providers have the right and sufficient resources to meet the OEM's design, manufacturing, and support needs. A problem that the EMS provider can resolve reduces the engineering load on the OEM, and the time saved reduces the product's time to market. Striking the right balance is needed since EMS services vary widely.

Design-Related Issues

Three issues pertaining to design have come to the forefront. First, as networking and telecommunications equipment manufacturers pare their workforces and sell their manufacturing operations to EMS providers, they are becoming more dependent on IC suppliers for help in designing their systems, the complexities of which are becoming difficult for contract manufacturers and distributors to handle effectively. Thus, IC designers will need to have a systems background.

Second, three levels of EMS design involvement that can affect design for manufacturability have been identified, going from almost none to complete involvement. These are (1) OEMs either give the company a fully developed product; (2) look for EMS engineering support in the middle of product development at or about the prototype phase; or (3) engage with the OEM at the conceptual design phase. Contract manufacturers are moving toward or already providing new product introduction (NPI) services that help get high-volume projects underway more quickly. The sooner the customer and CM can get together on a new project, the sooner such issues as design for manufacturability and design for test can be resolved. The closer the CM can get to product inception, the quicker the ramp to volume or time to market and the lower the overall product cost.

The third issue is that significant delays are being experienced by OEMs in their product development efforts. According to a report published by AMR Research Inc. (Boston, MA) in May 2001, product design times for OEMs in PC, peripherals, and telecommunications markets have increased by an average of 20% as a result of the growing outsourcing trend. The report identifies two troubling issues: the OEM design process is being frustrated by an increase in the number of participants collaborating in product development, and OEMs are losing their ability to design products that can be easily transferred to manufacturing.

In their quest for perfection, OEM design engineers create a design bill of materials that has to be translated into a manufacturing bill of materials at the EMS level because design engineers aren't familiar with how the EMS providers produce the product. They've lost the connection with manufacturing they had when manufacturing was internal. OEMs are designing their products one way, and EMS providers preparing the product for manufacturing have to rewrite (the design) to be compatible with the manufacturing process. The disconnect is in the translation time.

Because OEMs in many cases have relinquished production of printed wiring assemblies and other system-level components, their design teams have begun to move further away from the manufacturing process. When the OEMs did the manufacturing, their design engineers would talk to the manufacturing personnel and look at the process. Since the OEMs don't do manufacturing anymore, they can no longer do that. The result is increased effort between OEM design and EMS manufacturing. Time is being wasted both because the design has to be redone and then changes have to be made to make the product more manufacturable by the EMS provider.

The EMS Provider's Viewpoint

An OEM looking for the right EMS partner must recognize that the provider's manufacturing challenges differ from those the OEM would experience. First, the provider generally enjoys little or no control over the design itself (although, as mentioned, this is changing). Layout changes, adding or enhancing self-tests, etc. will not compensate for a design that is inherently difficult to manufacture in the required quantities. An experienced EMS provider will work with a customer during design to encourage manufacturability and testability and discourage the reverse, but ultimate design decisions rest with the company whose name goes on the product.

One primary concern of the CM in accepting a contract is to keep costs down and avoid unpleasant surprises that can make thin profit margins vanish completely. The OEMs need to be sensitive to this situation, remembering that the object is for everyone in the relationship to come out ahead. Keeping costs low requires flexibility. For the EMS provider, that might mean balancing manufacturing lines to accommodate volume requirements that may change without warning. A board assembled and tested on line A on one day may be assigned to a different line the next day because line A is running another product. Similarly, volume ramp-ups may demand an increase in the number of lines manufacturing a particular product. Achieving this level of flexibility with the least amount of pain requires that manufacturing plans, test fixtures, and test programs are identical and perform identically from one line to the next. Also, one make and model of a piece of manufacturing or test equipment on one manufacturing line must behave identically to the same make and model on another manufacturing line (repeatability and reproducibility), which is not always the case.

The CMs depend on OEMs for products to maintain high volumes on their manufacturing lines to maximize capacity and lower the overhead associated with maintaining state-of-the-art equipment that may not always be fully loaded (uti-

Manufacturing/Production Practices

lized). A contract manufacturer may send production from one geographical location to another for many reasons. Tax benefits and import restrictions such as local-content requirements may encourage relocating part or all of a manufacturing operation elsewhere in the world. The EMS provider may relocate manufacturing just to reduce costs. Shipping distances and other logistics may make spreading production over several sites in remote locations more attractive as well. Again, seamless strategy transfer from one place to another will reduce each location's startup time and costs. To offset the geographical time differences, crisp and open communications between the EMS provider and the OEM are required when problems and/or questions arise.

The first 6 to 12 months in a relationship with an EMS provider are critical. That period establishes procedures and defines work habits and communication paths. Planners/schedulers should allow for longer turnaround times to implement product and process changes than with in-house projects.

The Bottom Line

Connecting outsourcing to an OEM's business strategy, selecting the right opportunities and partners, and then supporting those relationships with a system designed to manage the risk and opportunities are the essential success factors for implementing an outsourcing business model. For an OEM, specific needs and technical expertise requirements must be evaluated to select an appropriate EMS provider. The selected CM's manufacturing and test strategy must match the OEM's. Identifying the right partner requires that a satisfactory tradeoff among quality, cost, and delivery factors be found. If the product reach is global, the EMS provider should have a worldwide presence as well. Because the product could ship from manufacturing lines anywhere in the world, the EMS provider should strive for consistency and uniformity in the choice of manufacturing and test equipment, thus minimizing the variation of produced product and the effort and cost of that wide implementation and distribution. In the final analysis, there are no canned solutions. One size cannot fit all.

Reference 1 discusses the process for selecting contract manufacturers, CM selection team responsibilities, qualifying CMs, integrating CMs into the OEM process flow, material and component supplier responsibilities and issues, and selecting a manufacturing strategy.

7.2 PRINTED WIRING ASSEMBLY TESTING

7.2.1 Introduction

Defects are inherent to the processes used in creating the products we make. Virtually no step in these processes has 100% yield—it just isn't possible. So effective manufacturing becomes an engineering problem in choosing how to minimize the conflicts and thus the potential defects. As a result, performing effective testing to identify and correct defects is as important to the manufacturing process as buying quality materials and installing the right equipment.

In fact, the manufacturing process has become part of what is considered as traditional test. Test is thus comprised of the cumulative results of a process that includes bare board (PCB) test; automated optical inspection (AOI); x-ray, flying probe, manufacturing defect analyzer (MDA), ICT, and functional test solutions. Electrical testing is an important part of manufacturing because visual inspection is not sufficient to ensure a good PWA.

The electrical design, the physical design, and the documentation of all impact testability. The types of defects identified at test vary from product, depending on the manufacturing line configuration and the types of component packages used. The difficult part of testing is the accumulation and processing of and action taken on defect data identified through the manufacturing test processes. The manufacturing defect spectrum for a given manufacturing process is a result of the specific limits of that manufacturing process. As such, what is an actionable item is ultimately process determined and varies by line design. A typical PWA manufacturing defect spectrum is shown in Table 3.

Opens and shorts are the most predominant failure mechanisms for PWAs. There is a difference in the failure mechanisms encountered between through-hole technology (THT) and surface-mount technology (SMT). Practically, it is difficult to create an open connection in THT unless the process flow is not reach-

TABLE 3 Typical Defect Spectrum

Process defects
    Missing components
    Wrong component orientation
    Wrong component value (resistors, capacitors)
    Shorts (process or PCB)
    Opens (process or PCB)
    Missing screws
    Solder issues (missing solder, insufficient solder, solder balls, insufficient heal)
    Wrong label
    Misalignment
    Wrong firmware
Component defects
    Defective components
    Part interaction defects (combined tolerance issues, timing faults)
Design and tolerance issues

Manufacturing/Production Practices

ing part of the assembly, there are contaminants in the assembly, or a lead on the part is bent during the insertion process. Otherwise the connection to the part is good. The bigger problem is that solder will bridge from one component lead to the next to create a short. When soldering using SMT, the most significant problem is likely to be an open connection. This typically results from insufficient reflow. Since it is more difficult for manufacturers to catch (detect) opens with ATE, some tune their assembly processes toward using additional solder to make the manufacturing process lean more toward shorts than it would if left in the neutral state.

Testing SMT or THT will find all of the shorts unless SMT doesn't allow full nodal coverage. Either technology allows electrical testing to find most opens. For example, when testing a resistor or capacitor, if there is an open, it will be detected when the part is measured. The issue is finding open connections in ICs. For the typical test, a mapping is made of the diode junctions present between IC pins and the power/ground rails of the IC under test. Normally this works fine, but in many cases an IC is connected to both ends of the trace. This looks like two parallel diodes to the test system. If one is missing due to an open connection in the IC, the tester will miss it since it will measure the other diode that is in parallel, not knowing the difference.

Every test failure at any deterministic process step must have data analyzed to determine the level of existing process control. Data can be compared to the manufacturing line's maximum capability or line calibration factor to achieve a relative measure of line performance quality. The result is a closed loop data system capable of reporting the manufacturing quality monitored by the various test systems. In general, it has been observed that 80% of the defects encountered at the various test stages are process related. Looking at the defect spectrum of Table 3 it is seen that the nature of the defects has not changed with technology. Rather it is the size of the defect spectrum that has changed.

7.2.2 Visual Inspection

After a printed circuit board has been tested and assembled (i.e., all components soldered) the assembly is first subjected to a visual inspection to look for gross manufacturing defects. These include defects in solder joints, assembly defects involving wrong or missing components, and solder defects under devices such as ball grid arrays (BGAs) via x-ray. Visual examination is efficient and cost effective in detecting manufacturing defects and alleviates the burden of a more expensive electrical test and time-consuming diagnostics. The objective is to find the defects at the lowest cost point.

Visual inspection employs both manual and automated techniques such as optical magnifiers, optical comparators, closed-circuit television (CCTV) magnified display, automated optical inspection, and x-ray. There is an increased trend toward the use of x-ray inspection using x-ray–based machine vision systems, which provide pin-level diagnostics. The reasons for this are

1. The shrinking sizes of passive components and finer linewidths of PCBs result in much denser PWAs than in the past. Thus, solder joints are much more critical as they become smaller and closer together, making human inspection more difficult and inaccurate.
2. The increased use of chip scale and ball grid array packages in which soldered connections around the periphery of an IC package are replaced by an underside array of solder balls that are used for electrical connections. The result is that the solder joint connections are hidden from view underneath the package and cannot be inspected by humans or conventional optical inspection or machine vision systems.
3. Densely populated PWAs are so complex that there is no way to access enough test nodes. While board density makes physical access difficult, electrical design considerations (radiofrequency shielding, for example) may make probing extremely difficult if not impossible. X-ray inspection provides an excellent solution for inspecting solder joints where limited access or board topography makes it impossible to probe.
4. A functional failure (via functional test) only identifies a segment of problem circuitry and the diagnostics are not robust.

7.2.3 Electrical Testing

Historically the first microview of the quality and predictability of the manufacturing process has been at PWA electrical test (either ICT or functional test). Even though a PWA may undergo a rigorous visual inspection, the only way to ensure that all components are functional, contain no electrical shorts or opens, and that the circuitry performs as designed is by conducting an electrical test.

Electrical test at the PWA, module, and system levels exists for the following reasons:

Verify that the PWA, module, and system operates as designed
Continuously improve the manufacturing process and thus the quality
Reduce costs
Ensure the level of quality and reliability demanded by the customer is met

Board test can be divided into two phases: manufacturing process test and functional test. Manufacturing process test verifies that the PWA has been assembled correctly by checking for correct component interconnection and component orientation. The PWA is tested for opens and shorts as well as resistance, capacitance, inductance, diode junctions, and active part operation. Manufacturing defect analyzers and in-circuit testers are used to perform this step. The MDA tester,

Manufacturing/Production Practices

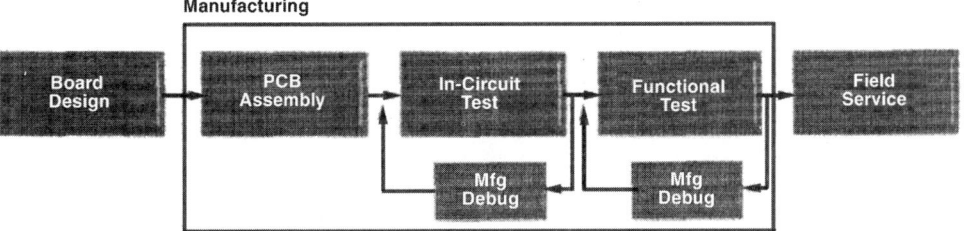

FIGURE 4 Typical electrical tests conducted during PWA manufacturing.

often called an analog ICT, assumes that the components are good and endeavors to find how they may be improperly installed on the PCB. In-circuit testing takes testing one step further by checking ICs in the circuit for operation. Going from MDA to full ICT improves the test coverage by a few percent, but at multiple times tester and test fixture development cost.

Functional test verifies that the PWA operates as designed. Does it perform the tasks (function) that it was designed for? Figure 4 is a block diagram of the electrical tests performed during manufacturing and the sequence in which they are performed, from simple and least expensive (visual inspection, not shown) to complex and most expensive (functional test).

In the 1960s and 1970s PWAs were simple in nature and edge connectors provided easy test access using functional test vectors. In the 1980s, increased PWA density led to the use of ICT or bed-of-nails testing to mechanically access hard-to-get-at nodes. In the 1990s, due to continually increasing PWA density and complexity, boundary scan, built-in self-test (BIST), vectorless tests, and unpowered opens testing gained increased importance with the decreasing use of ICT.

Table 4 summarizes and compares some of the key features of the various electrical test methods being used. These are further discussed in the following sections.

In-Circuit Testing

The most basic form of electrical test is by use of a manufacturing defects analyzer, which is a simple form of an in-circuit tester. The MDA identifies manufacturing defects such as solder bridges (shorts), missing components, wrong components, and components with the wrong polarity as well as verifies the correct value of resistors. Passive components and groups or clusters of components can be tested by the MDA. The PWA is connected to the MDA through a bed-of-nails fixture. The impedance between two points is measured and compared against the expected value. Some MDAs require programming for each component, while

TABLE 4 Comparison of Test Methods

Self test	Functional test	In-circuit test	Vectorless open test
Limited defect coverage	Catches wider spectrum of defects	Highest defect coverage	No models required
Limited diagnosis	Better diagnostics than self-test	Precise diagnostics	Overclamp and probes required
Inexpensive	Most expensive test	Very fast	Fast test development
Can help ICT (boundary scan)	Computer run	Least expensive test	High defect coverage
No test fixture required	Test fixture required	Requires test fixture and programming	Reversed capacitors detectable
	Long test development time and costs	Facilitates device programming	
	Easy data logging/analysis of results	Easy data logging/analysis of results	

Manufacturing/Production Practices 329

others can learn the required information by testing a known good PWA. The software and programming to conduct the tests is not overly complicated.

Conventional in-circuit testers are the workhorses of the industry. They have more detection capability than MDAs and extended analog measurement capability as well as the added feature of digital in-circuit test for driving integrated circuits. The programming and software required for conventional in-circuit testing is more complicated than for an MDA, but simpler than for functional testing.

An in-circuit test is performed to detect any defects related to the PWA assembly process and to pinpoint any defective components. The ICT searches for defects similar to those found by both visual inspection and by the MDA but with added capability; ICT defect detection includes shorts, opens, missing components, wrong components, wrong polarity orientation, faulty active devices such as nonfunctioning ICs, and wrong-value resistors, capacitors, and inductors.

To be effective, in-circuit test requires a high degree of nodal access, therefore the tester employs a bed-of-nails fixture for the underside of the PWA. The bed-of-nails fixture is an array of spring-loaded probes; one end contacts the PWA, and the other end is wired to the test system. The PWA-to-fixture contact pressure is vacuum activated and maintained throughout the test. The fixture contacts as many PWA pads, special test pads, or nodes as possible such that each net can be monitored. (A node is one circuit on the assembly such as GROUND or ADDR0.) With contact to each net, the tester can access every component on the PWA under test and find all shorts. The industry standard spring probe for many years was the 100-mil size, typical of through-hole technology probes for the vast majority of devices that had leads on 0.100-in. centers. 75-mil probing was developed in response to the needs of SMT, then 50-mil, then 38-mil; and now even smaller spacing probes are available.

A major concern in ICT is test fixture complexity. Fixtures for testing THT PWAs are generally inexpensive and reliable compared with the fixtures required for testing SMT PWAs if the latter are not designed with test in mind. This leads to high fixture costs, but also makes it extremely difficult to make highly reliable, long-lasting fixtures. Dual-sided probing allows access to both sides of a PWA (whether for ICT or functional text). Complex PWAs may require the use of clam-shell bed-of-nails fixturing to contact the component side as well as the back or underside of the PWA, but it adds significantly to the cost of the test fixture.

Also, sometimes not all nodes are accessible, compromising test coverage. In-circuit testing is constantly being refined to provide improved test coverage while minimizing the number of required test (probe) points. This is accomplished by including testability hooks in the design and layout of the PWA such as using boundary scan devices, clustering groups of components, being judicious in test point selection, adding circuitry to provide easier electrical access, and placing test points at locations that simplify test fixture design.

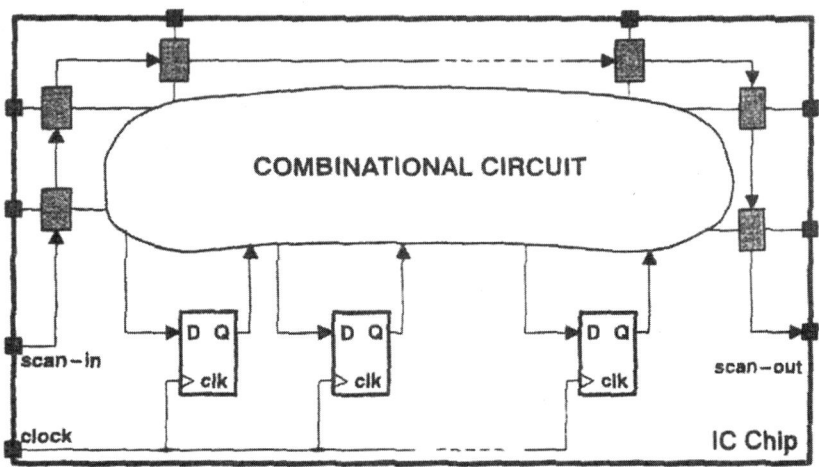

FIGURE 5 Boundary scan circuit showing latches added at input/output pins to make IC testable.

By way of summary from Chapter 3, the use of boundary scan has been effective in reducing the number of test points. In boundary scan the individual ICs have extra on-chip circuitry called boundary scan cells or latches at each input/output (see Fig. 5). These latches are activated externally to isolate the I/O (wire bond, IC lead, PWA pad, external net) from the internal chip circuitry (Fig. 6), thereby allowing ICT to verify physical connections (i.e., solder joints).

Boundary scan reduces the need for probe access to each I/O because an input signal to the latches serially connects the latches. This allows an electrical check of numerous solder joints and nets extending from device to device. The reality of implementing boundary scan is that most PWAs are a mix of ICs (both analog and digital functions) with and without boundary scan using a mix of ad hoc testing strategies and software tools. This creates a problem in having a readily testable PWA.

Conventional in-circuit testing can be divided into analog and digital testing. The analog test is similar to that performed by an MDA. Power is applied through the appropriate fixture probes to make low-level DC measurements for detecting shorts and the value of resistors. Any shorts detected must first be repaired before further testing. The flying probe tester, a recent innovation in ICT, performs electrical process test *without* using a bed-of-nails fixture interface between the tester and the board under test. Originally developed for bare board testing, flying probe testing—together with associated complex software and programming—can effectively perform analog in-circuit tests. These systems use multiple, motor-operated, fast-moving electrical probes that contact device leads

Manufacturing/Production Practices

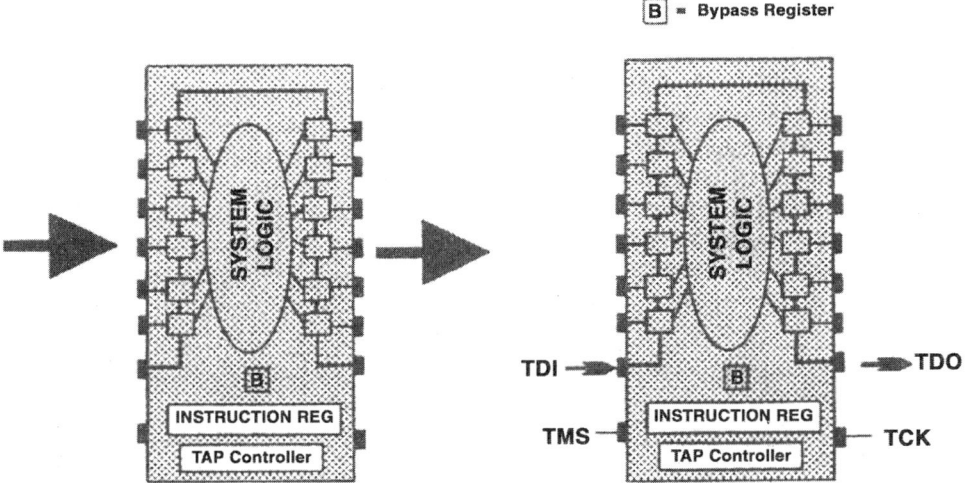

FIGURE 6 Integrated circuit with boundary scan in normal operating mode (left) and with boundary scan enabled for testing (right).

and vias and make measurements on the fly. The test heads (typically four or eight) move across the PWA under test at high speed as electrical probes located on each head make contact and test component vias and leads on the board, providing sequential access to the test points. Mechanical accuracy and repeatability are key issues in designing reliable flying probers, especially on dense PWAs with small lead pitches and trace widths.

Flying probers are often used during prototype and production ramp-up to validate PWA assembly line setup without the cost and cycle time associated with designing and building traditional bed-of-nails fixtures. In this application, flying probers provide fast turnaround and high fault coverage associated with ICT, but without test fixture cost. Flying probers have also been used for in-line applications such as sample test and for production test in low-volume, high-mix PWA assembly lines.

The second phase of analog test—an inherent capability of the conventional in-circuit tester—consists of applying low-stimulus AC voltages to measure phase-shifted currents. This allows the system to determine the values of reactive components such as capacitors and inductors. In taking measurements, each component is electrically isolated from others by a guarding process whereby selective probes are grounded.

In digital ICT, power is applied through selected probes to activate the ICs and the digital switching logic. Each IC's input and output is probed to verify proper switching. This type of test is made possible by the use of a technique called *back driving* in which an overriding voltage level is applied to the IC input to overcome the interfering voltages produced by upstream ICs. The back driving technique must be applied carefully and measurements taken quickly to avoid overheating the sensitive IC junctions and wire bonds (Fig. 7). An example of how this happens is as follows. Some companies routinely test each IC I/O pin for electrostatic discharge (ESD) susceptibility and curve trace them as well beginning at the PWA perimeter and working inward. They then move farther into the PWA back driving the just-tested ICs. These devices are weaker than the devices being tested (by virtue of the test) and fail.

In-circuit testing is not without its problems. Some of these include test fixture complexity, damage to PWAs due to mechanical force, inefficiency and impracticality of testing double-sided PWAs, and the possibility of overdriving ICs causing thermal damage. Because of these issues many companies have eliminated ICT. Those companies that have done so have experienced an increase in board yield and a disappearance of the previously mentioned problems.

Vectorless Opens Testing

Vectorless opens testing uses a special top probe over the IC under test in conjunction with the other standard probes already present. By providing a stimulus and measuring the coupling through the IC, open connections can be accurately detected. Vectorless test can be used effectively even when IC suppliers are changed. It doesn't require the expensive programming of full ICT techniques.

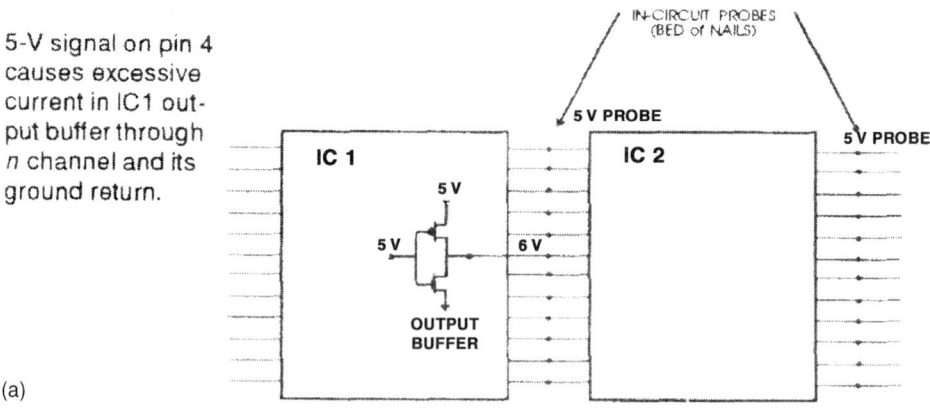

FIGURE 7 Back- (or over-) driving digital ICs can result in adjacent ICs being overdriven (a and c), resulting in overheating and temperature rise (b), leading to permanent damage.

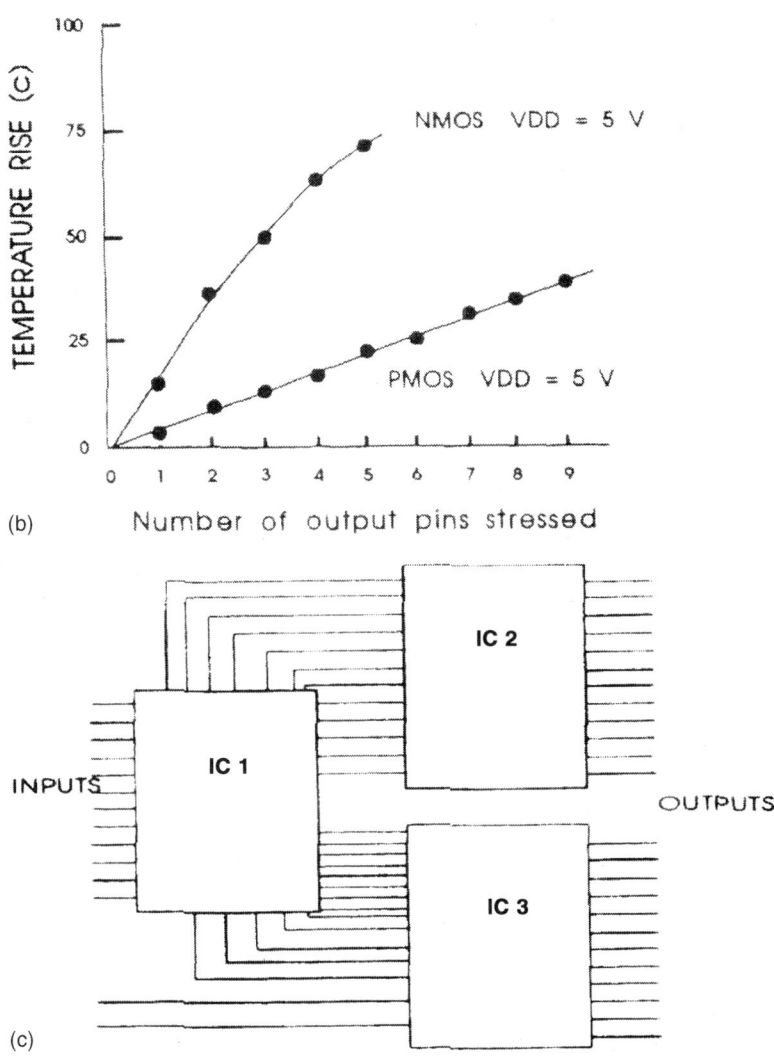

(b) Number of output pins stressed

(c)

Functional Testing

Although an in-circuit test is effective in finding assembly defects and faulty components, it cannot evaluate the PWA's ability to perform at clock speeds. A functional test is employed to ensure that the PWA performs according to its intended design function (correct output responses with proper inputs applied). For functional testing, the PWA is usually attached to the tester via an edge connector and powered up for operation similar to its end application. The PWA inputs are stimulated and the outputs monitored as required for amplitude, timing, frequency, and waveform.

Functional testers are fitted with a guided probe that is manually positioned by the operator to gain access to circuit nodes on the PWA. The probe is a troubleshooting tool for taking measurements at specific areas of the circuitry should a fault occur at the edge connector outputs. The probe is supported by the appropriate system software to assist in defect detection.

First-pass PWA yield at functional test is considerably higher when preceded by an in-circuit test. In addition, debugging and isolating defects at functional test requires more highly skilled personnel than that for ICT.

Cluster Testing

Cluster testing, in which several components are tested as a group, improves PWA testability and reduces the concerns with ICT. One begins with the components that are creating the testing problem and works outward adding neighboring components until a cluster block is defined. The cluster is accessed at nodes that are immune to overdrive. Address and data buses form natural boundaries for clusters. Cluster testing combines the features of both ICT and functional test.

Testing Microprocessor Based PWAs

The ubiquitous microprocessor-based PWAs present some unique testing challenges since they require some additional diagnostic software that actually runs on the PWA under test, in contrast to other board types. This software is used to exercise the PWA's circuits, such as performing read/write tests to memory, initializing I/O devices, verifying stimuli from external peripherals or instruments, generating stimuli for external peripherals or instruments, servicing interrupts, etc. There are a number of ways of loading the required test code onto the PWA under test.

1. With a built-in self-test, the tests are built into the board's boot code and are run every time the board is powered up, or they can be initiated by some simple circuit modification such as link removal.
2. With a test ROM, a special test ROM is loaded onto the PWA during test. On power up, this provides the necessary tests to fully exercise the board.
3. The required tests can be loaded from disk (in disk-based systems) after power up in a hot mock-up situation.
4. The tests can be loaded via an emulator. The emulator takes control of the board's main processor or boot ROM and loads the necessary test code by means of this connection.

When a microprocessor-based board is powered up, it begins running the code contained in its boot ROM. In a functional test environment this generally consists of various test programs to exercise all areas of the PWA under test. An emulator provides an alternative approach by taking control of the board after PWA power up and providing the boot code. Two types of emulation are used: processor emulation and ROM emulation.

Processor Emulation. Many microprocessor manufacturers incorporate special test circuitry within their microprocessor designs which is generally accessible via a simple three- to five-wire serial interface. Instructions can be sent through this interface to control the operation of the microprocessor. The typical functions available include the following:

Stop the microprocessor.
Read/write to memory.
Read/write to I/O.
Set breakpoints.
Single step the microprocessor.

Using these low-level features, higher level functions can be constructed that will assist in the development of functional test programs. These include

Download test program to microprocessor under test's memory.
Run and control operation of downloaded program.
Implement test program at a scripting language level by using the read/write memory or I/O features.
Recover detailed test results, such as what data line caused the memory to fail.
Control and optimize the sequence in which test programs run to improve test times.

Emulators are commercially available that already have these functions preprogrammed.

ROM Emulation. If microprocessor emulation is not available, ROM emulation can be a viable alternative. A ROM emulator replaces the boot ROM of the (DUT) device under test. This means that there must be some way of disabling or removing this. Once connected, the ROM emulator can be used in two different ways. In the first, the DUT is run from the ROM emulator code (after the test code has been downloaded to the ROM emulator), rather than from the PWA's own boot code. This removes the need to program the ROM with boot code and test code. Alternatively, the PWA's own boot ROM can be used to perform the initial testing and then switch to the ROM emulator to perform additional testing. In the second method, the emulator is preloaded with some preprogrammed generic tests (such as read/write to memory and RAM test, for example). These are controlled by means of a scripting language to quickly implement comprehensive test programs.

Most commercially available emulators come with preprogrammed diagnostic tests that include ROM CRC test; RAM test; I/O bus test; PCI/compact PCI/PMC bus test; and ISA bus test, to name several. In addition, tests can be written using simple scripting languages rather than C or assembly language.

Since PWA throughput is driven by test time, Emulators provide a reduction in test time, thus increasing PWA throughput. They also possess the advantage of having an easy-to-modify test code. This means that redundant test code can be removed, test sequencing can be optimized, and the test code compressed, leading to dramatic improvements in test time with minimal programmer input. The major disadvantage of using an emulator is cost. However, due to the benefit of improved test development times and PWA test throughput, this investment can be rapidly recouped.

Testing Challenges

Limited electrical access is the biggest technical challenge facing in-circuit test. A steady decline in accessible test points is occurring due to shrinking board geometries. The decline in electrical accessibility is due to both the use of smaller test targets and a reduction in the number of test targets. Designers driven to create smaller, denser PWAs view test targets as taking away board real estate that does not add value to the end product. The problem is compounded by the increased use of smaller parts such as 0402 capacitors and by denser digital designs employing micro-BGAs, chip scale packages, and chip-on-board (or direct chip attach) methods.

Smaller test targets mean more expensive and less reliable fixturing. Reduced test target numbers mean reduced test coverage. This typically forces test engineers to write more complex cluster tests at the expense of component level diagnostics.

Very-high-frequency PWA designs, such as for cellular phones, are also a challenge to in-circuit test. When PWAs operate at 1 GHz and higher, the component values become very small and difficult to measure, assuming that test points are available. At such frequencies, test points become small radiating antennas, causing designers to avoid their use. This is why cellular phone manufacturers traditionally have not embraced in-circuit technology as readily as in other markets, having chosen instead to use x-ray and optical inspection systems.

7.3 ENVIRONMENTAL STRESS SCREENING

Once a product has been released to production manufacturing, most of the component, design, and manufacturing issues should be resolved. Depending on market requirements and product and process maturity, environmental stress screening (ESS) can be used to quickly identify latent component, manufacturing, and

Manufacturing/Production Practices

workmanship issues that could later cause failure at the customer's site. Optimal ESS assumes that design defects and margin issues have been identified and corrected through implementation of accelerated stress testing at design.

Also called accelerated stress testing, ESS has been extensively used for product improvement, for product qualification, and for improving manufacturing yields for several decades. In the early 1980s, computer manufacturers performed various stress screens on their printed wiring assemblies. For example, in the 1981–1983 timeframe Apple Computer noticed that "a number of their boards would suddenly fail in the midst of manufacturing, slowing things down. It happened to other computer makers, too. In the industry it had been seen as an accepted bottleneck. It was the nature of the technology—certain boards have weak links buried in places almost impossible to find. Apple, working with outside equipment suppliers, designed a mass-production PWA burn-in system that would give the boards a quick, simulated three month test—a fast burn-in—the bad ones would surface" (3). The result was that this "burn-in system brought Apple a leap in product quality" (4).

Now a shift has taken place. As the electronics industry has matured, components have become more reliable. Thus, since the late 1980s and early 1990s the quality focus for electronic equipment has moved from individual components to the attachment of these components to the PCB. The focus of screening has changed as well, migrating to the PWA and product levels. The cause of failure today is now much more likely to be due to system failures, hardware–software interactions, workmanship and handling issues (mechanical defects and ESD, for example), and problems with other system components/modules such as connectors and power supplies. Stress screening is an efficient method of finding faults at the final assembly or product stage, using the ubiquitous stresses of temperature, vibration, and humidity, among others.

The lowest *cost of failure* point has moved from the component level to board or PWA test, as shown in Figure 7 of Chapter 4. This is currently where screening can have the greatest benefit in driving product improvement. As was the case for components, increasing reliability mitigates the necessity for screening. The decision process used to initiate or terminate screening should include technical and economic variables. Mature products will be less likely candidates for screening than new products using new technology. Given the decision to apply ESS to a new product and/or technology, as the product matures, ESS should be withdrawn, assuming a robust field data collection–failure analysis–corrective action process is in place. Figure 8 puts the entire issue of ESS into perspective by spanning the range of test from ICs to systems and showing the current use of each. Figure 1 of Ref. 5 shows a similar trend in reliability emphasis.

Much has been written in the technical literature over the past 10 years regarding accelerated stress testing of PWAs, modules, and power supplies. The

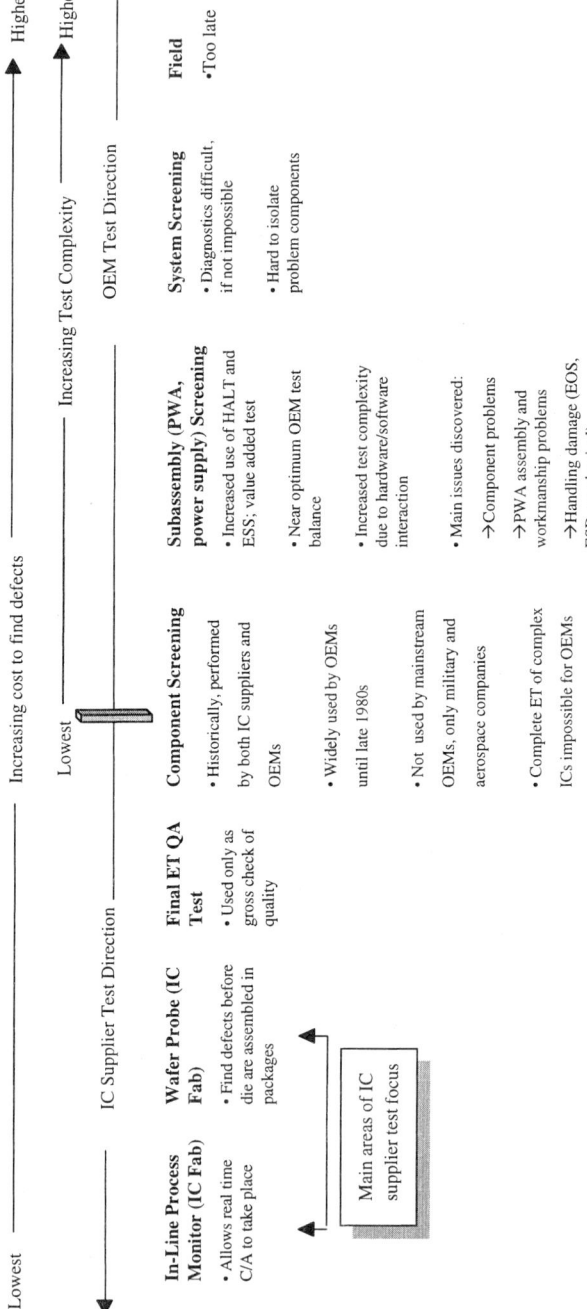

FIGURE 8 Environmental stress screening perspective. C/A, Corrective action; OEM, original equipment manufacturer; final ET QA, electrical test; EOS, electrical overstress; ESD, electrical static discharge.

Manufacturing/Production Practices

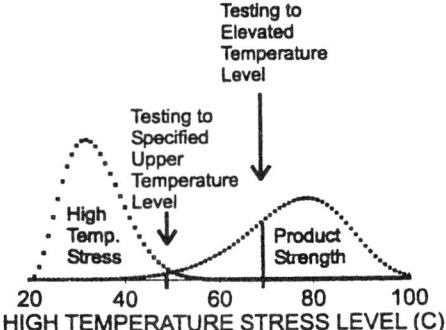

FIGURE 9 Accelerated stress testing uses substantially higher than normal specification limits.

philosophy of accelerated stress tests is best summarized by the following: *The system reliability of a complex electronic product can be improved by screening out those units most vulnerable to environmental stress.*

What Is Environmental Stress Screening?

Environmental stress screening applies stresses that are substantially higher than those experienced in normal product use and shipping (Fig. 9); product weaknesses and variations are better revealed at heightened stress levels. The technique may also include stresses that do not occur in normal use. The requirements on the stress stimuli are

1. They are severe enough to precipitate and detect relevant latent defects, i.e., the family of defects that show up in normal use.
2. They don't damage good products, causing early wearout and reduced life in the field.

The Institute of Environmental Sciences and Technology (IEST) definition of environmental stress screening is as follows: "Environmental stress screening is a process which involves the application of a specific type of environmental stress, on an accelerated basis, but within design capability, in an attempt to surface latent or incipient hardware flaws which, if undetected, would in all likelihood manifest themselves in the operational or field environment." Simply stated, *ESS is the application of environmental stimuli to precipitate latent defects into detectable failures.*

The stress screening process thus prevents defective products from being shipped to the field and offers the opportunity to discover and correct product weaknesses early in a product's life.

Why Perform Environmental Stress Screening?

Manufacturing processes and yields improve with the application of ESS because it precipitates out intermittent or latent defects that may not be caught by other forms of testing. It thus prevents these defects from causing long-term reliability problems following shipment to the customer. It identifies workmanship, manufacturing, and handling problems such as solder joint defects, poor interconnections (sockets and connectors), marginal adhesive bonding, and material defects.

Thus, effective ESS

1. Precipitates relevant defects at minimal cost and in minimal time
2. Initiates a closed-loop failure analysis and corrective action process for all defects found during screening
3. Increases field reliability by reducing infant mortalities or early life failures
4. Decreases the total cost of production, screening, maintenance, and warranty

Environmental Stress Screening Profiles

Environmental stress screening applies one or more types of stress including random vibration, temperature, temperature cycling, humidity, voltage stressing, and power cycling on an accelerated basis to expose latent defects. Table 5 provides a listing of the typical failure mechanisms accelerated and detected by application of thermal cycling, random vibration, and combined thermal cycling and

TABLE 5 Screening Environments Versus Typical PWA Failure Mechanisms Detected

Thermal cycling	Vibration	Thermal cycling and vibration
Component parameter drift	Particle contamination	Defective solder joints
PCB opens/shorts	Chafed/pinched wires	Loose hardware
Component incorrectly installed	Defective crystals	Defective components
Wrong component	Adjacent boards rubbing	Fasteners
Hermetic seal fracture	Components shorting	Broken part
Chemical contamination	Loose wire	Defective PCB etch
Defective harness termination	Poorly bonded part	
Poor bonding	Inadequately secured high-mass parts	
Hairline cracks in parts	Mechanical flaws	
Out-of-tolerance parts	Part mounting problems	
	Improperly seated connectors	

Manufacturing/Production Practices

TABLE 6 Combined Environment Profile Benefits and Accelerating Factors in Environmental Stress Screening

---

Combined environments are truer to actual use conditions.
Hot materials are looser/softer so vibration at high temperatures creates larger motions and displacements, accelerating material fatigue and interfacial separation.
Many materials (like rubber and plastics) are stiffer or brittle when cold; vibration stress at cold temperature can produce plasticity and brittle fracture cracks.
Six-axis degree of freedom vibration excites multiple structural resonant modes.
Overstress drives materials further along the S–N fatigue curve.
Thermal expansion/contraction relative movements during temperature cycling combined with relative movements from random vibration also accelerate material fatigue.

---

random vibration environments. The most effective screening process uses a combination of environmental stresses. Rapid thermal cycling and triaxial six–degree of freedom (omniaxial) random vibration have been found to be effective screens.

Rapid thermal cycling subjects a product to fast and large temperature variations, applying an equal amount of stress to all areas of the product. Failure mechanisms such as component parameter drift, PCB opens and shorts, defective solder joints, defective components, hermetic seals failure, and improperly made crimps are precipitated by thermal cycling. It has been found that the hot-to-cold temperature excursion during temperature cycling is most effective in precipitating early life failures.

Random vibration looks at a different set of problems than temperature or voltage stressing and is focused more on manufacturing and workmanship defects. The shift to surface mount technology and the increasing use of CMs make it important to monitor the manufacturing process. Industry experience shows that 20% more failures are detected when random vibration is added to thermal cycling. Random vibration should be performed before thermal cycling since this sequence has been found to be most effective in precipitating defects. Random vibration is also a good screen to see how the PWA withstands the normally encountered shipping and handling vibration stresses. Table 6 summarizes some benefits of a combined temperature cycling and six-axis (degree of freedom) random vibration ESS profile and lists ways in which the combined profile precipitates latent defects.

Once detected, the cause of defects must be eliminated through a failure analysis to root cause–corrective action implementation–verification of improvement process. More about this later.

The results of highly accelerated life testing (HALT) (Chapter 3), which was conducted during the product design phase, are used to determine the ESS profile for a given product, which is applied as part of the normal manufacturing

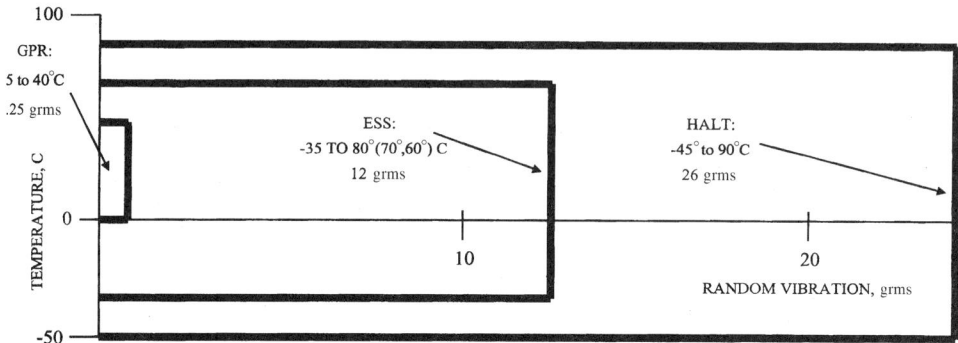

FIGURE 10 Environmental stress screening profile for high-end computer server. GPR (general product requirements) = operating environment limits.

process. The ESS profile selected for a given product must be based on a practical, common-sense approach to the failures encountered, usage environment expected, and costs incurred. The proper application of ESS will ensure that the product can be purged of latent defects that testing to product specifications will miss.

Figure 10 shows an example of a HALT profile leading to an ESS profile compared with the product design specification general product requirements (GPR) for a high-end computer server. From this figure you can see that the screen provides accelerated stress compared to the specified product environment. Recently compiled data show that at PWA functional test, 75% of defects detected are directly attributed to manufacturing workmanship issues. Only 14% of failures are caused by defective components or ESD handling issues, according to EIQC Benchmark.

Figure 11 shows how a selected ESS profile was changed for a specific CPU after evaluating the screening results. The dark curve represents the original ESS profile, and the portions or sections labeled A, B, and C identify various circuit sensitivities that were discovered during the iterative ESS process. For section A, the temperature sensitivity of the CPU chip was causing ESS failures that were judged unlikely to occur in the field. The screen temperature was lowered while an engineering solution could be implemented to correct the problem. In section B the vibration level was reduced to avoid failing tall hand-inserted capacitors whose soldering was inconsistent. Section C represented a memory chip that was marginal at low temperature. Each of the problems uncovered in these revisions was worked through a root cause/physics of failure–corrective action process, and a less stressful ESS screen was used until the identified improvements were implemented.

Manufacturing/Production Practices

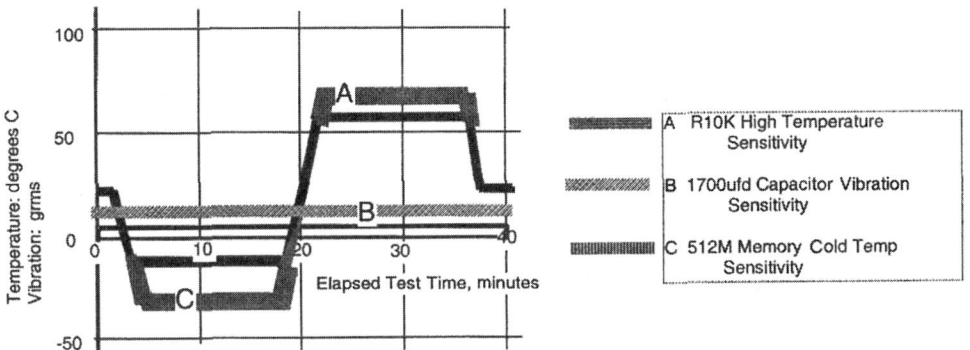

FIGURE 11 Adjusting the ESS profile.

Environmental Stress Screening Results

The results achieved by careful and thoughtful implementation of ESS are best illustrated by means of several case studies. The first three case studies presented here are somewhat sparse in the description of the ESS process (because the implementation of ESS at these companies was in its infancy), yet they serve to illustrate how the implementation of ESS led to positive product improvements. The two other case studies contain more detail and are written in a prose style because they come from a source where ESS has been a way of life for a long period of time.

Case Studies

Case Study 1: Computer Terminal.   Relevant details were as follows:

1000 components per module.
Two years production maturity.
75 modules screened.
Temperature cycling and random vibration screens were used.

Results:

Initial screens were ineffective.
Final screens exposed defects in 10–20% of the product.

Failure spectrum:

Parts: 60% (mainly capacitors)
Workmanship: 40% (solder defects)

Case Study 2: Power Supply. Relevant details were as follows:

29 manufacturing prototypes screened.
200 components per module.
Test, analyze, fix, and text (TAFT) approach was used to drive improvement.
Temperature, vibration, and humidity screens were used.

Results: Defects exposed in 55% of the product:

Workmanship: cold solder joints, bent heat sinks, and loose screws
Design issues: components selected and poor documentation

Case Study 3: Large Power Supply. Relevant details were as follows:

4500 components per power supply.
One unit consists of 75 modules.
Two years production maturity.
75 units were screened.
Temperature cycling was the screen of choice.

Results:

Initially, defects exposed in 25% of the product. This was reduced through the TAFT process to <3%.
Concurrently mean time between failures (MTBF) was improved from 40 hr at the beginning of screening to >20,000 hr.
Improvements were achieved through minimal design changes, improved manufacturing processes, and upgrading components used and controls exercised.

Case Study 4: Complex Communications PWA. A CPU board just entering production was subjected to the ESS profile shown in the lower portion of Figure 12. The upper portion shows the type and number of failures and when and where they occurred in the ESS cycle.

The failure categories in Figure 12 are derived from console error messages generated by the computer's operating system. Each error (fault) is traced to the component level; the defective component is replaced; and the board is retested. From Figure 12 we can see which applied stress tests result in what failures. Note that the Comm Logic error occurs exclusively during the 60°C temperature dwell. Other error groupings suggest that the initial temperature ramp-down and the first vibration segment are also effective precipitators of ESS failures.

The Comm Logic failure was corrected by replacing the Comm Logic ASIC; and then the board was retested. Using the data from the initial 10 boards tested at 80°C and the other results, a probability distribution function (PDF) for this fault as a function of temperature was constructed, as shown in Figure 13.

Manufacturing/Production Practices

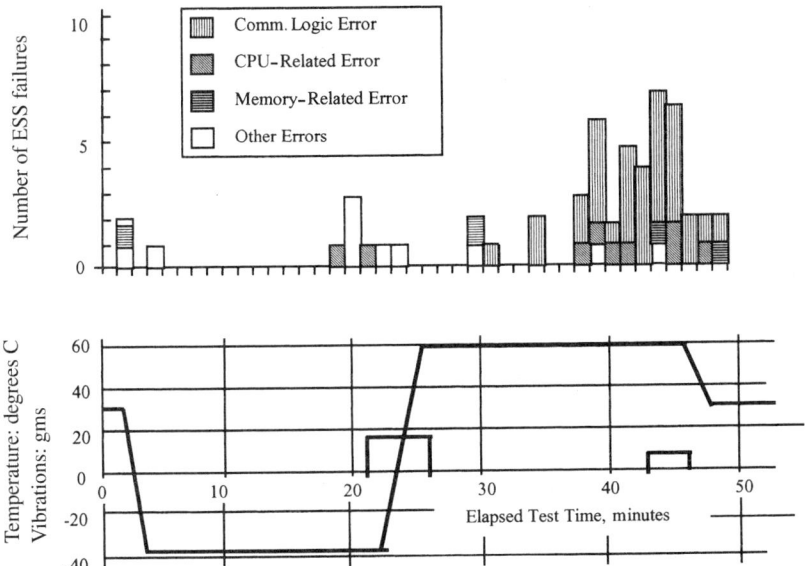

FIGURE 12 Environmental stress screening failures by type and screen time.

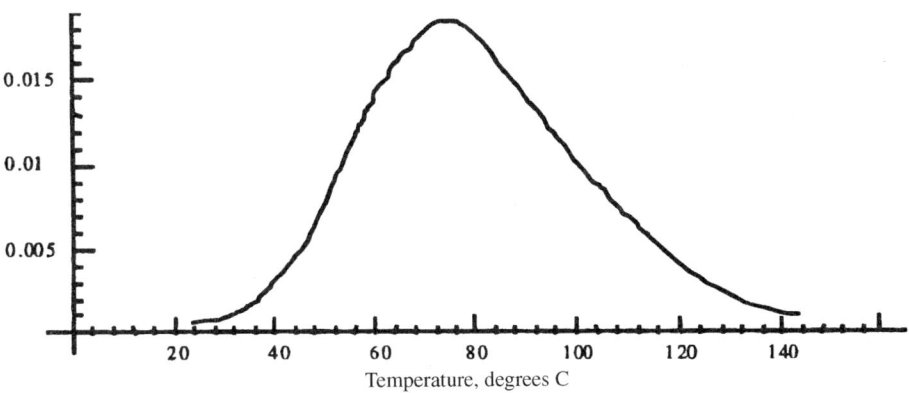

FIGURE 13 Failure probability as a function of temperature.

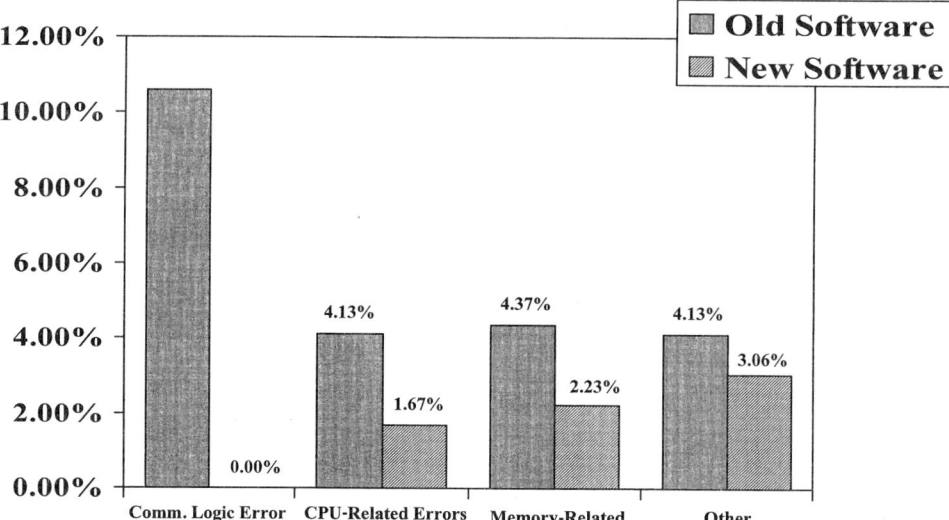

FIGURE 14   Failure rates for two software versions.

The results shown in Figure 13 were particularly disturbing since a small but significant probability of failure was predicted for temperatures in the normal operating region. The problem could be approached either by reworking the Comm Logic ASIC or by attempting a software workaround.

The final solution, which was successful, was a software revision. Figure 14 also shows the results obtained by upgrading system software from the old version to the corrected version. In Figure 14, the failures have been converted to rates to show direct comparison between the two software versions. Note that the Comm Logic error has been completely eliminated and that the remaining error rates have been significantly diminished. This result, completely unexpected (and a significant lesson learned), shows the interdependence of software and hardware in causing and correcting CPU errors.

Case Study 5: Family of CPU PWAs.   Figure 15 is a bar chart showing ESS manufacturing yields tracked on a quarterly basis. Each bar is a composite yield for five CPU products. Note that the ESS yield is fairly constant. As process and component problems were solved, new problems emerged and were addressed. In this case, given the complexity of the products, 100% ESS was required for the entire life of each product.

Figure 16 shows a detailed breakout of the 3Q97 ESS results shown in the last bar of Figure 15, and adds pre-ESS yields for the five products in production that make up the 3Q97 bar. This chart shows the value of conducting ESS in

Manufacturing/Production Practices

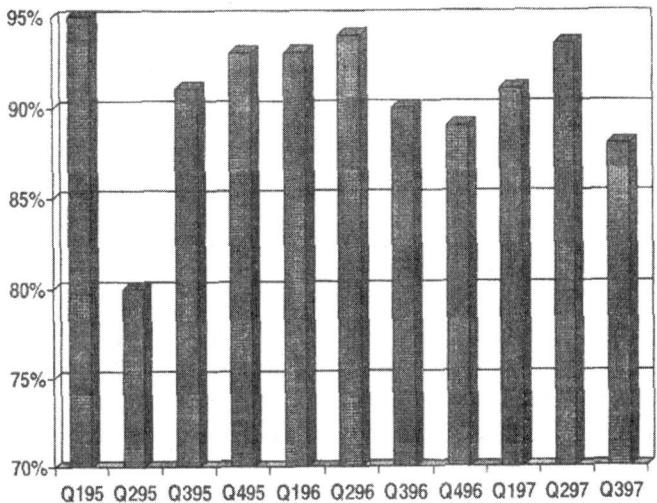

FIGURE 15 Printed wiring assembly ESS yields.

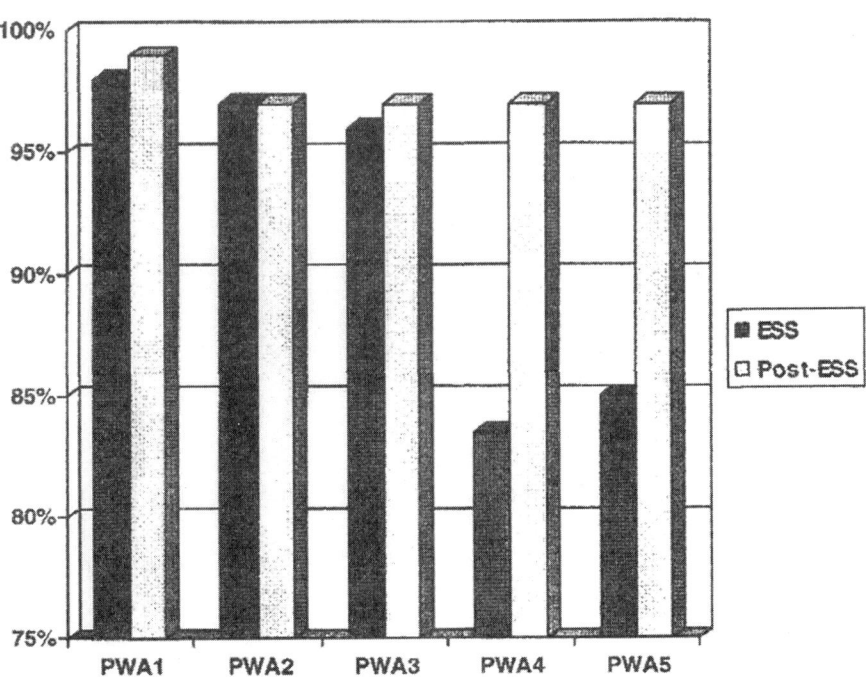

FIGURE 16 Manufacturing yield by PWA type for 3Q97.

production and the potential impact of loss in system test or the field if ESS were not conducted. Notice the high ESS yield of mature PWAs (numbers 1–3) but the low ESS yield of new boards (4 and 5), showing the benefit of ESS for new products. Also, note particularly that the post-ESS yields for both mature and immature products are equivalent, indicating that ESS is finding the latent defects. Figure 16 also shows that the value of ESS must be constantly evaluated. At some point in time when yield is stable and high, it may make sense to discontinue its use for that PWA/product. Potential candidates for terminating ESS are PWA numbers 1–3.

Figures 17–20 show the results of ESS applied to another group of CPU PWAs expressed in terms of manufacturing yield. Figures 17 and 18 show the ESS yields of mature PWAs, while Figures 19 and 20 show the yields for new CPU designs. From these figures, it can be seen that there is room for improvement for all CPUs, but noticeably so for the new CPU designs of Figures 19 and 20. Figures 17 and 18 raise the question of what yield is good enough before we cease ESS on a 100% basis and go to lot testing, skip lot testing, or cease testing altogether. The data also show that ESS has the opportunity to provide real product improvement.

The data presented in Figures 17 through 20 are for complex high-end CPUs. In the past, technology development and implementation were driven primarily by high-end applications. Today, another shift is taking place; technology is being driven by the need for miniaturization, short product development times

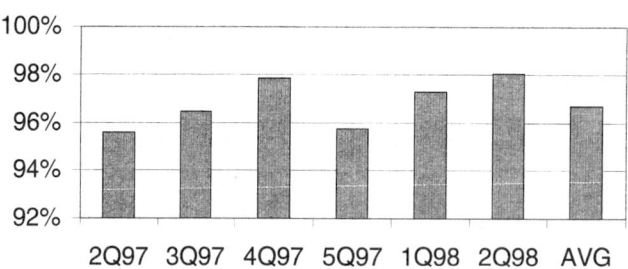

FIGURE 17 Yield data for CPU A (mature CPU PWA).

Manufacturing/Production Practices

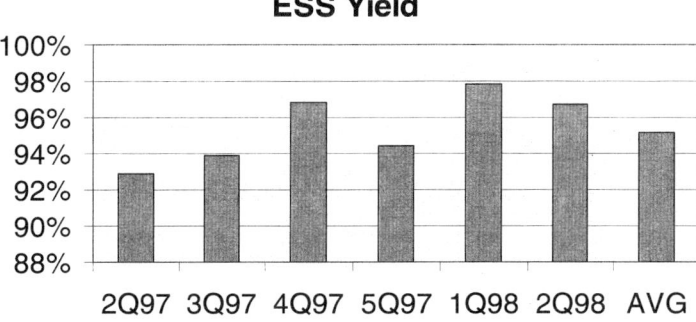

FIGURE 18 Yield data for CPU B (mature CPU PWA).

(6 months) and short product life cycles (<18 months), fast time to market, and consumer applications. Products are becoming more complex and use complex ICs. We have increasing hardware complexity and software complexity and their interactions. All products will exhibit design- or process-induced faults. The question that all product manufacturers must answer is how many of these will we allow to get to the field. Given all of this, an effective manufacturing defect test strategy as well as end-of-line functional checks are virtually mandated.

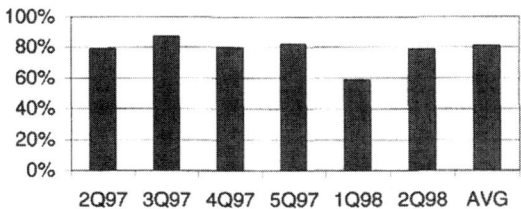

FIGURE 19 Yield data for CPU C (new CPU PWA design).

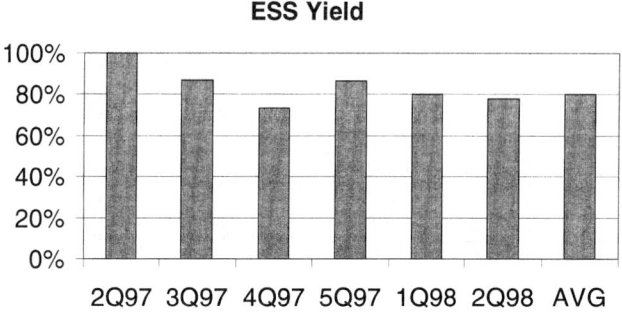

	2Q97	3Q97	4Q97	5Q97	1Q98	2Q98	AVG
YIELD	100.0%	87.0%	73.3%	86.3%	80.2%	77.7%	80.2%
FAILED	0	17	121	56	18	39	251
TESTED	8	131	453	410	91	175	1268

FIGURE 20   Yield data for CPU D (new CPU PWA design).

Environmental Stress Screening and the Bathtub Curve

The system reliability of a complex electronic product can be improved by screening out those units most vulnerable to various environmental stresses. The impact of environmental stress screening on product reliability is best seen by referring to the bathtub failure rate curve for electronic products reproduced in Figure 21. The improved quality and reliability of ICs has reduced the infant mortality (early

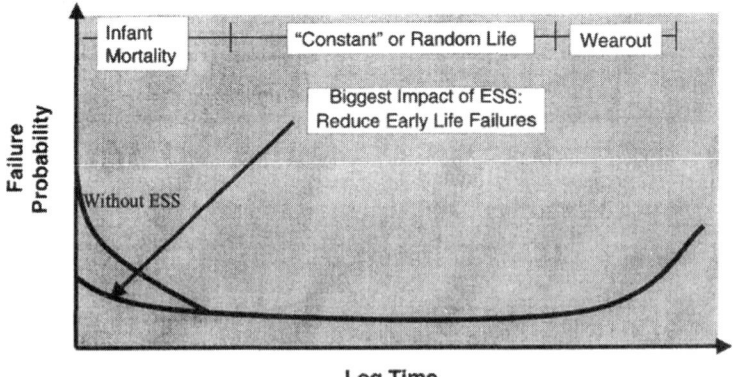

FIGURE 21   The bathtub failure rate curve.

life) failure rate in PWAs over the past 10–15 years. Application of ESS during manufacturing can further reduce early life failures, as shown in the figure.

The question that needs to be answered is how do failures from an ESS stress-to-failure distribution map onto the field failure time-to-failure distribution. Figure 22 graphically depicts this question. Failures during the useful life (often called the steady state) region can be reduced by proper application of the ESS–failure analysis to root cause–implement corrective action–verify improvement (or test–analyze–fix–test) process. The stress-to-fail graph of Figure 22 indicates that about 10–15% of the total population of PWAs subjected to ESS fail. The impact of this on time to failure (hazard rate) is shown in the graph on the right of that figure.

Implementing an ESS manufacturing strategy reduces (improves) the infant mortality rate. Field failure data corroborating this for a complex computer server CPU are shown in Figure 23. Data were gathered for identical CPUs (same design revision), half of which were shipped to customers without ESS and half with ESS. The reason for this is that an ESS manufacturing process was implemented in the middle of the manufacturing life of the product, so it was relatively easy to obtain comparison data holding all factors except ESS constant. The top curve shows failure data for PWAs not receiving ESS, the bottom curve for PWAs receiving ESS.

Figure 24 shows field failure data for five additional CPU products. These data support the first two regions of the generic failure curve in Figure 21. Figure 24 shows the improvements from one generation to succeeding generations of a product family (going from CPU A to CPU E) in part replacement rate (or failure rate) as a direct result of a twofold test strategy: HALT is utilized during the product design phase and an ESS strategy is used in manufacturing with the resultant lessons learned being applied to improve the product.

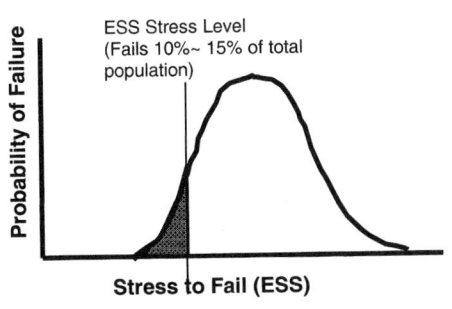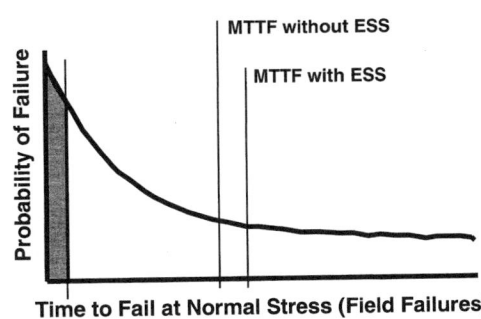

FIGURE 22 Mapping ESS and field failure distributions.

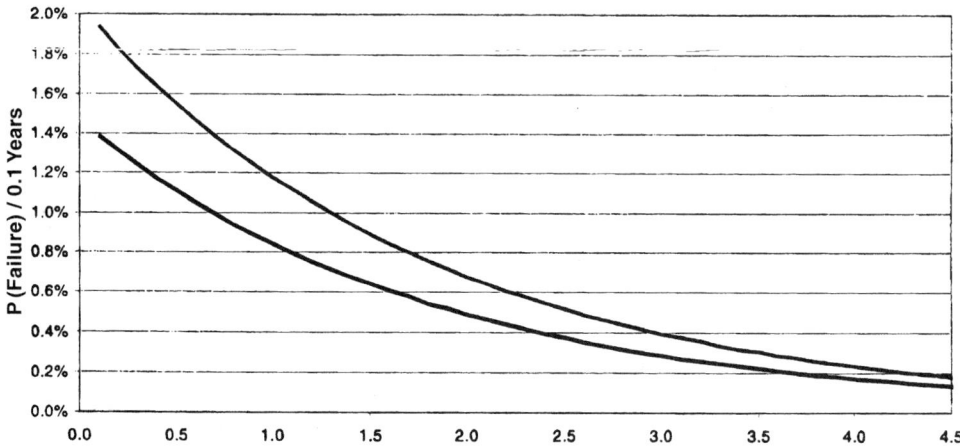

FIGURE 23 Early failure rate improvement achieved by using ESS.

One of the striking conclusions obtained from Figure 24 is that there is no apparent wearout in the useful lifetime of the products, based on 4 years of data. Thus the model for field performance shown in Figure 22 does not include a positive slope section corresponding to the wearout region of the bathtub curve.

Cost Effectiveness of Environmental Stress Screening

Environmental stress screening is effective for both state-of-the-art as well as mature PWAs. The "more bang for the buck" comes from manufacturing yield improvement afforded by applying 100% ESS to state-of-the-art PWAs. However, it has value for mature PWAs as well. In both cases this translates to less system test problems and field issues.

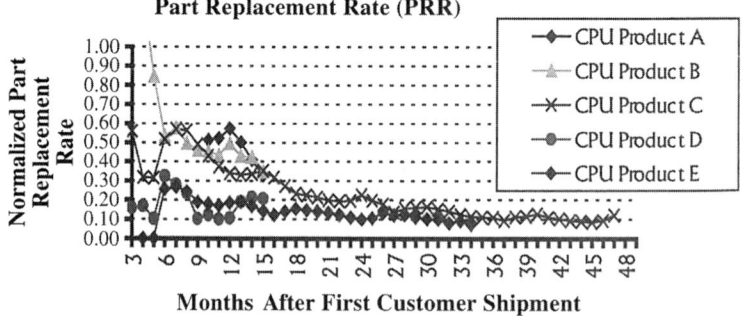

FIGURE 24 Field failure distributions of different CPU products.

Manufacturing/Production Practices

The previous section showed the effectiveness of 100% ESS. Pre-ESS manufacturing yields compared with post-ESS yields show improvement of shippable PWA quality achieved by conducting 100% ESS. This is directly translated into lower product cost, positive customer goodwill, and customer product rebuys.

In all of the preceding discussions, it is clear that a key to success in the field of accelerated stress testing is the ability to make decisions in the face of large uncertainties. Recent applications of normative decision analysis in this field show great promise. Figures 25 and 26 show the probability of achieving positive net present savings (NPS) when applying ESS to a mature CPU product and a new CPU product, respectively.

The NPS is net present savings per PWA screened in the ESS operation. It is the present value of all costs and all benefits of ESS for the useful lifetime of the PWA. A positive NPS is obtained when benefits exceed costs.

This approach to decisionmaking for ESS shows that there is always a possibility that screening will not produce the desired outcome of fielding more reliable products, thus reducing costs and saving money for the company. Many variables must be included in the analysis, both technical and financial. From the two distributions shown in Figures 25 and 26, it is seen that there is a 20% chance we will have a negative NPS for CPU product B and an 80% chance we will have a negative NPS for CPU product E. The decision indicated is to continue

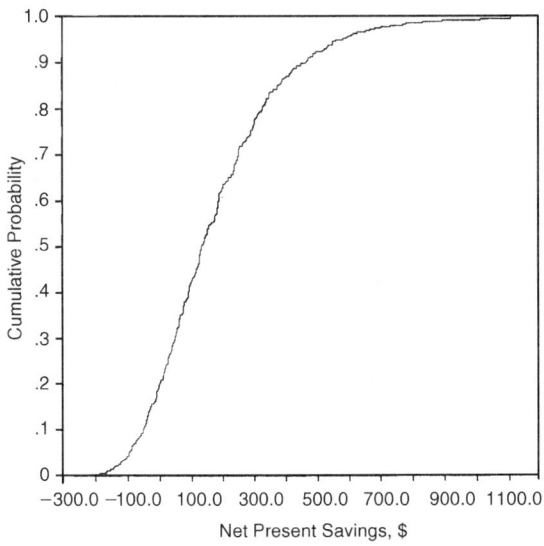

FIGURE 25 Probability distribution. Net present savings for ESS on a new CPU product (B in Fig. 24).

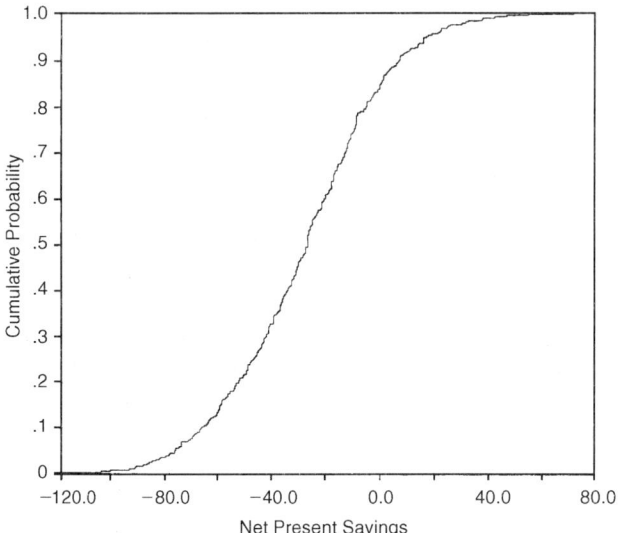

FIGURE 26 Probability distribution. Net present savings for ESS on a mature CPU product (E in Fig. 24).

ESS for CPU product B, and to stop ESS for CPU product E. The obvious next step is to analyze the decision of whether or not to perform ESS on a sample basis.

Here we see the historical trend repeating itself—as products become more reliable, there is less need to perform ESS. Just as in the case for components, the economics of screening become less favorable as the product becomes more reliable. Emphasis then shifts to sampling or audit ESS.

7.3.1 Lessons Learned Through Testing

A small percentage (typically about 10%) of PWAs that are electrically tested and exposed to ESS exhibit anomalies or problem conditions, resulting in manufacturing yield loss. These situations, depending on severity and frequency of occurrence, are typically investigated through a troubleshooting process to determine the cause of the problem or failure. The troubleshooting process requires that effective and efficient diagnostics are available to isolate the problem, usually to one or more components. These components are then removed from the PWA and tested separately on automated testing equipment (ATE) to verify that the PWA anomaly was due to a specific component (in reality only a small percentage of removed components are tested due to time and cost constraints). The ATE testing of the component(s) removed often results in a finding that there is nothing

Manufacturing/Production Practices

wrong with them (no trouble found—NTF), they meet all published specifications. In the electronics equipment/product industry 40–50% of the anomalies discovered are no trouble found. Much time in manufacturing is spent troubleshooting the NTFs to arrive at a root cause for the problem. Often the components are given to sustaining engineering for evaluation in a development system versus the artificial environment of automated test equipment. Investigating the cause of no trouble found/no defect found (NDF)/no problem found (NPF) components is a painstaking, costly, and time-consuming process. How far this investigation is taken depends on the product complexity, the cost of the product, the market served by the product, and the amount of risk and ramifications of that risk the OEM is prepared to accept. Listed in the following sections are lessons learned from this closed-loop investigative anomaly verification–corrective action process.

Lesson 1: PWAs are complex structures.

The "fallout" from electrical tests and ESS is due to the complex interactions between the components themselves (such as parametric distribution variation leading to margin pileup when all components are interconnected on the PWA, for example), between the components and the PWA materials, and between the hardware and software.

Lesson 2: Determine the root cause of the problem and act on the results.

Performing electrical testing or ESS by itself has no value. Its what is done with the outcome or results of the testing that counts. Problems, anomalies and failures discovered during testing must be investigated in terms of both short- and long-term risk. For the short-term, a containment and/or screening strategy needs to be developed to ensure that the defective products don't get to the field/customer. For the long term, a closed-loop corrective action process to preclude recurrence of the failures is critical to achieving lasting results. In either case the true root cause of the problem needs to be determined, usually through an intensive investigative process that includes failure analysis. This is all about risk evaluation, containment, and management. It is imperative that the containment strategies and corrective actions developed from problems or failures found during electrical testing and ESS are fed back to the Engineering and Manufacturing Departments and the component suppliers. To be effective in driving continuous improvement, the results of ESS must be

1. Fed back to Design Engineering to select a different supplier or improve a supplier's process or to make a design/layout change
2. Fed back to Design Engineering to select a different part if the problem was misapplication of a given part type or improper interfacing with other components on the PWA

3. Fed back to Design Engineering to modify the circuit design, i.e., use a mezzanine card, for example
4. Fed back to Manufacturing to make appropriate process changes, typically of a workmanship nature

Lesson 3: Troubleshooting takes time and requires a commitment of resources.

Resources required include skilled professionals along with the proper test and failure analysis tools. There is a great deal of difficulty in doing this because engineers would prefer to spend their time designing the latest and greatest product rather than support an existing production design. But the financial payback to a company can be huge in terms of reduced scrap and rework, increased revenues, and increased goodwill, customer satisfaction, and rebuys.

Lesson 4: Components are not the major cause of problems/anomalies.

Today the causes of product/equipment errors and problems are due to handling problems (mechanical damage and ESD), PCB attachment (solderability and workmanship) issues; misapplication/misuses of components (i.e., design application not compatible with component); connectors; power supplies; electrical overstress (EOS); system software "rev" versions; and system software–hardware interactions.

Lesson 5: The majority of problems are NTF/NDF/NPF.

In analyzing and separately testing individual components that were removed from many PWAs after the troubleshooting process, it was found that the removed components had no problems. In reality though, there is no such thing as an NDF/NTF/NPF; the problem has just not been found due either to insufficient time or resources being expended or incomplete and inaccurate diagnostics. Subsequent evaluation in the product or system by sustaining engineering has revealed that the causes for NTF/NDF/NPF result from

1. Shortcuts in the design process resulting in lower operating margins and yields and high NTF. This is due to the pressures of fast time to market and time to revenue.
2. Test and test correlation issues, including low test coverage and not using current (more effective) revision of test program and noncomprehensive PWA functional test software.
3. Incompatible test coverage of PWA to component.
4. Component lot-to-lot variation. Components may be manufactured at various process corners impacting parametric conditions such as bus hold, edge rate, and the like.
5. Design margining/tolerance stacking of components used due to para-

metric distributions (variation). For example, timing conditions can be violated when ICs with different parametric distributions are connected together; yet when tested individually, they operate properly within their specification limits.
6. Importance and variation of unspecified component parameters on system operation.
7. System/equipment software doesn't run the same way on each system when booted up. There are differences and variations in timing parameters such as refresh time. The system software causes different combinations of timing to occur and the conditions are not reproducible. Chip die shrinks, which speed up a component, tend to exacerbate timing issues/incompatibilities.
8. System noise and jitter; noisy power supplies; long cable interconnection lengths; microprocessor cycle slip (40% of microprocessor problems); and pattern/speed sensitivity.
9. Statistical or load-dependent failures—those failures that occur after the system (computer) has been run for a long time, such as cycle lurch and correctable memory errors.

I decided to follow up the NDF/NTF/NPF issue further and surveyed my colleagues and peers from Celestica, Compaq Computer Corp., Hewlett-Packard, IBM, Lucent Technologies, SGI, and Sun Microsystems to find out what their experience had been with NDFs. The results are summarized in Table 7.

There are several proposed approaches in dealing with NDF/NTF/NPF. These are presented in ascending order of application. I thank my colleagues at Celestica for providing these approaches.

Strategies for No Defect Found Investigation

Test Three Times. In order to verify the accuracy of the test equipment and the interconnection between the test equipment and the unit under test (UUT), it is recommended that the UUT be tested three times. This involves removing the UUT from the test fixture or disconnecting the test equipment from the UUT and reinstalling/reconnecting the UUT to the test equipment three times. If the UUT fails consistently, the accuracy of the test equipment or the interconnection is not likely to be the cause of the UUT failure. If, however, after disconnect/reconnect the UUT does not fail again, there is a possibility that the test equipment may play a part in the failure of the UUT, requiring further investigation. This method applies to both PWAs and individual components.

Remove/Replace Three Times. In order to verify the interconnection between the suspected failing component and the PWA, the suspected failing component should be disconnected from the PWA and reinstalled in the PWA three times (where it is technically feasible). If the suspected failing component does

TABLE 7 Computer Industry Responses Regarding No Defect Found Issues

The single largest defect detractor (>40%) is the no defect/no trouble found issue.
Of 40 problem PWA instances, two were due to manufacturing process issues and the balance to design issues.
Lucent expended resources to eliminate NDFs as a root cause and has found the real causes of so-called NDFs to be lack of training, poor specifications, inadequate diagnostics, different equipment used that didn't meet the interface standards, and the like.
No defect found is often the case of a shared function. Two ICs together constitute an electrical function (such as a receiver chip and a transmitter chip). If a problem occurs, it is difficult to determine which IC is the one with the defect because the inner functions are not readily observable.
Removing one or more ICs from a PWA obliterates the nearest neighbor, shared function, or poor solder joint effects that really caused the problem.
Replace a component on a PWA and the board becomes operational. Replacing the component shifted the PWA's parameters so that it works. (But the PWA was not defective.)
The most important and effective place to perform troubleshooting is at the customer site since this is where the problem occurred. However, there is a conflict here because the field service technician's job is to get the customer's system up and running as fast as possible. Thus, the technician is always shotgunning to resolve a field or system problem and removes two or three boards and/or replaces multiple components resulting in false pulls. The field service technician, however, is the least trained to do any troubleshooting and often ends up with a trunk full of components. Better diagnostics are needed.
Software–hardware interactions may happen once in a blue moon.
In a manufacturing, system, or field environment we are always attempting to isolate a problem or anomaly to a thing: PWA, component, etc. In many instances the real culprit is the design environment and application (how a component is used, wrong pull-up or pull-down resistor values, or a PWA or component doesn't work in a box that is heated up for example). Inadequate design techniques are big issues.
Testing a suspected IC on ATE often finds nothing wrong with the IC because the ATE is an artificial environment that is dictated by the ATE architecture, strobe placement, and timing conditions. Most often the suspected IC then needs to be placed in a development system using comprehensive diagnostics that are run by sustaining development engineers to determine if the IC has a problem or not. This ties in with Lessons 3 and 7.

Manufacturing/Production Practices 359

not fail consistently, this is a strong clue that the interconnection between the PWA and the suspected component contributes to the PWA failure and should be investigated further. If the failure does repeat consistently after removal and replacement three times, the suspected failing component should be used in a component swap, as described next.

Component Swap. Under the component swap technique, a suspected failing component is removed from the failing application and swapped into a similar, but passing application. Concurrently, the same component from the passing application is swapped into the failing application. If after the component swap, the failed PWA now passes and the passing PWA fails, that's a pretty strong clue that the component was the cause of the problem and failure analysis of the component can begin.

If after the swap, the failed PWA still fails and the passing PWA still passes, then the swapped component probably is not the cause of the problem and further failure investigation and fault isolation is required. If after the swap, both PWAs now pass, the cause probably has something to do with the interconnection between the component and the PWA (maybe a cold solder joint or fracture of a solder joint) and probably had nothing to do with the component. An NTF has been avoided.

The following examples illustrate the methodologies just discussed.

Example 1: A printed wiring assembly fails at system integration test. This failing PWA should be disconnected from the test hardware, reconnected and tested three times. If the PWA continues to consistently fail, proceed to component swap. In this swap environment, the failed PWA from the fail system is swapped with a passing PWA from a passing system. If the failed system now passes and the passing system now fails, there's a good chance that something is wrong with the PWA. The PWA should be returned to the PWA manufacturer for further analysis. If after the swap, the failed system still fails and the passing system still passes, then the suspect PWA isn't likely to be the root cause of the problem. If after the swap, both systems now pass, there may be some interconnect issue between the PWA and the system that should be assessed.

Example 2: A PWA manufacturer is testing a PWA at functional test and it fails. The PWA should be removed from the test fixture, replaced, and retested three times. If the PWA failure persists it is up to the failure analysis technician to isolate the failure cause to an electronic component. The suspected component should be removed and replaced (if technically feasible) and retested three times. If the failure still persists, the failed component should be removed from the failing PWA and swapped with a passing component from a passing PWA. Again, if after the swap, the failed PWA now passes and the passing PWA fails, it's reasonable to conclude that the component was the root cause of the problem and should be returned to the component supplier for root cause analysis and corrective action. If, however, after the swap, the failed PWA still fails and the passing PWA still passes, then the swapped component probably is not the cause of the problem. If after the swap, both PWAs now pass, the root cause may have something to do with the interconnection between the component and the PWA (such as a fractured solder joint) and nothing to do with the component.

Lesson 6: Accurate diagnostics and improved process mapping tests are required for problem verification and resolution.

Let's take an example of a production PWA undergoing electrical test to illustrate the point. Diagnostics point to two or three possible components that are causing a PWA operating anomaly. This gives a 50% and 33% chance of finding the problem component, respectively, and that means that the other one or two components are NTF. Or it may not be a component issue at all, in which case the NTF is 100%. Take another example. Five PWAs are removed from a system to find the one problem PWA, leaving four PWAs, or 80%, as NTF in the best case. This ties in with Lesson 5.

The point being made is that diagnostics that isolate a 50% NTF rate are unacceptable. Diagnostics are typically not well defined and need to be improved because PWAs are complex assemblies that are populated with complex components.

Lesson 7: Sustaining Engineering needs to have an active role in problem resolution.

Many of the anomalies encountered are traced to one or more potential problem components. The problem components need to be placed in the product/system to determine which, if any, component is truly bad. As such, sustaining engineering's active participation is required. This ties in with Lesson 3.

7.4 SYSTEM TESTING

The topic of system test begins with the definition of what a system test is. One definition of a system is "the final product that the customer expects to operate as a functioning unit; it has a product ID number." Another definition is "an item in which there is no next level of assembly."

Products or systems run the gamut from the simple to the complex—from digital watches, calculators, electronic games, and personal computers at one end of the spectrum to automobiles, nonstop mainframe computers, patient heart monitors, and deep space probes at the other end. Testing these products is directly related to the level of product complexity. Simple products are easier and less costly to test than are complex products.

System testing contains elements of both design and manufacturing. System test is used to ensure that the interconnected and integrated assemblies, PWAs, modules, and power supplies work together and that the entire system/product functions in accordance with the requirements established by the global specification and the customer. For example, a Windows NT server system could consist of basic computer hardware, Windows NT server and clustering software, a Veritas volume manager, a RAID box, a SCSI adapter card, and communications controllers and drivers.

Many of the same issues incurred in IC and PWA test cascade to the system (finished or complete product) level as well. Since products begin and end at the system level, a clear understanding of the problems expected must emanate from the customer and/or marketing-generated product description document. The system specifications must clearly state a test and diagnostic plan, anticipating difficult diagnostic issues (such as random failures that you know will occur); specify the expected defects and the specific test for these defects; and outline measurement strategies for defect coverage and the effectiveness of these tests.

Some of the issues and concerns that cause system test problems which must be considered in developing an effective system test strategy include

- Defects arise from PWA and lower levels of assembly as well as from system construction and propagate upward.
- Interconnections are a major concern. In microprocessor-based systems, after the data flow leaves the microprocessor it is often affected by glitches on the buses or defects due to other components.
- Timing compatability between the various PWAs, assemblies, and modules can lead to violating various timing states, such as bus contention and crosstalk issues, to name two.
- Effectiveness of software diagnostic debug and integration.
- Design for test (DFT) is specified at the top of the hierarchy, but implemented at the bottom: the IC designer provides cures for board and system test, and the board designer provides cures for system test.
- ICs, PWAs, assemblies, and modules have various levels of testability designed in (DFT) that may be sufficient for and at each of these levels. However, their interconnectivity and effectiveness when taken together can cause both timing and testing nightmares. Also the level of DFT implementation may be anywhere from the causal and careless to the comprehensive.
- System failures and intermittent issues can result in shortened component (IC) life.
- Computer systems are so complex that no one has figured out how to design out or test for all the potential timing and race conditions, unexpected interactions, and nonrepeatable transient states that occur in the real world.
- Interaction between hardware and software is a fundamental concern.

TABLE 8 System Test Defects

Connectors—opens, shorts, intermittent
Software integrity
Electrostatic discharge and electrical overstress
Errors in wiring
Timing errors due to subtle defects in timing paths
Leakage paths—solder slivers, foreign material
Marginal components
Improper or poor integrity connections
Incompatibility between subassemblies
Crosstalk
Cables pinched or cut
Jumper wires missing or in wrong configuration
Serial operation of components that individually are within specification but collectively cause failure
Integration of millions of gates (not concerned about individual components but interaction of all components and their variation)

A list of some typical system test defects is presented in Table 8. This list does not contain those manufacturing and workmanship defects that occur during PWA manufacturing and as detected by PWA visual inspection, ICT, and functional test: solder issues, missing and reversed components, and various PCB issues (vias and plated through-holes, for example). Since system test is so varied and complex, there is no way to address and do justice to the myriad unique issues that arise and must be addressed and solved. System testing includes not only electrical testing, but also run-in test—also called system burn-in (e.g., 72-hr run-in at 25°C, or 48-hr at 50°C)—that checks for product stability and facilitates exercising the product with the full diagnostic test set. The same lessons learned from PWA testing apply to systems test as well. Once a system/product successfully completes system test it can be shipped to the customer.

7.5 FIELD DATA COLLECTION AND ANALYSIS

System problems come from many sources that include environmental, application, individual component, and integration issues, for example. The environmental causes of failure in electronic systems include temperature, vibration, humidity, and dust, with more than half of the failures being attributed to thermal effects (Figure 27). Surface mount assemblies are particularly sensitive to thermal conditions whereby temperature cycling and power on/off cycles induce cycling strains and stresses in the solder interconnects. In general, vibration is less damaging to solder joints because high-frequency mechanical cycling does not allow enough time for creep strains to develop in solder joints.

Manufacturing/Production Practices

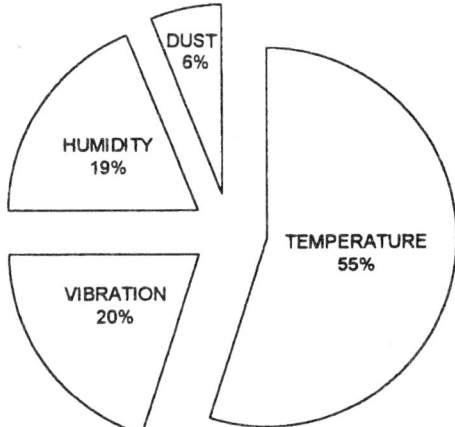

FIGURE 27 Different environmental conditions that cause failure in electronic systems/products.

Component failures are a concern and were briefly discussed in Chapter 4.

Then there is the issue of application-related problems that must be considered. Some examples include software–hardware integration and interaction with power supplies; disk drives and other equipment; the timing issues that occur when interconnecting multiple ICs with different setup and hold conditions; and the interaction of the different system PWAs with one another.

The best understanding of customer issues and thus product reliability comes from gathering and analyzing product field data. An analysis of field reliability data indicates

> The accuracy of the predictions made and the effectiveness of the reliability tests conducted
> Whether established goals are being met
> Whether product reliability is improving

In addition to measuring product reliability, field data are used to determine the effectiveness of the design and manufacturing processes, to correct problems in existing products, and to feed back corrective action into the design process for new products. Regular reports are issued disseminating both good and bad news from the field. The results are used to drive corrective actions in design, manufacturing, component procurement, and supplier performance.

A comprehensive data storage system is used to determine accurate operation times for each field replaceable unit (FRU) and allows identification of a field problem that results in a unit being replaced, as well as the diagnosis and

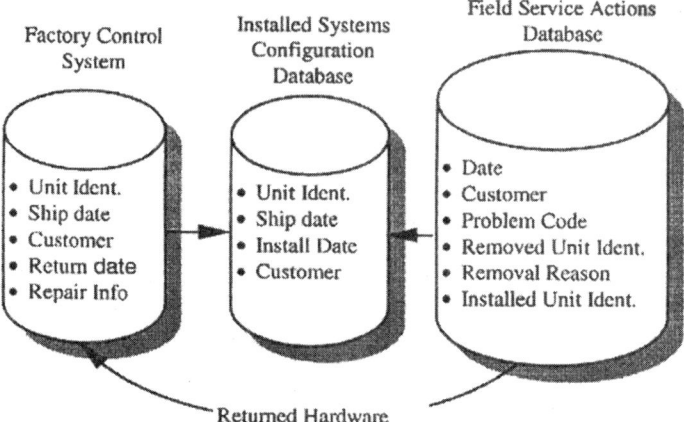

FIGURE 28 Example of field data collection system. (From Ref. 1. Courtesy of the Tandem Division of Compaq Computer Corporation Reliability Engineering Department.)

repair actions on the unit that caused the problem. Since the installation and removal dates are recorded for each unit, the analysis can be based on actual run times, instead of estimates based on ship and return dates.

At the Tandem Division of Compaq Computer Corp. data are extracted from three interlinked databases (Figure 28). Each individual FRU is tracked from cradle to grave, i.e., from its original ship date through installation in the field and, if removed from a system in the field, through removal date and repair. From these data, average removal rates, failure rates, and MTBF are computed. Individual time-to-fail data are extracted and plotted as multiply processed data on a Weibull hazard plot to expose symptoms of wearout.

Using the data collection system shown in Figure 28, a typical part replacement rate (PRR) is computed by combining installation data from the Installed Systems database and field removal data from the Field Service Actions database. The data are plotted to show actual field reliability performance as a function of time versus the design goal. If the product does not meet the goal, a root cause analysis process is initiated and appropriate corrective action is implemented. An example of the PRR for a disk controller is plotted versus time in Figure 29 using a 3-month rolling average.

Figure 30 shows a 3-month rolling average part replacement rate for a product that exhibited several failure mechanisms that had not been anticipated during the preproduction phase. Corrective action included respecification of a critical component, PWA layout improvement, and firmware updates. The new revision was tracked separately and the difference was dramatically demonstrated by the resultant field performance data. The old version continued to exhibit an

Manufacturing/Production Practices

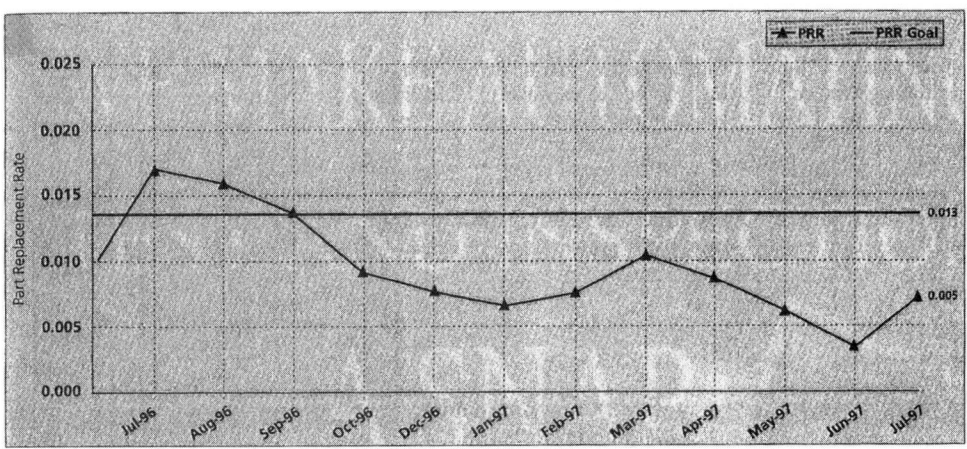

FIGURE 29 Example of a disk controller field analysis. (From Ref. 1. Courtesy of the Tandem Division of Compaq Computer Corporation Reliability Engineering Department.)

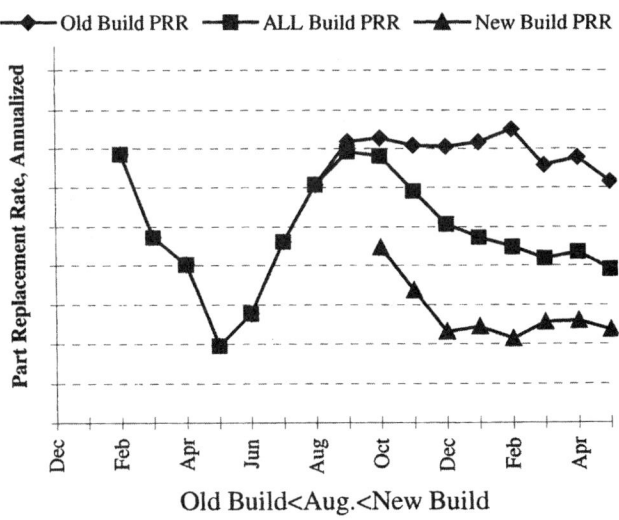

FIGURE 30 Improved PRR as a result of corrective action implemented in new product versions. (From Ref. 1. Courtesy of the Tandem Division of Compaq Computer Corporation Reliability Engineering Department.)

unsatisfactory failure rate, but the new version was immediately seen to be more reliable and quickly settled down to its steady state value, where it remained until the end of its life.

Power supplies are particularly troublesome modules. As such much data are gathered regarding their field performance. Figure 31 plots the quantity of a

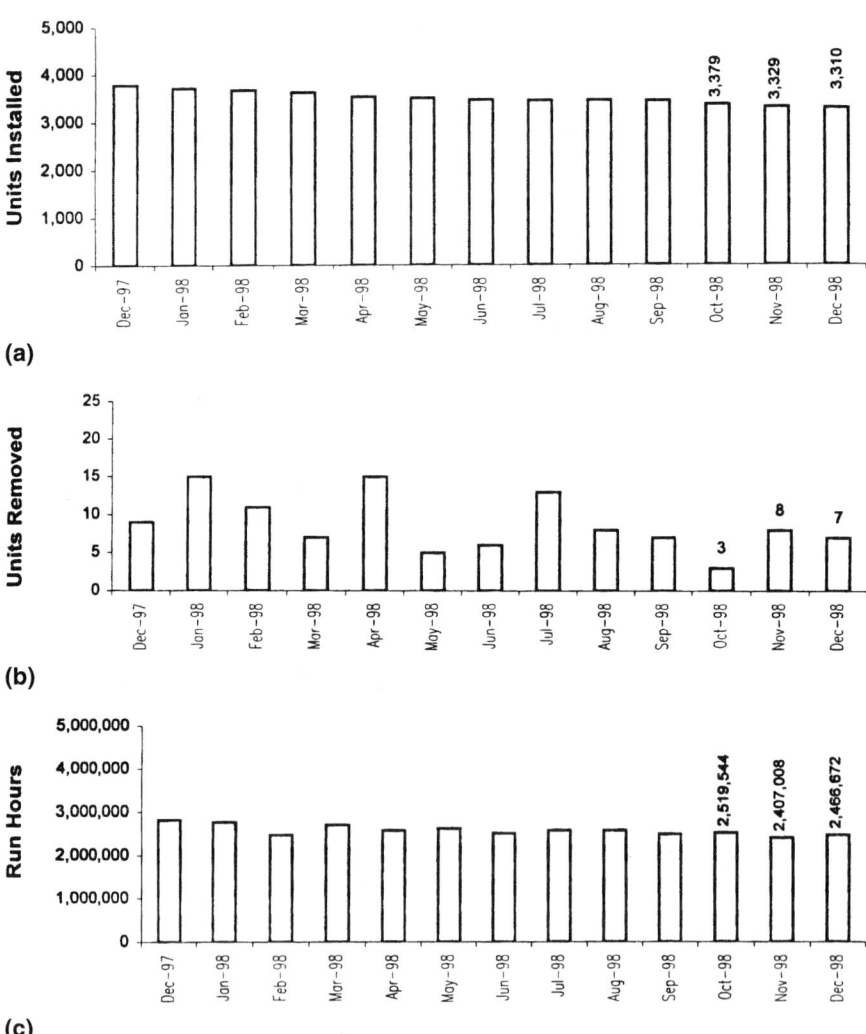

FIGURE 31 Power supply field data: (a) installed field base for a power supply; (b) field removal for the power supply; (c) part replacement rate/MTBF for the power supply.

Manufacturing/Production Practices

given power supply installed per month in a fielded mainframe computer (a), units removed per month from the field (b), and run hours in the field (c). The PRR and MTBF for the same power supply are plotted in Figure 32a and b, respectively.

Certain products such as power supplies, disk drives, and fans exhibit known wearout failure mechanisms. Weibull hazard analysis is performed on a regular basis for these types of products to detect signs of premature wearout. To perform a Weibull analysis, run times must be known for all survivors as

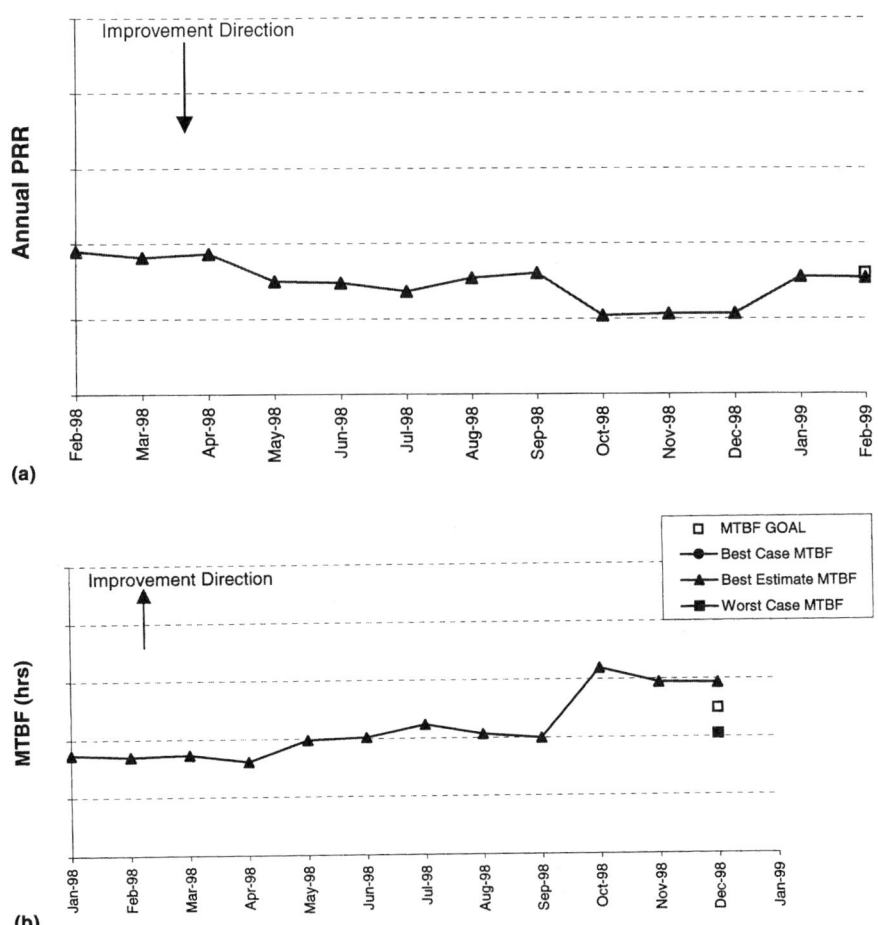

FIGURE 32 Example of 3-month rolling average plots of (a) PRR and (b) MTBF for power supply of Figure 31.

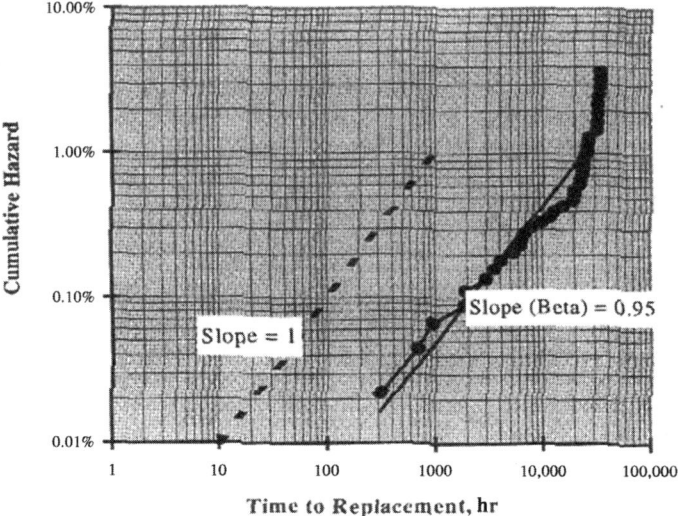

FIGURE 33 Weibull plot showing disk drive wearout. *Note*: The software's attempt to fit a straight line to these bimodal data illustrates the necessity of examining the graph and not blindly accepting calculations. (From Ref. 1. Courtesy of the Tandem Division of Compaq Computer Corporation Reliability Engineering Department.)

well as for removals. A spreadsheet macro can be used to compute the hazard rates and plot a log–log graph of cumulative hazard rate against run time. Figure 33 is a Weibull hazard plot of a particular disk drive, showing significant premature wearout. This disk drive began to show signs of wearout after 1 year (8760 hr) in the field, with the trend being obvious at about 20,000 hr. Field tracking confirmed the necessity for action and then verified that the corrective action implemented was effective.

Figure 34 is the Weibull hazard plot for the example power supply of Figures 31 and 32. This plot has a slope slightly less than 1, indicating a constant to decreasing failure rate for the power supply. Much effort was expended over an 18-month period in understanding the root cause of myriad problems with this supply and implementing appropriate corrective actions.

7.6 FAILURE ANALYSIS

Failures, which are an inherent part of the electronics industry as a result of rapidly growing IC and PWA complexity and fast time to market, can have a severe financial impact. Consequently, failures must be understood and corrective actions taken quickly. Figure 35 shows some of the constituent elements for the

Manufacturing/Production Practices

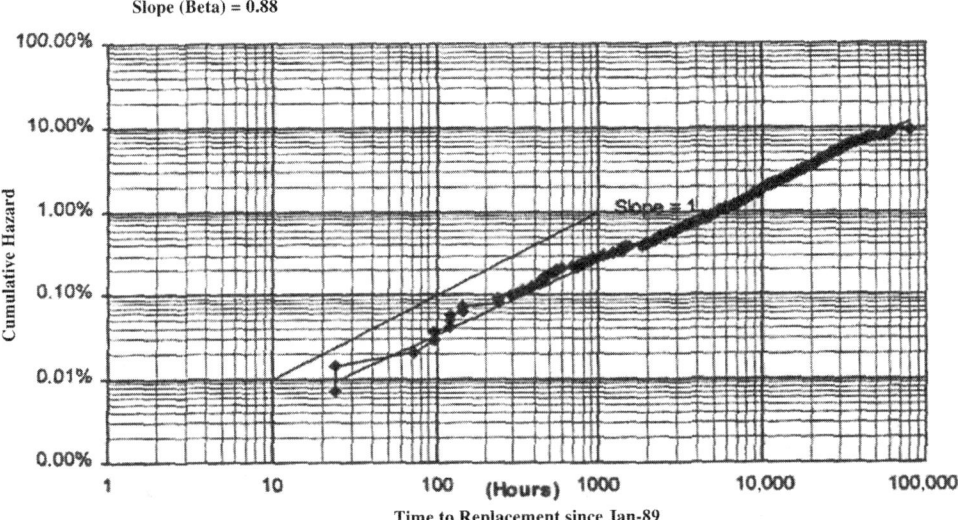

FIGURE 34 Weibull hazard plot for power supply of Figures 31 and 32. Slope >1: increasing failure rate. Slope = 1: constant failure rate. Slope <1: decreasing failure rate. (From Ref. 1. Courtesy of the Tandem Division of Compaq Computer Corporation Reliability Engineering Department.)

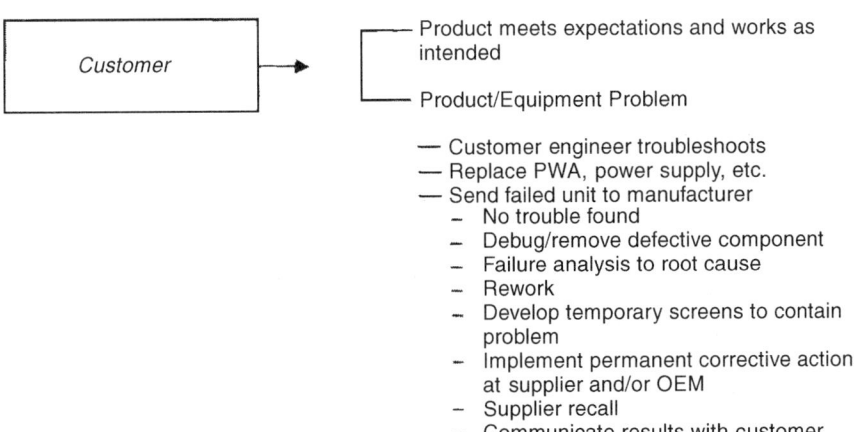

FIGURE 35 Customer PWA problem resolution.

last block of Figure 1 from Chapter 1—the customer—as relates to problems experienced with a product/system. In the field or during system test, the customer support engineer/field service engineer or manufacturing/hardware engineer, respectively, wastes no time in getting a system back online by replacing a PWA, power supply, or disk drive, for example. So typically the real investigative work of problem resolution starts at the PWA level. In some cases, because of the complexity of the ICs, it is the circuit designers themselves who must perform failure analysis to identify design and manufacturing problems.

Section 7.2.5 on lessons learned initiated the discussion of the investigative troubleshooting problem resolution process. The point was made that this process is all about risk evaluation, containment, and management. Problems that surface during development, manufacturing, and ESS as well as in the field need to be analyzed to determine the appropriate corrective action with this in mind. A metaphor for this process is a homicide crime scene. The police detective at the crime scene takes ownership and follows the case from the crime scene to interviewing witnesses to the autopsy to the various labs to court. This is much like what occurs in resolving a product anomaly. Another metaphor is that of an onion. The entire problem resolution process is like taking the layers off of an onion, working from the outside layer, or highest level, to the core, or root cause. This is depicted in Table 9.

If the PWA problem is traced to a component, then it first must be determined if the component is defective. If the component is defective, a failure analysis is conducted. Failure analysis is the investigation of components that fail during manufacturing, testing, or in the field to determine the root cause of the failure. This is an important point because corrective action cannot be effective without information as to the root cause of the failure. Failure analysis locates specific failure sites and determines if a design mistake, process problem, material defect, or some type of induced damage (such as operating the devices outside the maximum specification limits or other misapplication) caused it. A shortcoming of failure analysis to the component is that the process does not provide any information regarding adjacent components on the PWA.

A mandatory part of the failure analysis process is the prompt (timely) feedback of the analysis results to determine an appropriate course of action. Depending on the outcome of the failure analysis, the supplier, the OEM, the CM, or all three working together must implement a course of containment and corrective action to prevent recurrence. Proper attention to this process results in improved manufacturing yields, fewer field problems, and increased reliability.

Failure analysis involves more than simply opening the package and looking inside. Failure mechanisms are complex and varied, so it is necessary to perform a logical sequence of operations and examinations to discover the origin of the problem. The failure analysis of an IC is accomplished by combining a series of electrical and physical steps aimed at localizing and identifying the

Manufacturing/Production Practices

TABLE 9 Layers of the IC Problem Resolution Process

	Example

Complaint/symptom
  What happened?
  How does the problem manifest itself?
	Data line error

Failure mode
  Diagnostic or test result (customer site and IC supplier incoming test)?
  How is the problem isolated to a specific IC (customer site)?
  What is the measurement or characterization of the problem?
	Leakage on pin 31

Failure mechanism
  Physical defect or nonconformity of the IC?
  What is the actual anomaly that correlates to the failure mode?
	Oxide rupture

Failure cause
  Explanation of the direct origin or source of the defect.
  How was the defect created?
  What event promoted or enabled this defect?
	ESD

Root cause
  Description of the initial circumstances that can be attached to this problem.
  Why did this problem happen?
	Improper wrist strap use

ESD, electrostatic discharge.

ultimate cause of failure (see Fig. 36). The process of Figure 36 is shown in serial form for simplicity. However, due to the widely varying nature of components, failures, and defect mechanisms, a typical analysis involves many iterative loops between the steps shown. Identifying the failure mechanism requires an understanding of IC manufacturing and analysis techniques and a sound knowledge of the technology, physics, and chemistry of the devices plus a knowledge of the working conditions during use. Let's look at each of the steps of Figure 36 in greater detail.

Fault Localization

The first and most critical step in the failure analysis process is fault localization. Without knowing where to look on a complex VLSI component, the odds against locating and identifying a defect mechanism are astronomical. The problem is like the familiar needle in the haystack.

Because of the size and complexity of modern VLSI and ULSI components, along with the nanometric size of defects, it is imperative to accurately localize faults prior to any destructive analysis. Defects can be localized to the nearest logic block or circuit net or directly to the physical location of the responsible

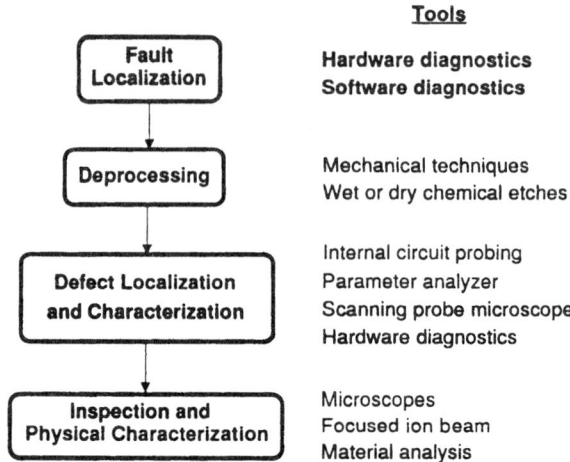

FIGURE 36 Simplified diagram of IC failure analysis process.

defect. There are two primary methods of fault localization: hardware-based diagnostics using physical parameters like light, heat, or electron-beam radiation, and software-based diagnostics using simulation and electrical tester (ATE) data.

Hardware diagnostic techniques are classified in two broad categories. The first is the direct observation of a physical phenomenon associated with the defect and its effects on the chip's operation. The second is the measurement of the chip's response to an outside physical stimulus, which correlates to the instantaneous location of that stimulus at the time of response. While a fault can sometimes be isolated directly to the defect site, there are two primary limitations of hardware diagnostics.

The first is that the techniques are defect dependent. Not all defects emit light or cause localized heating. Some are not light sensitive nor will they cause a signal change that can be imaged with an electron beam. As such it is often necessary to apply a series of techniques, not knowing ahead of time what the defect mechanism is. Because of this, it can often take considerable time to localize a defect.

The second and most serious limitation is the necessity for access to the chip's transistors and internal wiring. In every case, the appropriate detection equipment or stimulation beam must be able to view or irradiate the site of interest, respectively. With the increasing number of metal interconnect layers and the corresponding dielectric layers and the use of flip chip packaging, the only way to get to the individual transistor is through the backside of the die.

Software diagnostics are techniques that rely on the combination of fault simulation results and chip design data to determine probable fault locations. While it is possible to do this by manually analyzing failure patterns, it is impractical for ICs of even moderate complexity. Software diagnostics are generally categorized in two groups that both involve simulation of faults and test results: precalculated fault dictionaries and posttest fault simulation.

Deprocessing

Once the fault has been localized as accurately as possible the sample must be prepared for further characterization and inspection. At this stage the chip usually needs to first be removed from its package. Depending on the accuracy of fault localization and the nature of the failure, perhaps multiple levels of the interlevel insulating films and metal wiring may need to be sequentially inspected and removed. The process continues until the defect is electrically and physically isolated to where it is best identified and characterized.

To a great extent deprocessing is a reversal of the manufacturing process; films are removed in reverse order of application. Many of the same chemicals and processes used in manufacturing to define shapes and structures are also used in the failure analysis laboratory, such as mechanical polishing, plasma or dry etching, and wet chemical etching.

Defect Localization

Again, depending on the accuracy of fault localization and the nature of the failure, a second localization step or characterization of the defect may be necessary. At this point the defect may be localized to a circuit block like a NAND gate, latch, or memory cell. By characterizing the effects of the defect on the circuit's performance it may be possible to further pinpoint its location. Because the subsequent steps are irreversible it is important to gather as much information as possible about the defect and its location before proceeding with the failure analysis.

A number of tools and techniques exist to facilitate defect localization and characterization. Both optical source and micrometer-driven positioners with ultrafine probes (with tips having diameters of approximately 0.2 µm) are used to inject and measure signals on conductors of interest. High-resolution optical microscopes with long working-distance objectives are required to observe and position the probes. Signals can be DC or AC. Measurement resolution of tens of millivolts or picoamperes is often required. Because of shrinking linewidths it has become necessary to use a focused ion beam (FIB) tool to create localized probe pads on the nodes of interest. A scanning probe microscope (SPM) may be used to measure the effects of the defect on electrostatic force, atomic force, or capacitance. A number of other techniques are used for fault localization based on the specific situation and need. These are based on the use of light, heat, or electron-beam radiation.

Inspection and Defect Characterization

After exhausting all appropriate means to localize and characterize a fault, the sample is inspected for a physical defect. Once identified the physical defect must often be characterized for its material properties to provide the IC manufacturing line with enough information to determine its source.

Depending on the accuracy of localization, the fail site is inspected using one of three widely used techniques: optical, scanning-electron, or scanning-probe microscopy. Optical microscopy scans for anomalies on relatively long wires or large circuit blocks (latches, SRAM cells, etc.). While relatively inadequate for high-magnification imaging, optical microscopy is superior for its ability to simultaneously image numerous vertical levels. Nanometer-scale resolution is attained with scanning electron microscopy (SEM). In addition to its high magnification capabilities, SEM can evaluate material properties such as atomic weight and chemical content. However, it is limited to surface imaging, requiring delayering of films between inspection steps. For defects localized to extremely small areas (individual transistors, dynamic memory cell capacitors, etc.) a scanning probe microscope (SPM) can be used. This technique offers atomic-scale resolution and can characterize electrostatic potential, capacitance, or atomic force across small areas.

When these techniques cannot determine the material composition of the defect or are unable to locate a defect altogether, a suite of chemical and material analysis tools are utilized, e.g., transmission electron microscopy (TEM), auger electron spectroscopy (AES), and electron spectroscopy for chemical analysis (ESCA), to name several.

Reiterating, once the defect and its cause are identified, the information is fed back up the entire electronics design/manufacture chain to allow all members to identify risk and determine containment and a workaround solution (often implementing some sort of screening) until permanent corrective action is instituted. The need for this feedback and open communication cannot be stressed enough.

Integrated circuit failure analysis is a time-consuming and costly process that is not applicable or required for all products or markets. For lower cost consumer and/or disposable products or for personal computers, failure analysis is not appropriate. For mainframe computers, military/aerospace, and automotive applications, failure analysis of problem or failing components is mandatory.

ACKNOWLEDGMENTS

Portions of Section 7.1.2 excerpted from Ref. 1.
Much of the material for Section 7.2.4 comes from Refs. 2, 3 and 6.

Manufacturing/Production Practices

Portions of Section 7.3 excerpted from Ref. 7, courtesy of the Tandem Division of the Compaq Computer Corporation Reliability Engineering Department. Portions of Section 7.3.1 excerpted from Ref. 8.

REFERENCES

1. Hnatek ER, Russeau JB. PWA contract manufacturer selection and qualification, or the care and feeding of contract manufacturers. Proceedings of the Military/Aerospace COTs Conference, Albuquerque, NM, 1998, pp 61–77.
2. Hnatek ER, Kyser EL. Straight facts about accelerated stress testing (HALT and ESS)—lessons learned. Proceedings of The Institute of Environmental Sciences and Technology, 1998, pp 275–282.
3. Hnatek ER, Kyser EL. Practical Lessons Learned from Overstress Testing—a Historical Perspective, EEP Vol. 26-2: Advances in Electronic Packaging—1999. ASME 1999, pp 1173–1180.
4. Magaziner I, Patinkin M. The Silent War, Random House Publishers, 1989.
5. Lalli V. Space-system reliability: a historical perspective. IEEE Trans Reliability 47(3), 1998.
6. Roettgering M, Kyser E. A Decision Process for Accelerated Stress Testing, EEP Vol. 26-2: Advances in Electronic Packaging—1999, Vol. 2. ASME, 1999, pp 1213–1219.
7. Elerath JG et al. Reliability management and engineering in a commercial computer environment. Proceedings of the International Symposium on Product Quality and Integrity, pp 323–329.
8. Vallet D. An overview of CMOS VLSI/failure analysis and the importance of test and diagnostics. International Test Conference, ITC Lecture Series II, October 22, 1996.

FURTHER READING

1. Albee A. Backdrive current-sensing techniques provide ICT benefits. Evaluation Engineering Magazine, February 2002.
2. Antony J et al. 10 steps to optimal production. Quality, September 2001.
3. Carbone J. Involve buyers. Purchasing, March 21, 2002.
4. IEEE Components, Packaging and Manufacturing Technology Society Workshops on Accelerated Stress Testing Proceedings.
5. International Symposium for Testing and Failure Analysis Proceedings.
6. Kierkus M and Suttie R. Combining x-ray and ICT strategies lowers costs. Evaluation Engineering, September 2002.
7. LeBlond C. Combining AOI and AXI, the best of both worlds. SMT, March 2002.
8. Prasad R. AOI, Test and repair: waste of money? SMT, April 2002.
9. Radiation-induced soft errors in silicon components and computer systems. International Reliability Symposium Tutorial, 2002.

10. Ross RJ et al. Microelectronic Failure Analysis Desk Reference. ASM International, 1999 and 2001 Supplement.
11. Scheiber S. The economics of x-rays. Test & Measurement World, February 2001.
12. Serant E, Sullivan L. EMS taking up demand creation role. Electronic Buyers News, October 1, 2001.
13. Sexton J. Accepting the PCB test and inspection challenge. SMT, April 2001.
14. Verma A, Hannon P. Changing times in test strategy development. Electronic Packaging and Production, May 2002.

8
Software

8.1 INTRODUCTION

As stated at the outset, I am a hardware person. However, to paint a complete picture of reliability, I think it is important to mention some of the issues involved in developing reliable software. The point is that for microprocessor-based products, hardware and software are inextricably interrelated and codependent in fielding a reliable product to the customer base.

In today's systems, the majority of issues/problems that crop up are attributed to software rather than hardware or the interaction between hardware and software. For complex systems like high-end servers, oftentimes software fixes are made to address hardware problems because software changes can be made more rapidly than hardware changes or redesign. There appears to be a larger gap between customer expectations and satisfaction as relates to software than there is for hardware. Common software shortfalls from a system perspective include reliability, responsiveness to solving anomalies, ease of ownership, and quality of new versions.

An effective software process must be predictable; cost estimates and schedule commitments must be met with reasonable consistency; and the resulting products (software) should meet user's functional and quality expectations. The software process is the set of tools, methods, and practices that are used to produce a software product. The objectives of software process manage-

377

ment are to produce products according to plan while simultaneously improving an organization's capability to produce better products. During software development a creative tension exists between productivity (in terms of number of lines of codes written) and quality. The basic principles are those of statistical control, which is based on measurement. Unless the attributes of a process can be measured and expressed numerically, one cannot begin to bring about improvement. But (as with hardware) one cannot use numbers arbitrarily to control things. The numbers must accurately represent the process being controlled, and they must be sufficiently well defined and verified to provide a reliable basis for action. While process measurements are essential for improvement, planning and preparation are required; otherwise the results will be disappointing.

At the inception of a software project, a commitment must be made to quality as reflected in a quality plan that is produced during the initial project planning cycle. The plan is documented, reviewed, tracked, and compared to prior actual experience. The steps involved in developing and implementing a quality plan include

1. Senior management establishes aggressive and explicit numerical quality goals. Without numerical measures, the quality effort will be just another motivational program with little lasting impact.
2. The quality measures used are objective, requiring a minimum of human judgment.
3. These measures are precisely defined and documented so computer programs can be written to gather and process them.
4. A quality plan is produced at the beginning of each project. This plan commits to specific numerical targets, and it is updated at every significant project change and milestone.
5. These plans are reviewed for compliance with management quality goals. Where noncompliance is found, replanning or exception approval is required.
6. Quality performance is tracked and publicized. When performance falls short of the plan, corrective action is required.
7. Since no single measure can adequately represent a complex product, the quality measures are treated as indicators of overall performance. These indicators are validated whenever possible through early user involvement as well as simulated and/or actual operational testing.

8.2 HARDWARE/SOFTWARE DEVELOPMENT COMPARISON

There are both similarities and differences between the processes used to develop hardware solutions and those used to create software. Several comparisons are presented to make the point.

Software

First, a typical hardware design team for a high-end server might have 20 designers, whereas the software project, to support the server hardware design might involve 100 software engineers.

Second, in the hardware world there is a cost to pay for interconnects. Therefore, the goal is to minimize the number of interconnects. With software code, on the other hand, there is no penalty or associated cost for connections (GO TO statements). The lack of such a penalty for the use of connections can add complexity through the generation of "spaghetti code" and thus many opportunities for error. So an enforced coding policy is required that limits the use of GO TO statements to minimize defects.

Also like hardware design, software development can be segmented or divided into manageable parts. Each software developer writes what is called a unit of code. All of the units of code written for a project are combined and integrated to become the system software. It is easier to test and debug a unit of software code than it is to test and debug the entire system.

Hardware designers will often copy a mundane workhorse portion of a circuit and embed it in the new circuit design (Designers like to spend their time designing with the latest microprocessor, memory, DSP, or whatever, rather than designing circuitry they consider to be mundane and not challenging). This approach often causes interface and timing problems that are not found until printed wiring assembly (PWA) or system test. The results of software designers copying a previously written unit of code and inserting it in a new software development project without any forethought could post similarly dangerous results.

Another significant difference between hardware design and software code development is that a unit of software may contain a bug and still function (i.e., it works), but it does the wrong thing. However, if a hardware circuit contains a defect (the equivalent of a software bug), it generally will not function.

As a result of these similarities and differences, hardware developers and their software counterparts can learn from each other as to what methods and processes work best to build a robust and reliable product.

8.3 SOFTWARE AVAILABILITY

For most development projects the customer is concerned with the system's (equipment's) ability to perform the intended function whenever needed. this is called *availability*. Such measures are particularly important for system and communication programs that are fundamental to overall system operation. These generally include the control program, database manager, job scheduler, user interface, communication control, network manager, and the input/output system. The key to including any program in this list is whether its failure will bring down the critical applications. If so, its availability must be considered.

Availability cannot be measured directly but must be calculated from such probabilistic measures as the mean time between failure (MTBF) and the mean time required to repair and restore the system to full operation (MTTR). Assuming the system is required to be continuously available, availability is the percent of total time that the system is available for use:

$$\text{Availability} = \frac{\text{MTTR}}{\text{MTTR} + \text{MTBF}} \times 100$$

Availability is a useful measure of the operational quality of some systems. Unfortunately, it is very difficult to project prior to operational testing.

8.4 SOFTWARE QUALITY

8.4.1 Software Quality Estimate

An estimate of software quality includes the following steps:

1. Specify the new development project quality goals [normally included in the marketing requirements list (MRL)], which represent customer needs.
2. Document and review lessons learned. Review recent projects completed by the organization to identify the ones most similar to the proposed product. Where the data warrant, this is done for each product element or unit of code. Often a design team disbands after a project is complete (software code is written) and moves on to the next project. The same thing happens with hardware designs as well. This prevents iterative learning from taking place.
3. Examine available quantitative quality data for these projects to establish a basis for the quality estimate.
4. Determine the significant product and process differences and estimate their potential effects.
5. Based on these historical data and the planned process changes, project the anticipated quality for the new product development process.
6. Compare this projection with the goals stated in the MRL. Highlight differences and identify needed process improvements to overcome the differences and meet the goals; this leads to the creation of a project quality profile.
7. Produce a development plan that specifies the process to be used to achieve this quality profile.

In making a software quality estimate it is important to remember that every project is different. While all estimates should be based on historical experience, good estimating also requires an intuitive understanding of the special characteristics of the product involved. Some examples of the factors to consider are

What is the anticipated rate of customer installation for this type of product? A high installation rate generally causes a sharp early peak in defect rate with a rapid subsequent decline. Typically, programs install most rapidly when they require minimal conversion and when they do not affect overall system operation. Compiler and utility programs are common examples of rapidly installed products.

What is the product release history? A subsequent release may install quickly if it corrects serious deficiencies in the prior release. This, of course, requires that the earliest experience with the new version is positive. If not, it may get a bad reputation and be poorly accepted.

What is the distribution plan? Will the product be shipped to all buyers immediately; will initial availability be limited; or is there to be a preliminary trial period?

Is the service system established? Regardless of the product quality, will the users be motivated and able to submit defect reports? If not, the defect data will not be sufficiently reliable to validate the development process.

To make an accurate quality estimate it is essential to have a quality model. One of the special challenges of software engineering is that it is an intellectual process that produces software programs that do not obey the laws of nature. The models needed for estimation purposes must reflect the way people actually write programs, and thus don't lend themselves to mathematical formulas. It is these unique, nonuniform characteristics that make the software engineering process manageable. Some of these characteristics are

Program module quality will vary, with a relatively few modules containing the bulk of the errors.

The remaining modules will likely contain a few randomly distributed defects that must be individually found and removed.

The distribution of defect types will also be highly skewed, with a relatively few types covering a large proportion of the defects.

Since programming changes are highly error-prone; all changes should be viewed as potential sources of defect injection.

While this characterization does not qualify as a model in any formal sense, it does provide focus and a framework for quality planning.

8.4.2 Software Defects and Statistical Process Control

There is a strong parallel between hardware and software in the following areas:

Defects.

Defect detection [for an IC one can't find all of the defect/bugs. Stuck at fault (SAF) coverage ranges from 85–98%. However, delay, bridging, opens, and other defects cannot be found with SAF. 100% test coverage is impossible to obtain. Software bug coverage is 50%.] root cause analysis and resolution
Defect prevention
Importance of peer design reviews. Code review should be conducted, preferably without the author being present. If the author is present, he or she cannot speak. The reviewers will try to see if they can understand the author's thought process in constructing the software. This can be an eye-opening educational experience for the software developer.
Need for testing (automated testing, unit testing).
Use of statistical methodology (plan–do–check–act).
Need for software quality programs with clear metrics.
Quality systems.

Statistical process control (SPC) has been used extensively in hardware development and manufacturing. One SPC tool is the fishbone diagram (also called an Ishikawa or a cause–effect diagram), which helps one explore the reasons, or causes, for a particular problem or effect. Figure 1 shows a fishbone diagram for register allocation defects in a compiler. This diagram enables one

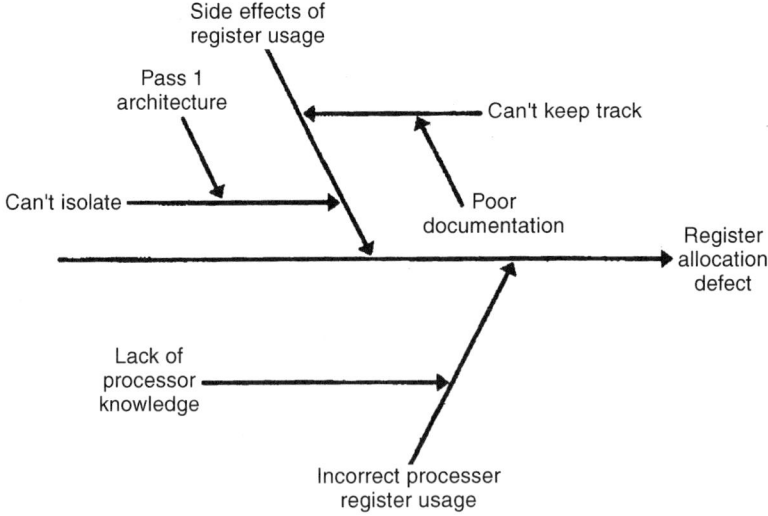

FIGURE 1 Typical cause–effect diagram.

to graphically identify all the potential causes of a problem and the relationship with the effect, but it does not illustrate the magnitude of a particular cause's effect on the problem.

Pareto diagrams complement cause–effect diagrams by illustrating which causes have the greatest effect on the problem. This information is then used to determine where one's problem-solving efforts should be directed. Table 1 is a Pareto distribution of software module defect densities or defect types. The defects are ranked from most prevalent to least. Normally, a frequency of occurrence, expressed either numerically or in percentage, is listed for each defect to show which defects are responsible for most of the problems. Table 2 is a Pareto

TABLE 1 Pareto List of Software Defects

User interface interaction
 1. User needs additional data fields.
 2. Existing data need to be organized/presented differently.
 3. Edits on data values are too restrictive.
 4. Edits on data values are too loose.
 5. Inadequate system controls or audit trails.
 6. Unclear instructions or responses.
 7. New function or different processing required.

Programming defect
 1. Data incorrectly or inconsistently defined.
 2. Initialization problems.
 3. Database processing incorrect.
 4. Screen processing incorrect.
 5. Incorrect language instruction.
 6. Incorrect parameter passing.
 7. Unanticipated error condition.
 8. Operating system file handling incorrect.
 9. Incorrect program control flow.
 10. Incorrect processing logic or algorithm.
 11. Processing requirement overlooked or not defined.
 12. Changes required to conform to standards.

Operating environment
 1. Terminal differences.
 2. Printer differences.
 3. Different versions of systems software.
 4. Incorrect job control language.
 5. Incorrect account structure or capabilities.
 6. Unforeseen local system requirements.
 7. Prototyping language problem.

TABLE 2   Pareto List of Software Errors

Error category	Frequency of occurrence	Percent
Incomplete/erroneous specification	349	28
Intentional deviation from specification	145	12
Violation of programming standards	118	10
Erroneous data accessing	120	10
Erroneous decision logic or sequencing	139	12
Erroneous arithmetic computations	113	9
Invalid timing	44	4
Improper handling of interrupts	46	4
Wrong constants and data values	41	3
Inaccurate documentation	96	8
Total	1202	100

listing of software errors both numerically and as a percentage of the total. By focusing process improvement efforts on the most prevalent defects, significant quality improvements can be achieved.

The real value of SPC is to effectively define areas for software quality improvement and the resultant actions that must be taken.

8.4.3   Software Quality Measures

The specific quality measures must be selected by each organization based on the data available or easily gathered, but the prime emphasis should be on those that are likely to cause customer problems. Program interrupt rate is one useful measure of quality that is important to customers. Many organizations restrict the definition of valid defects to those that require code changes. If clear criteria can be established, however, documentation changes also should be included.

Both software defect and test data should be used as quality indicators. Defect data should be used since it is all that most software development organizations can obtain before system shipment. Then, too, if defects are not measured, the software professionals will not take any other measures very seriously. They know from experience that the program has defects that must be identified and fixed before the product can be shipped. Everything else will take second priority.

It is important to perform software testing. This should begin as the software is being developed, i.e., unit testing. It is also worthwhile to do some early testing in either a simulated or real operational environment prior to final delivery. Even with the most thorough plans and a highly capable development team, the operational environment always seems to present some unexpected problems.

These tests also provide a means to validate the earlier installation and operational plans and tests, make early availability measurements, and debug the installation, operation, and support procedures.

Once the defect types have been established, normalization for program size is generally required. Defects per 1000 lines of source code is generally the simplest and most practical measure for most organizations. This measure, however, requires that the line-of-code definition be established. Here the cumulative number of defects received each month are plotted and used as the basis for corrective action.

The next issue is determining what specific defects should be measured and over what period of time they should be measured. This again depends on the quality program objectives. Quality measures are needed during development, test, and customer use. The development measures provide a timely indicator of software code performance; the test measures then provide an early validation; and the customer-use data complete the quality evaluation. With this full spectrum of data it is possible to calibrate the effectiveness of development and test at finding and fixing defects. This requires long-term product tracking during customer use and some means to identify each defect with its point of introduction. Errors can then be separated by release, and those caused by maintenance activity can be distinguished.

When such long-term tracking is done, it is possible to evaluate many software process activities. By tracking the inspection and test history of the complete product, for example, it is possible to see how effective each of these actions was at finding and removing the product defects. This evaluation can be especially relevant at the module level, where it provides an objective way to compare task effectiveness.

8.4.4 Defect Prevention

Defect prevention begins with (1) a clear understanding of how software defects (bugs) occur and (2) a clear management commitment to quality. Investigation of the factors that result in high-quality software being written revealed an interesting finding: the quality of the written software code is directly related to the number of interruptions that the software developer experiences while writing the code, such as someone stopping by for a chat. The interruptions cause the developer to lose the mental construct being formulated, which is continually updating mentally, to write the code. On the other hand, software code quality was found to be independent of such things as the developer's previous experience, the most recent project the developer was involved with, the level of education, and the like. So as much as possible, an interruption-free and noise-free environment needs to be provided for software developers (use white noise generators to block out all noise sources).

The management commitment must be explicit, and all members of the organization must know that quality comes first. Management must direct code "walk-throughs" and reviews. Until management delays or redirects a project to meet the established quality goals and thus ensures that the software development process has the right focus, people will not really believe it. Even then, the point must be reemphasized and the software engineers urged to propose quality improvement actions, even at the potential cost of schedule delays. In spite of what the schedule says, when quality problems are fixed early, both time and resources are saved. When management really believes this and continually conveys and reinforces the point to the software developers, the right quality attitudes will finally develop. It is important to identify defects and ways to prevent them from occurring. As with hardware,

> The cost of finding and repairing defects increases exponentially the later they are found in the process.
> Preventing defects is generally less expensive than finding and repairing them, even early in the process.

Finding and fixing errors accounts for much of the cost of software development and maintenance. When one includes the costs of inspections, testing, and rework, as much as half or more of the typical development bill is spent in detecting and removing errors. What is more, the process of fixing defects is even more error-prone than original software creation. Thus with a low-quality process, the error rate spiral will continue to escalate.

Hewlett-Packard found that more than a third of their software errors were due to poor understanding of interface requirements. By establishing an extensive prototyping and design review program, the number of defects found after release was sharply reduced.

A development project at another company used defect prevention methods to achieve a 50% reduction in defects found during development and a 78% reduction in errors shipped. This is a factor of 2 improvement in injected errors, and a 4-to-1 improvement in shipped quality.

Finding and identifying defects is necessary but not sufficient. The most important reason for instituting defect prevention is to provide a continuing focus for process improvement. Unless some mechanism drives process change, it will not happen in an orderly or consistent way. A defect prevention mindset focuses on those process areas that are the greatest source of trouble, whether methods, technology, procedures, or training.

The fundamental objective of software defect prevention is to make sure that errors, once identified and addressed, do not occur again. Defect prevention cannot be done by one or two people, and it cannot be done sporadically. Everyone must participate. As with any other skill, it takes time to learn defect prevention well, but if everyone on the project participates, it can transform an organization.

Most software developers spend much of their working lives reacting to defects. They know that each individual defect can be fixed, but that its near twin will happen again and again and again. To prevent these endless repetitions, we need to understand what causes these errors and take a conscious action to prevent them. We must then obtain data on what we do, analyze it, and act on what it tells us. This is called the Deming or Shewhart cycle:

1. *Defect reporting.* This includes sufficient information to categorize each defect and determine its cause.
2. *Cause analysis.* The causes of the most prevalent defects are determined.
3. *Action plan development.* Action teams are established to devise preventions for the most prevalent problems.
4. *Action implementation.* Once the actions have been determined, they must be implemented. This generally involves all parts of the organization in a concerted improvement effort.
5. *Performance tracking.* Performance data are gathered, and all action items are tracked to completion.
6. *Starting over.* Do it all again; this time focus on the most prevalent of the remaining defects.

As an example let's take cause analysis and delve into how this should be done. Cause analysis should be conducted as early as possible after a defect is found. For a given product, some useful guidelines on holding these sessions are

1. Shortly after the time that all product modules have completed detailed design, cause analysis sessions should be held for all problems identified during the design inspections/reviews.
2. Shortly after the last module has completed each test phase, cause analysis meetings should be held for all problems found for the first few modules as soon as possible after they have completed each development inspection or test phase.
3. To be of most value to the later modules, cause analysis meetings should be held for all problems found for the first few modules as soon as possible after they have completed each development inspection or test phase.
4. Cause analysis reviews should be held on a product after a reasonable number of user problems have been found after customer release. Such reviews should be held at least annually after release, even if the defect rate is relatively low. These reviews should be continued until no significant number of new defects is reported.

5. The cause analysis meetings should be held often enough to permit completion in several hours. Longer meetings will be less effective and harder to schedule. Short cause analysis meetings are generally most productive.

The objective of the cause analysis meeting is to determine the following:

1. What caused each of the defects found to date?
2. What are the major cause categories?
3. What steps are recommended for preventing these errors in the future?
4. What priorities are suggested for accomplishing these actions?

Defect identification and improvement have been discussed, but the real solution is to learn from the past and apply it to present software development projects to prevent defects in the first place. The principles of software defect prevention are

1. *The programmers must evaluate their own errors.* Not only are they the best people to do so, but they are most interested and will learn the most from the process.
2. *Feedback is an essential part of defect prevention.* People cannot consistently improve what they are doing if there is not timely reinforcement of their actions.
3. *There is no single cure-all that will solve all the problems.* Improvement of the software process requires that error causes be removed one at a time. Since there are at least as many error causes as there are error types, this is clearly a long-term job. The initiation of many small improvements, however, will generally achieve far more than any one-shot breakthrough.
4. *Process improvement must be an integral part of the process.* As the volume of process change grows, as much effort and discipline should be invested in defect prevention as is used on defect detection and repair. This requires that the process is architected and designed, inspections and tests are conducted, baselines are established, problem reports written, and all changes tracked and controlled.
5. *Process improvement takes time to learn.* When dealing with human frailties, we must proceed slowly. A focus on process improvement is healthy, but it must also recognize the programmers' need for a reasonably stable and familiar working environment. This requires a properly paced and managed program. By maintaining a consistent, long-term focus on process improvement, disruption can be avoided and steady progress will likely be achieved.

8.4.5 The SEI Capability Maturity Model

Because of the dominance and importance of software to a product's success, there is an increased focus on software reliability. This is accomplished by improving how software is developed, i.e., software development organizations. The following list provides the steps that lead to an improved software development organization:

1. Hire someone with previous software development project management experience to lead the software development process. If you've never done a software development project or led a development project, "you don't know what you don't know."
2. Understand the current status of the development process.
3. Develop a vision of the desired process.
4. Establish a list of required process improvement actions in order of priority.
5. Produce a plan to accomplish the required actions.
6. Commit the resources to execute the plan.
7. Start over at step 2.

The effective development of software is limited by several factors—such as an ill-defined process, inconsistent implementation, and poor process management—and hinges on experienced competent software developers (who have a track record). A model was devised to assess and grade a company's software development process, to address these factors, and to drive continuous improvement to the next level of sophistication or maturity by the Software Engineering Institute of Carnegie Mellon University. The SEI Capability Maturity Model (CMM) structure addresses the seven steps by characterizing the software development process into five maturity levels to facilitate high-reliability software program code being written. These levels are

Level 1: *Initial.* Until the process is under statistical control, no orderly progress in process improvement is possible.

Level 2: *Repeatable.* The organization has achieved a stable process with a repeatable level of statistical control by initiating rigorous project management of commitments, costs, schedules, and changes.

Level 3: *Defined.* The organization has defined the process. This helps ensure consistent implementation and provides a basis for a better understanding of the process.

Level 4: *Managed.* The organization has initiated comprehensive process measurements and analyses beyond those of cost and schedule performance.

Level 5: Optimizing. The organization now has a foundation for continuing improvement and optimization of the process.

The optimizing process helps people to be effective in several ways:

It helps managers understand where help is needed and how best to provide the people with the support they require.
It lets the software developers communicate in concise, quantitative terms.
It provides the framework for the software developers to understand their work performance and to see how to improve it.

A graphic depiction of the SEI Capability Maturity Model is presented in Figure 2. Notice that as an organization moves from Level I (Initial) to Level 5 (Optimizing) the risk decreases and the productivity, quality, and reliability increase. Surveys have shown that most companies that are audited per SEI criteria score between 1.0 and 2.5. There are very few 3s and 4s and perhaps only one 5 in the world.

These levels have been selected because they

Reasonably represent the actual historical phases of evolutionary improvement of real software organizations

Maturity Levels	Key Software Process Areas (Institutionalized)	Results
Optimizing *focus on:* *empowering individuals*	• Process change management • Technology innovation • Defect prevention	Productivity & Quality
Managed *focus on:* *empowering projects*	• Quality management • Process measurement and analysis	
Defined *focus on:* *project organization*	• Peer reviews • Intergroup coordination • Software product engineering • Integrated software management • Training program • Organization process definition • Organization process focus	
Repeatable *focus on:* *individual project*	• Software configuration management • Software quality assurance • Software subcontract management • Software project tracking and oversight • Software project planning • Software requirements management	
Initial *focus:* *individual*	• None	Risk

FIGURE 2 The five levels of the SEI capability maturity model.

Represent a measure of improvement that is reasonable to achieve from the prior level

Suggest interim improvement goals and progress measures

Easily identify a set of immediate improvement priorities once an organization's status in this framework is known

While there are many other elements to these maturity level transitions, the primary objective is to achieve a controlled and measured process as the foundation for continuing improvement.

The process maturity structure is used in conjunction with an assessment methodology and a management system to help an organization identify its specific maturity status and to establish a structure for implementing the priority improvement actions, respectively. Once its position in this maturity structure is defined, the organization can concentrate on those items that will help it advance to the next level. Currently, the majority of organizations assessed with the SEI methodology are at CMM Level 1, indicating that much work needs to be done to improve software development.

Level 1: The Initial Process

The initial process level could properly be described as ad hoc, and it is often even chaotic. At this level the organization typically operates without formalized procedures, cost estimates, or project plans. Tools are neither well integrated with the process nor uniformly applied. Change control is lax, and there is little senior management exposure or understanding of the problems and issues. Since many problems are deferred or even forgotten, software installation and maintenance often present serious problems.

While organizations at this level may have formal procedures for planning and tracking their work, there is no management mechanism to ensure that they are used. The best test is to observe how such an organization behaves in a crisis. If it abandons established procedures and essentially reverts to coding and testing, it is likely to be at the initial process level. After all, if the techniques and methods are appropriate, then they should be used in a crisis; if they are not appropriate in a crisis, they should not be used at all. One reason why organizations behave in this fashion is that they have not experienced the benefits of a mature process and thus do not understand the consequences of their chaotic behavior. Because many effective software actions (such as design and code inspections or test data analysis) do not appear to directly support shipping the product, they seem expendable.

In software, coding and testing seem like progress, but they are often only wheel spinning. While they must be done, there is always the danger of going in the wrong direction. Without a sound plan and a thoughtful analysis of the problems, there is no way to know.

Level 2: The Repeatable Process Level

Organizations at the initial process level can improve their performance by instituting basic project controls. The most important are project management, management oversight, quality assurance, and change control.

The fundamental role of a *project management* system is to ensure effective control of commitments. This requires adequate preparation, clear responsibility, a public declaration, and a dedication to performance. For software, project management starts with an understanding of the job's magnitude. In all but the simplest projects, a plan must then be developed to determine the best schedule and the anticipated resources required. In the absence of such an orderly plan, no commitment can be better than an educated guess.

A suitably disciplined software development organization must have *senior management oversight*. This includes review and approval of all major development plans prior to the official commitment. Also, a quarterly review should be conducted of facility-wide process compliance, installed quality performance, schedule tracking, cost trends, computing service, and quality and productivity goals by project. The lack of such reviews typically results in uneven and generally inadequate implementation of the process as well as frequent overcommitments and cost surprises.

A *quality assurance* group is charged with assuring management that software work is done the way it is supposed to be done. To be effective, the assurance organization must have an independent reporting line to senior management and sufficient resources to monitor performance of all key planning, implementation, and verification activities.

Change control for software is fundamental to business and financial control as well as to technical stability. To develop quality software on a predictable schedule, requirements must be established and maintained with reasonable stability throughout the development cycle. While requirements changes are often needed, historical evidence demonstrates that many can be deferred and incorporated later. Design and code changes must be made to correct problems found in development and test, but these must be carefully introduced. If changes are not controlled, then orderly design, implementation and test is impossible and no quality plan can be effective.

The repeatable process level has one other important strength that the initial process does not: it provides control over the way the organization establishes its plans and commitments. This control provides such an improvement over the initial process level that people in the organization tend to believe they have mastered the software problem. They have achieved a degree of statistical control through learning to make and meet their estimates and plans. This strength, however, stems from their prior experience at doing similar work.

Some of the key practices for software project planning are

The project's software development plan is developed according to a documented procedure.

Estimates for the size of the software work products are derived according to a documented procedure.

The software risks associated with the cost, resource, schedule, and technical aspects of the project are identified, assessed, and documented.

Organizations at the repeatable process level thus face major risks when they are presented with new challenges. Examples of the changes that represent the highest risk at this level are the following:

Unless they are introduced with great care, new tools and methods will affect the process, thus destroying the relevance of the intuitive historical base on which the organization relies. Without a defined process framework in which to address these risks, it is even possible for a new technology to do more harm than good.

When the organization must develop a new kind of product, it is entering new territory. For example, a software group that has experience developing compilers will likely have design, scheduling, and estimating problems when assigned to write a real-time control program. Similarly, a group that has developed small self-contained programs will not understand the interface and integration issues involved in large-scale projects. These changes again destroy the relevance of the intuitive historical basis for the organization's process.

Major organizational changes can also be highly disruptive. At the repeatable process level, a new manager has no orderly basis for understanding the organization's operation, and new team members must learn the ropes through word of mouth.

Level 3: The Defined Process

The key actions required to advance from the repeatable level to the next stage, the defined process level, are to establish a process group, establish a development process architecture, and introduce a family of software engineering methods and technologies.

Establish a process group. A process group is a technical resource that focuses exclusively on improving the software process. In software organizations at early maturity levels, all the people are generally devoted to product work. Until some people are given full-time assignments to work on the process, little orderly progress can be made in improving it.

The responsibilities of process groups include defining the development process, identifying technology needs and opportunities, advising the projects, and conducting quarterly management reviews of process status and performance. Because of the need for a variety of skills, groups smaller than about four professionals are unlikely to be fully effective. Small organizations that lack the experience base to form a process group should address these issues by using specially formed committees of experienced professionals or by retaining consultants.

The assurance group is focused on enforcing the current process, while the process group is directed at improving it. In a sense, they are almost opposites: assurance covers audit and compliance, and the process group deals with support and change.

Establish a software development process architecture. Also called a development life cycle, this describes the technical and management activities required for proper execution of the development process. This process must be attuned to the specific needs of the organization, and it will vary depending on the size and importance of the project as well as the technical nature of the work itself. The architecture is a structural description of the development cycle specifying tasks, each of which has a defined set of prerequisites, functional descriptions, verification procedures, and task completion specifications. The process continues until each defined task is performed by an individual or single management unit.

If they are not already in place, a family of software engineering methods and technologies should be introduced. These include design and code inspections, formal design methods, library control systems, and comprehensive testing methods. Prototyping should also be considered, together with the adoption of modern implementation languages.

At the defined process level, the organization has achieved the foundation for major and continuing progress. For example, the software teams when faced with a crisis will likely continue to use the process that has been defined. The foundation has now been established for examining the process and deciding how to improve it.

As powerful as the Defined Process is, it is still only qualitative: there are few data generated to indicate how much is accomplished or how effective the process is. There is considerable debate about the value of software measurements and the best ones to use. This uncertainty generally stems from a lack of process definition and the consequent confusion about the specific items to be measured. With a defined process, an organization can focus the measurements on specific tasks. The process architecture is thus an essential prerequisite to effective measurement.

Level 4: The Managed Process

The key steps required to advance from the defined process to the next level are

1. Establish a minimum basic set of process measurements to identify the quality and cost parameters of each process step. The objective is to

quantify the relative costs and benefits of each major process activity, such as the cost and yield of error detection and correction methods.
2. Establish a process database and the resources to manage and maintain it. Cost and yield data should be maintained centrally to guard against loss, to make it available for all projects, and to facilitate process quality and productivity analysis.
3. Provide sufficient process resources to gather and maintain this process database and to advise project members on its use. Assign skilled professionals to monitor the quality of the data before entry in the database and to provide guidance on analysis methods and interpretation.
4. Assess the relative quality of each product and inform management where quality targets are not being met. An independent quality assurance group should assess the quality actions of each project and track its progress against its quality plan. When this progress is compared with the historical experience on similar projects, an informed assessment can generally be made.

In advancing from the initial process through the repeatable and defined processes to the managed process, software organizations should expect to make substantial quality improvements. The greatest potential problem with the managed process level is the cost of gathering data. There is an enormous number of potentially valuable measures of the software process, but such data are expensive to gather and to maintain.

Data gathering should be approached with care, and each piece of data should be precisely defined in advance. Productivity data are essentially meaningless unless explicitly defined. Several examples serve to illustrate this point:

The simple measure of lines of source code per expended development month can vary by 100 times or more, depending on the interpretation of the parameters. The code count could include only new and changed code or all shipped instructions. For modified programs, this can cause variations of a factor of 10.

Noncomment nonblank lines, executable instructions, or equivalent assembler instructions can be counted with variations of up to seven times.

Management, test, documentation, and support personnel may or may not be counted when calculating labor months expended, with the variations running at least as high as a factor of 7.

When different groups gather data but do not use identical definitions, the results are not comparable, even if it makes sense to compare them. It is rare that two projects are comparable by any simple measures. The variations in task complexity caused by different product types can exceed 5 to 1. Similarly, the cost per line of code of small modifications is often two to three times that for new programs. The degree of requirements change can make an enormous difference, as can the design status of the base program in the case of enhancements.

Process data must not be used to compare projects or individuals. Its purpose is to illuminate the product being developed and to provide an informed basis for improving the process. When such data are used by management to evaluate individuals or teams, the reliability of the data itself will deteriorate.

Level 5: The Optimizing Process

The two fundamental requirements for advancing from the managed process to optimizing process level are

1. Support automatic gathering of process data. All data are subject to error and omission, some data cannot be gathered by hand, and the accuracy of manually gathered data is often poor.
2. Use process data both to analyze and to modify the process to prevent problems and improve efficiency.

Process optimization goes on at all levels of process maturity. However, with the step from the managed to the optimizing process there is a major change. Up to this point software development managers have largely focused on their products and typically gather and analyze only data that directly relate to product improvement. In the optimizing process, the data are available to tune the process itself. With a little experience, management will soon see that process optimization can produce major quality and productivity benefits.

For example, many types of errors can be identified and fixed far more economically by design or code inspections than by testing. A typically used rule of thumb states that it takes one to four working hours to find and fix a bug through inspections and about 15 to 20 working hours to find and fix a bug in function or system test. To the extent that organizations find that these numbers apply to their situations, they should consider placing less reliance on testing as the primary way to find and fix bugs.

However, some kinds of errors are either uneconomical to detect or almost impossible to find except by machine. Examples are errors involving spelling and syntax, interfaces, performance, human factors, and error recovery. It would be unwise to eliminate testing completely since it provides a useful check against human frailties.

The data that are available with the optimizing process give a new perspective on testing. For most projects, a little analysis shows that there are two distinct activities involved: the removal of defects and the assessment of program quality. To reduce the cost of removing defects, inspections should be emphasized, together with any other cost-effective techniques. The role of functional and system testing should then be changed to one of gathering quality data on the programs.

This involves studying each bug to see if it is an isolated problem or if it indicates design problems that require more comprehensive analysis.

With the optimizing process, the organization has the means to identify the weakest elements of the process and to fix them. At this point in process improvement, data are available to justify the application of technology to various critical tasks, and numerical evidence is available on the effectiveness with which the process has been applied to any given product. It is then possible to have confidence in the quality of the resulting products.

Clearly, any software process is dependent on the quality of the people who implement it. There are never enough good people, and even when you have them there is a limit to what they can accomplish. When they are already working 50 to 60 hr a week, it is hard to see how they could handle the vastly greater challenges of the future.

The optimizing process enhances the talents of quality people in several ways. It helps managers understand where help is needed and how best to provide people with the support they require. It lets the software developers communicate in concise, quantitative terms. This facilitates the transfer of knowledge and minimizes the likelihood of wasting time on problems that have already been solved. It provides a framework for the developers to understand their work performance and to see how to improve it. This results in a highly professional environment and substantial productivity benefits, and it avoids the enormous amount of effort that is generally expended in fixing and patching other peoples' mistakes.

The optimizing process provides a disciplined environment for software development. Process discipline must be handled with care, however, for it can easily become regimentation. The difference between a disciplined environment and a regimented one is that discipline controls the environment and methods to specific standards, while regimentation applies to the actual conduct of the work.

Discipline is required in large software projects to ensure, for example, that the many people involved use the same conventions, don't damage each other's products, and properly synchronize their work. Discipline thus enables creativity by freeing the most talented software developers from the many crises that others have created.

Unless we dramatically improve software error rates, the increased volume of code to be generated will mean increased risk of error. At the same time, the complexity of our systems is increasing, which will make the systems progressively more difficult to test. In combination these trends expose us to greater risks of damaging errors as we attempt to use software in increasingly critical applications. These risks will thus continue to increase as we become more efficient at producing volumes of new code.

As well as being a management issue, quality is an economic one. It is always possible to do more reviews or to run more tests, but it costs both time and money to do so. It is only with the optimizing process that the data are available to understand the costs and benefits of such work. The optimizing process provides the foundation for significant advances in software quality and simultaneous improvements in productivity.

There are few data on how long it takes for software organizations to advance through the maturity levels toward the optimizing process. What can be said is that there is an urgent need for better and more effective software organizations. To meet this need, software managers and developers must establish the goal of moving to the optimizing process.

Example of Software Process Assessment

This section is an excerpt from an SEI software process assessment (including actual company and assessor dialog) that was conducted for a company that develops computer operating system software. The material is presented to facilitate learning, to identify the pertinent issues in software development up close, to provide a perspective on how an organization deals with the issues/items assessed, and to see the organization's views (interpretation) of these items.

Item: Process Focus and Definition.

Assessment: *We do not have confidence in formal processes and we resist the introduction of new ones.*

Process focus establishes the responsibility for managing the organization's software process activities.

Process definition involves designing, documenting, implementing, maintaining and, enforcing the organization's standard software development process.

The standard software development process defines the phases and deliverables of the software life cycle and the role of each responsible organization in the life cycle. It defines criteria for completion of each phase and standards for project documents. The standard process is flexible and adaptable, yet it is followed and enforced. The standard process is updated when necessary and improvements are implemented systematically (e.g., through controlled pilot tests).

Process definition also involves managing other "assets" related to the standard software process, such as guidelines and criteria for tailoring the standard process to individual project needs and a library of software process–related documentation. In addition, process related data are collected and analyzed for the purpose of continuous process improvement.

1. *We have had experiences with inflexible processes that constricted our ability to do good work.* Inflexible processes waste time, lower productivity, involve unnecessary paperwork, and keep people from doing the right thing. People in our company have dealt with inflexible processes and therefore resist the introduction of new processes.

2. *Processes are not managed, defined, implemented, improved, and enforced consistently.* Like products, key software processes are assets that must be managed. At our company, only some processes have owners and many of our processes are not fully documented. As such, people find it difficult to implement and/or improve these processes.
3. *Processes (e.g., the use of coding standards) that are written down are not public, not consistently understood, not applied in any standard way, and not consistently enforced.* Even when we define standards, we do not publicize them nor do we train people in the proper use of the standard. As a result, standards are applied inconsistently and we cannot monitor their use or make improvements.
4. *We always go for the "big fix" rather than incremental improvements.* People find the big fix (replacement of a process with a completely different process) to be disruptive and not necessarily an improvement. People are frustrated with our apparent inability to make systematic improvements to processes.
5. *The same (but different) process is reinvented by many different groups because there is no controlling framework.* Some software development groups and individual contributors follow excellent practices. These methods and practices (even though not documented) are followed consistently within the group. However, since each project and development group develops its own methodology, the methods and practices are ad hoc and vary from group to group and project to project. For example, the change control process exists in many different forms, and people across our company use different platforms (Macintosh, SUN, PC, etc.) and different formats for documentation.
6. *There is no incentive to improve processes because you get beaten up.* People who try to implement or improve processes are not supported by their peers and management. People do not work on process improvement or share best practices for several reasons, including
    There is no reward for doing so.
    Such activity is regarded by peers and managers as not being real work.
    "Not invented here" resistance.
    Why bother; it is too frustrating.
    Management rewards the big fix.
7. *Our core values reinforce the value of the individual over the process.* We encourage individual endeavor and invention but this is often interpreted as being at odds with the use of standard practices.

People in our company believe that formal processes hinder an individual's ability to do good work, rather than see processes as a way to be more productive, enabling them to produce higher quality products. Therefore, we tend to do things "my way" rather than use standard practices. "Processes are okay; you can always work around them" is a typical sentiment.

The lack of process focus and definition affects customer perception of our company as well as our internal activities.

From an internal perspective, we cannot reliably repeat our successes because we do not reuse our processes. Instead of improving our processes, we spend time reinventing our processes; this reduces our productivity and impacts schedules. When people move within our company they spend time trying to learn the unique practices of the new group. When we hire new people, we cannot train them in processes that are not defined.

From the customer perspective, process audits provide the assurance that we can produce quality products. Lack of process definition means that our company may not be able to provide a strong response to such audits leading to the perception that we are not a state-of-the-art organization and that we need better control over our software development and management practices. This has a direct effect on sales to our current and future customers.

Item: Project Commitment.

Assessment: *At every level we commit to do more work than is possible.*

Software project planning involves developing estimates for the work to be performed, producing a plan (including a schedule) to perform the work, negotiating commitments, and then tracking progress against the plan as the project proceeds. Over the course of the project, as circumstances change, it may be necessary to repeat these planning steps to produce a revised project plan.

The major problem with our project planning is that everyone, including individual contributors, first-line and second-line managers, and executives, commits to more work than is possible given the time and resources available.

Several factors contribute to this problem. Our initial estimates and schedules are often poor, causing us to underestimate the time and resources required to accomplish a task. Furthermore, we do not analyze the impact nor do we renegotiate schedules when requirements or resources change. These practices are perpetuated because we reward people for producing short schedules rather than accurate schedules.

1. *Our planning assumptions are wrong.* We make a number of mistakes in our planning assumptions. For example,
    We do not allocate enough time for planning. We are often "too busy to plan." We schedule one high-priority project after another with-

out allowing time for postmortems on the previous project and planning for the next project.

We do not allow enough time in our schedules for good engineering practices such as design inspections.

We schedule ourselves and the people who work for us at 100% instead of allowing for meetings, mail, education, vacation, illness, minor interruptions, time to keep up with the industry, time to help other groups, etc.

We do not keep project history so we have no solid data from which to estimate new projects. We rely almost entirely on "swags."

We "time-slice" people too many ways. Ten people working 10% of their time on a project will not produce one person's worth of work because of loss of focus and time to switch context.

2. *Commitments are not renegotiated when things change.* In our business, priorities change and critical interrupts must be serviced. However, when requirements change and when we reassign resources (people, machines), we do not evaluate the impact of these changes on project schedules. Then we are surprised when the original project schedule slips.

3. *It is not okay to have a 4-year end date, but it is okay to have a 2-year end date that slips 2 years.* A "good" schedule is defined as a short schedule, not an accurate schedule. There is management pressure to cut schedules. People are rewarded for presenting a short schedule even if the project subsequently slips. There is fear that a project will be canceled if it has a long schedule.

4. *We rely on heroic efforts to get regular work done.* Heroic effort means long hours, pulling people off their assigned projects, overusing key people, etc. There will always be exceptional circumstances where these measures are required; however, this is our normal mode of operation. We count on both managers and developers putting in these heroic efforts to fix problems, keep regular projects moving forward, work with other groups, write appraisals, do planning, etc.

5. *We have a history of unreliable project estimates.* Most of our projects miss their schedules and exceed their staffing estimates. It is not uncommon for a project to overrun its schedule by 100 to 200%. It is common for a project to start, be put on hold, and not be restarted for several years.

Unreliable project estimates are taken for granted. Very rarely is there any formal analysis and recognition of the problem, for example, in the form of a project postmortem.

6. *There are no company standards or training for project plan-*

ning. Lack of organizational standards makes it difficult to compare or combine multiple project schedules, which in turn makes it difficult to recognize overcommitment. Lack of training perpetuates bad planning practices.

Our market is becoming more and more competitive, with much shorter product development cycles. When we miss our schedules because of overcommitment, we lose sales because we fail to deliver products within the narrow market window. We also lose credibility with our customers by making promises that we cannot keep. Customers complain that they do not know what new features and products are coming and when.

Overcommitment also impacts our internal operations (which indirectly affects our customers again). Our software development organization operates in crisis mode most of the time. We are so busy handling interrupts and trying to recover from the effects of those interrupts that we sacrifice essential software engineering and planning practices. We are "too busy to plan." We are also too busy to analyze our past mistakes and learn from them.

Overcommitment also has a serious effect on people. Both managers and developers get burned out, and morale suffers as they context switch and try to meet unrealistic schedules. They never feel like they've done a good job, they do not have time to keep up their professional skills, and creativity suffers. Low morale also means lower productivity now and higher turnover in the future.

Item: Software Engineering Practices.

Assessment: *We do not invest enough time and effort in defect prevention in the front end of the development cycle.*

Software engineering practices are those activities that are essential for the reliable and timely development of high-quality software. Some examples of software engineering practices are design documentation, design reviews, and code inspections. It is possible to develop software without such practices. However, software engineering practices, when used correctly, are not only effective in preventing defects in all phases of the software development life cycle, but also improve an organization's ability to develop and maintain many large and complex software products and to deliver products which meet customer requirements.

The phases of our software development life cycle are requirements definition, design, code, unit testing, product quality assurance (QA) testing, and integration testing. We spend a lot of time on and devote many resources to the back-end product QA and integration testing trying to verify that products do not have defects. Much less effort is devoted to ensuring that defects are not introduced into products in the first place. In other words, we try to test in quality instead of designing in quality, even though we know that a defect discovered late in the product development life cycle is much more expensive to correct than one detected earlier. It would be far more cost effective to avoid defects in the first place and to detect them as early as possible.

1. *We are attempting to get quality by increasing product testing.* Customers are demanding higher quality; we are responding by increasing our testing efforts. But increased product testing has diminishing returns. As noted, it is far more cost effective to avoid defects altogether or at least detect defects prior to product QA and integration testing.

2. *We do not recognize design as a necessary phase in the life cycle.* Few groups do design reviews and very few groups use formal design methodologies or tools. Code inspections are often difficult because there is no internal design document. Consequently, code inspections often turn into design reviews. This approach is misguided; it assumes that code is the only deliverable. Design work is an essential activity and design inspections have distinct benefits and are necessary in addition to code inspection. Moreover, design is a necessary phase of the life cycle.

3. *Our reward system does not encourage the use of good software engineering practices.* Developers are primarily held responsible for delivery of code (software) on time. This is their primary goal; all else is secondary. Managers typically ask about code and test progress, not about progress in design activities.

4. *We recognize people for heroic firefighting; we do not recognize the people who avoid these crises.* Code inspections are generally practiced, but may not be used and are sometimes ineffective. Often, people are not given adequate time to prepare for inspections, especially when code freeze dates are imminent. Also, we do not use the inspection data for anything useful, so many people do not bother entering these data.

    Unit testing by developers is not a general practice; it is not required, and is only done at the developer's discretion. Unit testing is sometimes omitted because of tight schedules and many developers rely on product QA testing to find the bugs.

5. *Few well-defined processes or state-of-the-art tools exist.* Most groups have their own processes; these tend to be ad hoc and are often undocumented. There are very few processes which are well documented and used throughout software development.

Our company has no software design tools other than whiteboards or paper and pencil and has no standards for the design phase. Developers have complained about the lack of state-of-the-art tools for a long time and are dismayed at the continued lack of such tools.

6. *No formal software engineering training is required.* Most developers know very little about software engineering. A few developers have learned about software engineering on their own; some of these have promoted software engineering practices within their own group. We don't offer any formal software engineering training.

7. *We have very little internal design documentation.* There are few internal design documents because there are no requirements for producing internal design documentation. When internal design documents exist, they are often out of date soon after code has been written because we do not have any processes for ensuring that the documentation is kept current. Furthermore, there is a lack of traceability between requirements, design, and code.

   When there is no internal design documentation, developers have two choices: either spend a lot of time doing software archeology (i.e., reading a lot of code trying to figure out why a specific implementation was chosen) or take a chance that they understood the design and rely on testing to prove that they didn't break it.

   Lack of internal design documentation makes code inspection much more difficult; code inspections often become design reviews.

When a defect is found, it means that all of the effort subsequent to the phase where the defect was introduced must be repeated. If a defect is introduced in the design phase and is found by integration testing, then fixing the defect requires redesign, recoding, reinspecting the changed code, rereleasing the fix, retesting by QA, and retesting by the integration testing staff—all this effort is collectively known as rework.

The earlier a defect is introduced and the farther the product progresses through the life cycle before the defect is found, the more work is done by various groups. Industry studies have shown that 83% of defects are introduced into products before coding begins; there is no reason to believe that we are significantly better than the industry average in this respect. Anything we do to prevent or detect defects before writing code will have a significant payoff.

Rework may be necessary, but it adds no value to the product. Rework is also very expensive; it costs us about $4 per share in earnings. Ideally, if we produced defect-free software, we would not have to pay for any rework.

It is difficult to hire and train people. People are not eager to have careers in software maintenance, yet that is what we do most of the time. We cannot have a standard training program if every development group uses different software development practices.

Increased testing extends our release cycle. We get diminishing returns on money spent doing testing.

Item: Requirements Management.

Assessment: *We do a dreadful job of developing and managing requirements.*

The purpose of requirements management is to define product requirements in order to proceed with software development, and then to control the flow of requirements as software development proceeds.

In the first phase of the software development life cycle it is important to identify the needs of the intended customers or market and establish a clear understanding of what we want to accomplish in a software product. These requirements must be stated precisely enough to allow for traceability and validation at all phases of the life cycle and must include technical details, delivery dates, and supportability specifications, among other things. Various functional organizations can then use the requirements as the basis for planning, developing, and managing the project throughout the life cycle.

It is common for requirements to change while a product is under development. When this happens, the changes must be controlled, reviewed, and agreed to by the affected organizations. The changes must also be reflected in project plans, documents, and development activities. Having a core initial set of requirements means that any changes during the life cycle can be better controlled and managed.

Our profitability and market variability depends on developing the right product to meet customer needs. Requirements identify these needs. When we do not manage requirements, it affects our profitability and internal operations.

If we do not know what our customers want, we end up developing and shipping a product that may not meet the marketplace needs; we miss the market window; and customers do not get what they are willing to pay for. We may produce a product that lacks necessary functionality, is not compatible with some standard, does not look like or work well with our other products, is too slow, or is not easy to support.

Because we do not do primary market research, we do not have a sense of what the marketplace will need. We gather the bulk of the requirements from customers who mostly identify their immediate needs; we can seldom meet these needs because it takes time to produce the product.

Even when we know the requirements, we continue to add new requirements and to recycle the product through the development process. This is one reason why our products take a long time to reach the marketplace.

An industry-conducted study shows that 56% of bugs are introduced because of bad requirements; as stated earlier, bugs introduced early in the life cycle cause a lot of expensive rework. We have no reason to believe that we do a better job than the rest of the industry in managing requirements.

Poorly understood and constantly changing requirements mean that our plans are always changing. Development has to change its project plans and resources; sales volume predictions have to change; and staffing levels for all the supporting organizations such as education, documentation, logistics, and field engineering to go through constant change. All of these organizations rely on clear requirements and a product that meets those requirements in order to be able to respond in a timely and profitable manner.

Item: Cross-Group Coordination and Teamwork.

Assessment: *Coordination is difficult between projects and between functional areas within a single project.*

Cross-group coordination involves a software development group's participation with other software development groups to address program-level requirements, objectives, and issues.

Cross-group coordination is also required between the functional groups within a single project, that is, between Product Management, Software Development, Quality Assurance, Release, Training, and Support.

At our company cross-group coordination is typically very informal and relies on individual developers establishing relationships with other developers and QA people. Consequently, intergroup communication with cooperation often does not happen, and groups which should be working together sometimes end up being adversaries.

1. *There is no good process for mediating between software development teams.* Software development teams sometimes compete for resources; there is no standard procedure for resolving these conflicts.
2. *It is left to individual effort to make cross-group coordination happen.* There is usually very little planning or resource allocation for cross-group coordination. It is often left to the individual developer to decide when it is necessary and to initiate and maintain communication with other groups.
3. *"The only reward for working on a team is a T-shirt or plaque."* Individual effort is often recognized and rewarded. However, with few exceptions, there is very little reward or recognition for teamwork.

4. *Development and release groups do not communicate well with each other.* To many developers, the release group is perceived as an obstacle to releasing their product. People in the release group think that they are at the tail end of a bad process—they think that developers do not know how to correctly release their product and release defective software all the time. At times, the release group is not aware of development plans or changes to plans.
5. *There is a brick wall between some development and QA groups.* In some groups, development and QA people have little sense of being on the same team working toward a common goal.
6. *Development does not know or understand what product managers do.* Most developers do not know what the product management organization is supposed to do. They see no useful output from their product manager. In many cases, developers don't even know the product manager for their product.
7. *There is too much dependency on broadcast e-mail.* There are a lot of mail messages of the form "the ABC files are being moved to . . ." or "the XYZ interface is being changed in the TZP release. If you use it, you will need to . . ." These are sent to all developers in the hope that the information will reach those individuals who are affected. This use of broadcast e-mail requires people to read many messages that do not affect them. Also, if someone does not read such mail messages, they run the risk of not knowing about something that does affect them. This is a poor method of communication which results from people not knowing who their clients are.

Lack of coordination between projects has resulted in products that do not have similar user interfaces, do not use the same terminology for the same functions, do not look like they came from the same company, or simply do not work well together. This often results in rework, which increases our cost of development.

Poor coordination reduces productivity and hurts morale. Lack of effective coordination usually results in one group waiting for another to complete development of a software product; this causes the first group to be idle and the second group to be under pressure to deliver. It also causes some developers to work overtime to meet their deadline, only to discover that the other group is not ready to use the product.

SEI CMM Level Analysis.

Assessment: *Our company's practices were evaluated at Level 1 of CMM.*

The improvement areas that we have targeted are not all CMM Level 2 activities. Rather, these areas represent the most significant problem areas in our company and are crucial to resolve. We will, therefore, continue to use the SEI assessment and the CMM to guide us and not necessarily follow the model in sequence. Resolving the problem areas will certainly help us achieve at least Level 2 capability.

The assessment results offer no surprises nor do they offer a silver bullet. The assessment was a first step in identifying the highest priority areas for improvement. Knowing these priorities will help us target our efforts correctly—to identify and implement solutions in these high-priority areas. The assessment also generated enthusiasm and a high level of participation all across the company, which is encouraging and makes us believe that as an organization we want to improve our software development processes.

The SEI assessment was a collaborative effort between all organizations. Future success of this project, and any process improvement effort, will depend on sustaining this collaboration. Some development groups are using excellent software development practices and we want to leverage their work. They can help propagate the practices that work for them throughout the company by participating in all phases of this project.

The SEI assessment evaluates processes, not products. The findings, therefore, are relevant to the processes used to develop products. As a company we produce successful products with high quality. However, we have a tremendous cost associated with producing high-quality products. The assessment and follow-up activities will help us improve our processes. This in turn will have a significant impact on cost and productivity, leading to higher profits for us and higher reliability and lower cost of ownership for our customers.

REFERENCE

1. Humphrey WS. Managing the Software Process. Addison-Wesley Publishing, 1989.

Appendix A: An Example of Part Derating Guidelines

INTRODUCTION

These guidelines contain basic information for the derating of electrical and electromechanical components for the purpose of improving the system reliability. However, the designer should evaluate all components according to the requirements of the application.

Resistor Derating

Part type	Parameter	Derate to	Notes
Fixed, carbon composition RCR styles	Power	50%	Voltage should not exceed 75% of the published rating. These resistors should not be used in high humidity or high temperature environments. This resistor style has high voltage and temperature coefficients. Do not allow hot spot temperature to exceed 50% of maximum specification.

Part type	Parameter	Derate to	Notes
Fixed, metal film RNR styles	Power	60%	Slight inductive reactance at high frequency. Hot spot temperature should not be more than 60% of maximum specifications.
Fixed, film-insulated RLR styles	Power	60%	Hot spot temperature should not be more than 60% of maximum specification.
Fixed, precision wirewound RBR styles	Power	50%	Inductive effect must be considered. High resistance values use small-diameter windings (0.0007), which are failure prone in high humidity and high power environments. Hot spot temperature should be not more than 50% of maximum specification.
Fixed, power wirewound (axial lead) RWR styles	Power	50%	Inductive effects must be considered. Noninductively wound resistors are available. Resistance wire diameters limited to 1 mm with some exceptions for high values. Do not locate near temperature-sensitive devices. Do not allow hot spot temperature to exceed 50% of maximum specification.
Fixed, power wirewound (chassis mounted) RER styles	Power	50%	Inductive effects must be considered. Derate further when operated at temperatures exceeding 25°C. Use appropriate heat sinks for these resistors. Do not locate near temperature-sensitive devices. Do not allow hot spot temperature to exceed 50% of maximum specification.
Variable wirewound (trimming RTR styles)	Power	50%	Inductive and capacitive effects must be considered. Current should be limited to 70% of rated value to minimize contact resistance.
Variable nonwirewound (trimming)	Power	50%	Voltage resolution is finer than in wirewound types. Low noise. Current should be limited to 70% of rated value to minimize contact resistance.

Resistor Application Notes

1. *General considerations.* The primary rating consideration for resistors is the average power dissipation. A secondary stress rating is maximum impressed voltage, which can cause failure modes such as insulation breakdown or flashover. For pulse applications, the average power calculated from pulse magnitude, duration, and repetition frequency should be used to establish power derating requirements. Pulse magnitude should be used to establish voltage derating requirements.
2. *Failure modes and mechanisms.* The principal failure mechanism for resistors is the destruction of the resistive element by burning, melting, evaporating, or loss of an insulating coating from the resistance material. Failure will occur as a result of these phenomena at the hottest spot on the resistor element. In order to increase the resistor reliability, the circuit application should not generate more heat than can be eliminated by conduction, radiation, and convection. The amount of heat generated is I^2R (I = current, R = resistance), and all three methods of heat removal are typically utilized in a resistor installation.
3. *Design tolerance.* Because of the cumulative effects of temperature coefficient, voltage coefficient (load stress), and life, each application of a resistor must withstand a broader tolerance on the resistance value than merely the initial tolerance (purchase tolerance). The worst case tolerance that will not affect design performance and will accommodate long life reliability must be determined by the designer.

Capacitor Derating

Part type	Parameter	Derate to	Notes
Capacitor	DC working voltage	70%	Of WVDC, all capacitors
	Temperature	See notes, next section	Limit case temperature to 70% of maximum rated temperature
	Surge voltage	100%	Of WVDC or 80% surge rating, whichever is less
	RMS voltage	70%	Of WVDC, ceramic, glass or mica capacitors only; AC operation
	Peak voltage	See notes, next section	DC plus peak AC (ripple voltage) limited to DC derated voltage (*polarized capacitors*)
	Reverse voltage	0 V	All polarized capacitors except solid tantalum

Part type	Parameter	Derate to	Notes
	Reverse voltage	3%	Of WVDC, solid tantalum capacitors only
	Ripple voltage	70%	Of specification value
	Ripple current	70%	Of specification value
	Series impedance (external)	3 Ω/V	*Solid tantalum capacitors only*, external circuit impedance less than 3 Ω/V

Capacitor Application Notes

1. *General considerations.* Failure rate for capacitors is a function of temperature, applied voltage, and circuit impedance. Increased reliability may be obtained by derating the temperature and applied voltage and increasing circuit impedance. For all styles, the sum of the applied DC voltage and the peak AC ripple voltage shall not be greater than 70% of rated voltage.
2. *Failure modes and mechanisms.* Catastrophic failure of capacitors is usually caused by dielectric failure. Dielectric failure is typically a chemical effect, and for well-sealed units is a function of time, temperature, and voltage. The time–temperature relationship is well expressed by assuming that the chemical activity, and therefore degradation, preceeds at a doubled rate for each 10°C rise in temperature; e.g., a capacitor operating at 100°C will have half the life of a similar one operating at 90°C.
3. *Capacitor misuse.* A capacitor may fail when subjected to environmental or operational conditions for which the capacitor was not designed or manufactured. The designer must have a clear picture of the safety factors built into the units and of the numerous effects of circuit and environmental conditions on the parameters. It is not enough to know only the capacitance and the voltage rating. It is important to know to what extent the capacitance varies with the environment; how much the internal resistance of the capacitor varies with temperature, current, voltage, or frequency; and of the effects of all of these factors on insulation resistance, breakdown voltage, and other basic capacitor characteristics that are not essential to the circuit but which do invariably accompany the necessary capacitance.
4. *Design tolerance.* Because of the cumulative effects of temperature, environment, and age, each capacitor application must consider a tolerance that is independent of the initial (or purchase) tolerance. In other

words, the design must tolerate some changes in the capacitance and other device parameters in order to assure long life reliability.
5. *Permissible ripple voltage.* Alternating current (or ripple) voltages create heat in electrolytic capacitors in proportion to the dissipation factor of the devices. Maximum ripple voltage is a function of the frequency of the applied voltage, the ambient operating temperature, and the capacitance value of the part.
6. *Surge current limitation.* A common failure mode of solid tantalum electrolytic capacitors is the catastrophic short. The probability of this failure mode occurring is increased as the "circuit" surge current capability and the circuit operating voltage are increased. When sufficient circuit impedance is provided, the current is limited to a level that will permit the dielectric to heal itself. The designer should provide a minimum of 3 ohms of impedance for each volt of potential ($3\Omega/V$) applied to the capacitor.
7. *Peak reverse voltage.* The peak reverse AC voltage applied to solid tantalum electrolytic capacitors should not exceed 3% of the forward voltage rating. Nonsolid tantalum electrolytic capacitors must never be exposed to reverse voltages of any magnitude. A diode will not protect them. Reverse voltages cause degradation of the dielectric system.

Discrete Semiconductor Derating

Part type	Parameter	Derate to	Notes
Transistors	Junction temperature	See note 3	Critical parameter
	Voltage	70%	Of maximum rating at 25°C ambient
	Current	70%	Of maximum rating at 25°C ambient
	Power	70% See note 5	Of maximum rating at 25°C ambient
Diodes, rectifier and switching	Junction temperature	See note 3	Critical parameter
	Forward current (I_F)	60%	Of maximum rated value at 25°C
	Reverse voltage (V_R)	80%	Of maximum rated value at 25°C
	Power	70%	Of maximum rating at 25°C ambient
	Surge current, peak	75%	Of maximum rated value

Part type	Parameter	Derate to	Notes
Diodes, zener	Junction temperature	See note 3	Critical parameter
	Zener current	70%	Of maximum rated value
	Surge current, peak	75%	Of maximum rated value
	Power	50%	Of maximum rated value
Silicon-controlled rectifiers	Junction temperature	See note 3	Critical parameter
	Forward current (RMS)	70%	Of maximum rated value
	Peak inverse voltage	70%	Of maximum rated value
FETs	Junction temperature	See note 3	
	Voltage	70%	Of maximum rated value

Discrete Semiconductor Application Notes

1. *General considerations.* Average power is the primary stress consideration for transistor derating. An important secondary stress rating is the maximum impressed voltage. High junction temperature (T_j) is a major contributor to transistor failure and is proportional to power stress.

    Average current is the primary stress consideration for diode derating. An important secondary stress rating is the maximum reverse voltage which can cause pn junction breakdown. The voltage–current characteristic in the forward direction is the important functional characteristic of all diodes except for Zener diodes, where the voltage–current characteristic in the reverse direction (avalanche region) is utilized for voltage regulation.

2. *Worst-case derating.* In derating all transistors, diodes, and silicon-controlled rectifiers, the worst-case combination of DC, AC, and transient voltages should be no greater than the stress percentages in the preceding table.

3. *Derated junction temperature.* The recommended derated junction temperature for semiconductor components is

Appendix A: An Example of Part Derating Guidelines

Mfg spec. T_j max.	Derated T_j
125°C	95°C
150°C	105°C
175°C	115°C
200°C	125°C

4. *Junction temperature calculation.* The junction temperature of a device can be calculated using either of the following formulas:

   $$T_j = T_a + 0_{ja} Pd$$
   $$T_j = T_c + 0_{j-c} Pd$$

   where

   T_a = Ambient temperature
   T_c = Case temperature
   0_{ja} = Thermal resistance of junction to ambient air, °C/W
   0_{jc} = Thermal resistance of junction to case, °C/W
   Pd = Power dissipation, W

5. *Safe operating area.* Bipolar transistors are known to experience a failure mode called *secondary breakdown* even though the operating currents and voltages are well within the permissible limits set by thermal dissipation requirements. Secondary breakdown results in a collapse of the device's voltage-sustaining abilities with the accompanying sharp rise in collector current. This usually results in permanent damage, almost always a collector-to-emitter short. Transistor manufacturers have dealt with this breakdown problem by publishing ratings for each type of device. These ratings are commonly referred to as "Safe Operating Area Ratings," or SOAR, and usually appear in a graphic form. All power transistors and many small signal transistors have SOAR ratings. It is essential that circuit designers refer to SOAR graphs to complete a reliable design.

6. *Thermal cycling.* Silicon power transistors are subject to thermal fatigue wearout due to the different coefficients of thermal expansion of the materials making up the transistor. Thermal cycling of power transistors can be controlled by reducing the case temperature change due to power dissipation and by the use of an adequate heat sink.

Microcircuit Derating

Part type	Parameter	Derate to	Notes
Integrated circuit, linear/analog	Supply voltage	80%	In some devices, supply voltage and current are the major contributors to increased junction temperature.
	Junction temperature	See note 2	Critical parameter.
Integrated circuit, regulator	Input voltage max.	80%	
	Input voltage min.	$V_{min} + 10\%$	Do not allow V_{in} to go below 110% of V_{min} specification to assure proper regulation.
	Output current	80%	
	Junction temperature	See note 2	Critical parameter.
Integrated circuit, digital	Fan out	80%	Limit input and output source and sink currents to 80% of maximum specification.
	Junction temperature	See note 2	Power dissipation is intrinsic to each device and, to a large degree, cannot be controlled. But junction temperature can be reduced by proper cooling, airflow, lower switching speeds, and low fanout. Low power devices are recommended where speed is not a critical factor.

Integrated Circuit Application Notes

1. *General considerations.* The critical derating characteristics for semiconductors is junction temperature. The junction temperature of a device is a function of power dissipation, package style, air flow, ambient temperature, and package characteristics such as material, die size, and process techniques.
2. *Derated junction temperature.* The recommended derated junction temperature for semiconductor components is

Appendix A: An Example of Part Derating Guidelines

Mfg spec. T_j max.	Derated T_j
125°C	95°C
150°C	105°C
175°C	115°C
200°C	125°C

3. *Junction temperature calculation.* The junction temperature of a device can be calculated using either of the following formulas:

$$T_j = T_a + 0_{ja}Pd$$
$$T_j = T_c + 0_{jc}Pd$$

where

T_a = Ambient temperature
T_c = Case temperature
0_{ja} = Thermal resistance of junction to ambient air, °C/W
0_{jc} = Thermal resistance of junction to case, °C/W
Pd = Power dissipation, W

4. *Thermal resistance.* The following is a list of thermal resistances of typical low pin count IC package styles:

IC package pins	Case style	0_{j-c}	0_{j-a}
3	T0-220	35	160
8	Plastic	45	125
8/10	Metal can	25	150
14	Plastic	44–65	108–120
14	Cerdip	25–40	80–130
16/18	Plastic	44–75	108–160
16/18	Cerdip	22–40	66–150
20	Plastic	32–44	81–123
20	Cerdip	16–35	68–122
24/28	Plastic	43–57	85–125
24/28	Cerdip	10–30	28–65
40	Plastic	27–50	62–95
40	Cerdip	7–25	36–60
64	Side braze	7	20

For device configurations not listed, specific thermal resistances should be obtained from the IC manufacturer.

The thermal resistance information was drawn from available vendors' data (TI, Signetics, Motorola, Fairchild, National, Intel). Thermal resistance values will vary depending on the vendor, die size, linear airflow around the device, etc.

5. *Operational type specifications.* For operational specifications (i.e., supply voltages, offset, transients, etc.), designs should perform within minimum or maximum specifications, whichever causes worst-case circuit operation. Typical specifications do not guarantee performance.

Remember to consider noise and thermal characteristics over the desired operating range.

Magnetic Devices Derating

Part type	Parameter	Derate to	Notes
Inductor, general	Current	70%	Of maximum rated current
Transformer, audio	Current	70%	Of maximum DC current
Transformer, power	Current	60%	Of maximum DC current
Transformer, pulse (low power)	Current	60%	Of maximum DC current

Magnetic Devices Application Notes

1. *General considerations.* The reliability of inductive devices is a function of operating hot spot temperature as it relates to the insulation capability. Derating of inductive devices is accomplished by assuming that the hot spot temperature does not exceed a selected value which is less than the maximum rated hot spot temperature. The hot spot temperature is the sum of the ambient temperature and the internal temperature rise at the hottest spot in the windings. The temperature rise under operating conditions can be determined by direct measurement or by analysis.
2. *Maximum hot spot temperature.* The maximum permissible hot spot temperature for standard inductors shall be 25°C less than the specified temperature.
3. *Failure modes and mechanisms.* The failures in inductive devices generally fall into two categories: (1) insulation breakdown and (2) open conductor.
    3.1 *Insulation breakdown—transformers.* Insulation breakdown can result from moisture penetration; poor insulation on conductors, between layers or to the magnetic core; or as a prolonged

Appendix A: An Example of Part Derating Guidelines 419

"hot spot" or sustained overload. Conduction from one turn to another results in a "shorted turn" effect, causing high power consumption. This additional power is dissipated in the transformer, causing further heat to be generated.

3.2 *Insulation breakdown—inductors and coils.* Insulation breakdown in these devices results in degraded performance, since the impedance is reduced but the power consumption is usually not increased enough to cause open conductors. If the device is a filter inductor, ripple will increase.

3.3 *Open conductor.* This type of failure is almost always catastrophic in nature. Transformers will have no output and coils or inductors will have infinite impedance and resistance.

4. *Frequency.* Power transformers and inductors are designed to operate efficiently over a limited frequency range. Operation outside this range, particularly at lower frequencies, will result in overheating.

5. *Saturation.* Power inductors used as filters usually carry a large direct current component. If this component exceeds the value specified, the inductance can be reduced because of core saturations.

6. *Custom magnetics.* When special devices are designed for a particular circuit application, the derating and application factors noted above for standard parts must be observed. The constraints of the application (e.g., ambient operating temperature, current and voltage characteristics, etc.) must be determined beforehand and applied as factors in the design of the device. For example, the expected ambient operating temperature and the hot spot temperature must be determined, and the class of insulation for the windings then selected to meet the derating guidelines.

Connector Derating

Part type	Parameter	Derate to	Notes
Coaxial	Current	50%	
Multipin	Current	50%	
	Surge current	100%	Of DC current rating
	Body surface temperature	70%	Of rated temperature
Cable	Current	50%	
	Surge current	100%	Of DC current rating
	Body surface temperature	70%	Of rated temperature

Connector Application Notes

1. *General considerations* The primary derating consideration for connectors is the temperature rise of the material due to power transmis-

sion through the contacts. High temperatures will accelerate aging, resulting in increased resistance of the contacts and drastically shortening life.
2. *Failure modes and mechanisms.* Friction and wear are the main failure mechanisms that will cause deterioration of connector reliability. Galvanic corrosion and fretting corrosion initiated by slip on connector contact roughness will lead to a direct attack on the connector contact interfaces by oxygen, water, and contaminants. The slip action, which can be produced by vibrations or cyclic heating and cooling, causes an acceleration in the formation of oxide films. Lubricants used to reduce friction and wear play an active part in the failure mechanisms and add to the complexities of potential film formations.
3. *Contact materials.* Gold contact material is recommended for connector reliability. Gold–tin mating is not recommended because of long-term reliability problems. The gold–tin bimetallic junction becomes subject to galvanic corrosion which leads to high contact resistance and eventual failure. Tin-plated contacts must not be used to make or break current; arcing quickly destroys the tin plating.
4. *Parallel pins.* When pins are connected in parallel to increase the current capacity, allow for at least a 25% surplus of pins over that required to meet the 50% derating for each pin, assuming equal current in each. The currents will not divide equally due to differences in contact resistance.
5. *IC sockets.* Integrated circuit sockets should not be used unless absolutely necessary. In most cases, the failure rate of the socket exceeds that of the IC plugged into it.

Switch Derating

Part type	Load type	Parameter	Derate to
Switch or relay	Capacitive	Contact current	50%
	Resistive	Contact current	50%
	Inductive	Contact current	40%
	Motor	Contact current	20%
	Filament	Contact current	10%

Switch Application Notes

1. *General considerations.* The derating of switches is centered around electrical and physical protection of the contacts. Electrical protection is achieved by derating the load that the contact must handle. Physical protection is mainly achieved through contact material selection and environmental protection.

Appendix A: An Example of Part Derating Guidelines

2. *Peak in-rush current.* Peak in-rush current should not exceed 50% of the maximum surge current rating.
3. *Arc supression.* Provide for arc suppression of all switched inductive loads.
4. *Contacts.* Operate contacts in parallel for redundancy only and never to "increase the current rating." Do not operate contacts in series to "increase voltage rating."
5. *Switch/relay mounting.* Proper support should be used to prevent any deflection of the switch due to shock, vibration, or acceleration. The direction of motion of the contacts should not be coincident with the expected direction of shock.
6. *Relay coils.* Coils are designed for specific voltage ratings to ensure good contacting. Applied coil voltage should always be within $\pm 10\%$ of the nominal rating.

Fuse Derating

Current Rating

Most fuses are rated so that at 110% of the rating, the temperature rise (from room temperature, 21°C or 70°F) at the hottest points of the fuse will not exceed 70°C. Under other conditions, the rating must be adjusted up or down (uprating or derating) as indicated on the graph below.

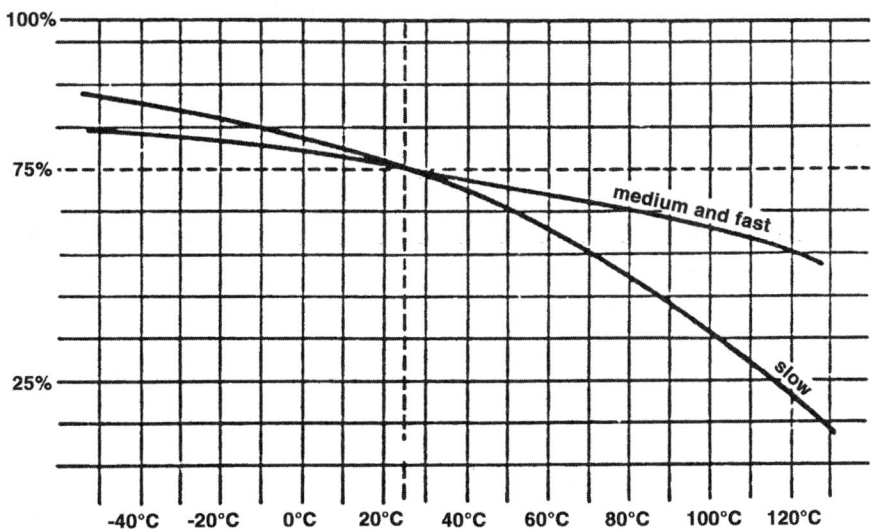

Maximum Ambient Temperature

Voltage Rating

Circuit voltage should not exceed fuse rating.

Fuse Application Notes

1. *Derating or uprating a fuse.* The best practice is to select a fuse rating which allows for a 25% derating. In other words, a fuse should normally be operated at about 75% of its rating in an ambient temperature of 25°C.
2. *Blow time.* Blow time is dependent upon the percent of rated current and on the thermal inertia of the fuse. The following figures are typical of the blowing time ranges when tested at 200% of rating.
   Fast   <1 s
   Medium   1–10 s
   Slow   >10 s
   The chart below shows the relation between percent of rated current and blowing time:

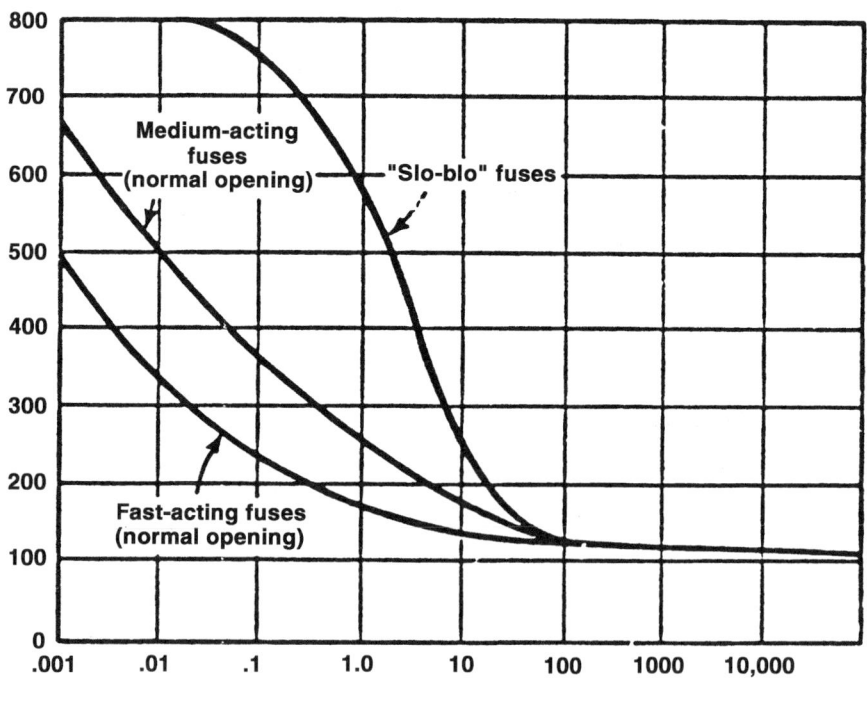

Blowing time, sec

Appendix B: FMEA Example for a Memory Module

Appendix B

Parts FMEA   Project: 1 Memory Module   Date: 12/9/02

ID #	Part name	Part functions	Failure modes (negative of the function)	Causes of failures	Occurrence of cause	Escape detection	Effect of failure (customer's perspective)	Severity of effect	RPN	Current controls	Recommended corrective action and status
1	PCB/PWA	Electric connection	Open/short	Solder short	1	1	Data corruption; function failure; none or intermittent	7	7	Net list test	
			Resistive	Crack	2	7		7	98	Supplier feedback	
				Scratch	1	1		7	7	Rating program	
				Contamination	1	10		7	70	Document rework procedures	
				Void	1	10		7	70		
				Overstress (tantalum)	3	5		7	105	Certified operator	Review tooling and thermal profile.
				Overetch	3	10		1	30		
				Bad rework	1	3		7	21		
				Overcurrent	1	1		7	7		
		Carrying components/ mechanical substrate	Parts loose or detached	Vibe-flex	3	7		7	147	Preventive maintenance scheduled design/ analysis	Austin, TX, to review system packaging, specifically that the module cover bracket and chassis tolerances are effective to control vibration.
				Wrong profile	1	3		7	21		
				Wrong paste	3	3		1	9		
				Plugged screen	3	3		7	63		
		Connector system	Intermittent open short	Fretting	3	10		1	30	Lubrication board cover	
				Wrong thickness, slot width/location	1	1		7	7		
				Oxidation	1	10		1	10		Tandem to review Micron's workmanship standards.
				Contamination	3	10		5	150		Review ACI process Cpk for thickness and purity.
				Poor plating	3	7		5	150		
		Present components air stream cool system	Burn-out	Airflow shadow	1	3		5	15		
		Entrap EMI/RFI carry lable for module		Airflow shadow	3	5		7	7	DVT measurement	Correct design errors if found.

Appendix B: FMEA Example for a Memory Module

#	Item	Function	Failure Mode	Effect	S	O	D	RPN	Actions/Controls
2	Connector	Electric connection	No solder joint	System not functional	3	3	7	63	Tolerances
			Fractured contact		3	3	7	63	
			Contact broken or damaged	Data corruption, no signal or data VCC to ground short					
		Poor connection	Poor solder joint	System not functional	3	5	5	75	
			Contamination	Data corruption	1	10	5	50	
			Alignment mismatch		1	1	1	1	
		Mechanical connection	See above						
			See above						
3	Chip caps (.1 uF, 1000 pF)	Decoupling HF	No connection						Quality of thin chip caps
			Poor connection						Future designs may consider series caps
			Short						
			Low capacitance/open	Fire	7	10	70		
			Open	Long-term reliability					
			Solderability	Noise/intermittent operation	1	3	7	21	Visual inspect
			Mismarked components	Signal integrity	1	10	7	70	FIFO inventory / Supplier quality
4	Tantalum cap (2 to 10 uF)	Decoupling LF	Short		1	1	7	7	Machine cals and PM
			Low capacitance/open		3	5	7	105	Supplier feedback audits
			Reverse polarity		3	3	1	9	Electrical test
					3	3	1	9	P&P programs from CA
			Solderability		3	3	1	9	Visual inspection / Online documents
5	Reflow	Reflow solder paste	Poor connection		3	7	7	49	Visual storage control
			Cold solder		1	1	7	21	Visual inspection
			Flux inactive	Poor signal	3	7	7	147	Process procedure/documentation
			Shorts						
			Solder paste slump	Functional failure	1	1	1	1	Paste selection / Functional test / Process automation
6	Kitting	Provide manufacturing with required components	Functional failure	Functional failure	3	3	5	45	Part verification Checklist
			Wrong parts received						Signed off by Inventory and Production
7	Rework	Repair ECs	Introduce secondary failure	Functional failure	3	1	1	3	Visual, functional test.
			Operator error						Review Micron selection and qualification process of paste suppliers (data and quality report).

Appendix B

Parts FMEA Project: 1 Memory Module Date: 12/9/02

ID #	Part name	Part functions	Failure modes (negative of the function)	Causes of failures	Occurrence of cause	Escape detection	Effect of failure (customer's perspective)	Severity of effect	RPN	Current controls	Recommended corrective action and status
8	Stencil printing	Apply solder paste	No paste	Misregistration	1	5	Functional failure	7	35	Visual specification	
			Excessive paste	Misregistration	1	1		1	1	Visual inspect, 1sm, etc.	
			Wrong paste	Operator error	3	1		3	9	Process documentation	Continuous improvement of paste quality.
			Low volume solder joint	Insufficient paste	3	7		7	147	1sm, visual inspection	Micron will supply qualification specification and report. Review stencil printing process specification.
9	Placement	Place components	Parts in wrong place	Part loaded wrong	1	1	No function	7	7	Feeder load list, visual	
				Program error	1	1		7	7	Documentation and revision control	
			Noncoplanar leads	Defective component	3	7	Erratic function	7	147	Visual, ICT	Insist parts on tape and reel. PM of equipment. Tandem purchasing to give better forecasting.
			No part placed	Pass-through mode	1	1		7	7	Visual	
			Part orientation	Programming error, t/r wrong	3	7		7	147	Visual, documentation	Tandem to give feedback on 10 prototypes to Micron.
					1	7		7	49	Visual, feeder PM	
			Damaged parts	Feeder malfunction	1	3		7	21		
				Defective components	3	7		7	147	Visual ICT	Supplier component qualification. Return all components for FA and RCA to supplier. Review Micron MRB F/A specification.
10	Cleaning	Remove flux residue	Ionic contamination	Dendrite growth	1	10	Long-term reliability	7	70	Process controls, flux quality	
				Corrosion							

Appendix B: FMEA Example for a Memory Module

#	Part	Function	Failure Mode	Cause	Effect	Sev	Occ	Det	RPN	Controls	Recommended Action
11	Visual	Catch solder balls	Solder short	Misprinted board	Functional failure	1	1	7	7	Visual, electrical test, supplier audit	
		Verify conformity to workmanship standards	Reversed component	Operator fatigue		1	1	7	7	Breaks (mandatory every 2 h), rotation	
			Misplaced component	Incorrect documentation		1	3	10	**300**	No control	Tandem reviews component CAD library for polarized parts. Provide mechanical sample if possible.
12	Chip resistor	Signal integrity	Missing component	Operator training		1	1	1	1	In-house training, operator certification	
			Poor signal	Crack	Poor performance	1	5	7	35	Supplier qualification, visual inspection	
				Out of tolerance		5	1	7	35	Supplier qualification	
				Poor solderability		1	1	7	7	Supplier qualification	
				Wrong value/mismarked		5	1	7	35	Visual inspection, supplier qualification	
13	4K SEEROM	Machine readable FRU ID record (firmware)	Wrong data written/read	ESD	Wrong record	7	1	5	35	Design/process	
				Latch-up	Record corruption	1	1	1	1	Design/process	
				Bad bit/row/column		3	1	5	15	Process/burn-in	
				Open/short addresses		3	1	5	15	Process/burn-in	
14	Trace cut	Provide ID code	Short/intermittent short	Not complete cut	Wrong ID code				0	Process, visual, test	Will be no cut CN in process.
			Leakage	Contamination	Wrong ID code	1	1	3	3	Process control	
			Corrosion	Contamination	Wrong ID code	1	1	3	3	Process control	
15	DRAM (4M×4)	Data storage	Wrong data written/read	Soft error	Parity error	3	1	1	3	Process/design	
				ESD		7	7	1	49	Process/design	
				Latch-up		1	1	7	7	Process/design	
				Timing drift		3	1	5	15	Burn-in	
				Wrong speed		3	1	5	15	Electrical test	
				Bad bit/row/column		3	1	7	21	Burn-in/electrical test	
				Open/short address		3	1	7	21	Burn-in/electrical test	
				Pattern sensitivity		3	1	3	9	Electrical test	
16	LVT244 Driver	Control signal buffer	Wrong data written/read	ESD	Parity error	7	1	7	49	Design/process	
				Latch-up		1	1	1	1	Design/process	
				Slow speed		3	3	10	90	Electrical test	
				Timing drift		3	1	10	30	Process/design	
			Part will not tristate	Missing pull-down		1	1	1	1	Process/design	
			Part remains in tristate	Open enable pin		1	1	1	1	Process/design	

Appendix B

Parts FMEA Project: 1 Memory Module Date: 12/9/02

ID #	Part name	Part functions	Failure modes (negative of the function)	Causes of failures	Occurrence of cause	Escape detection	Effect of failure (customer's perspective)	Severity of effect	RPN	Current controls	Recommended corrective action and status
17	Test	Verify functionality	Bad data	Ineffective test codes	1	5	Functional fail	7	35	Incoming specifications Board grader (coverage report of nodes) Design for test Verification checklist	←
				Defective hardware	3	5		7	105	Verify test with known good product (if available) Verify with grounded probe PM schedule	Micron to review and establish regular PM schedule on test hardware.
				Inadequate requirements	1	1		7	7	Standard program in place	→
		Verify timing	Functional failure	Operator error	1	7		7	49	Operator training	
				No coverage at present	1	7		7	49	DRAM supplier burn-in	
18	Driver (16244)	ADDR Buffer	Wrong data written/read	Address aliasing ESD Latch-up Slow speed Ground bounce Timing drift			Parity error			Pattern test Design Design Test Design/test Process/design	
			ADDR or data unavailable								
		Tristate for test	Part will not tristate	Defective driver Missing pull-down Defective pull-down			Interference during memory test			Process/design	
			Part stuck in tristate	Open enable pin Solder joint			Won't write system data			Process/design	

Appendix B: FMEA Example for a Memory Module

19	Packaging/shipping	Prepare product for shipment/storage	ESD damage	No ESD bag	1	7	Functional failure	7	49	ESD training ESD products
				ESD bags wear out						
		Provide marking (label)	Product damage	Operator error	1	7		7	49	Packaging procedures
			TANDEM needs to define criteria							
20	Depanel	Separate individual PCBs from manufacturing array	ESS damage	Ground failure	1	10		7	70	Monthly ESD audit Plastic exhaust replace.
				ESD discharge from plexiglass safety cover	1	10		7	70	Weekly PM Online documentation
			Physical damage	Improper mounting	3	1		1	3	Online documentation
			Wrong outline	Program error	3	1		1	3	Tooling ID Improve tooling plate.
21	Solder paste	Mechanical connection	Solder not reflowed	Improper profile	1	1	Data corruption	1	1	One profile on PN program
		Electrical connection		Powder oxidized	1	1	Functional failure	1	1	CFC, supplier audit
				Wrong flux	1	1	Shorts	1	1	CFC, supplier audit

Memory module action items list resulting from FMEA (action item assignees names suppressed):

1. PCB/PWA: Review tooling and thermal profile.
2. PCB/PWA: Review system packaging. Specifically, that the module cover bracket and chassis tolerances are effective to control vibration.
3. PCB/PWA: Review supplier's workmanship standard. Review ACI process Cpk for thickness and purity.
4. Reflow: Review selection and qualification process of paste suppliers (data and qual report).
5. Stencil printing: Continuous improvement of paste quality. Supplier will provide qualification specification and report. Review stencil printing process specification.
6. Placement: Insist that parts are on tape and reel.
7. Visual: OEM to review CAD library for polarized parts. Provide mechanical sample, if possible.
8. Trace cuts: There will be no cuts; CN is process.
9. Test: Supplier to review and establish regular preventive maintenance schedule.

CA, California; CN, change notice; FA, failure analysis; FRU, field replaceable unit; PM, preventive maintenance; RCA, root cause analysis.

Epilogue

The amount of reliability testing and analysis that a company does depends on its unique operating and design principles, on the type of products produced, their cost, the served market, and the customers' expectations. A balance is also required between what reliability customers want and what they are willing to pay for. Most companies perform some sort of analysis during the design and manufacturing of a product. Many of them analyze the problems that occur during the development phase. But few do so for those problems that occur during manufacturing and in the field, unless a trend is observed.

The approach for providing a reliable electronic product, as presented in this book, is from a very practical but high-end product perspective. Not all companies need or should have as comprehensive a reliability process as presented, nor are they all able to afford the resources to implement such a process in total. A complete implementation of these processes requires that a company have deep pockets and extensive internal resources. Nor is it always possible, due to market conditions and pressures, to do everything presented with the depth and thoroughness discussed. My goal at the outset was to make the reader aware of the different techniques and elements used and available in the industry. It is up to individual companies to assess their needs and choose and implement those elements or few essential items that make sense for them to ensure that their products meet their reliability goals. Thus, reliability is very unique to a company, a product type, and the served market.

The ability of a company to perform comprehensive design simulations, HALT, DVT, regulatory (safety and EMC) testing, ESS, reliability prediction, analysis and testing, and failure analysis in-house requires an enormous investment in specialized capital equipment and highly trained personnel. Most companies don't have the investment dollars and technical resources necessary to support and staff a complete facility for all of these reliability elements. None but the very largest companies can afford to own such resources themselves. Because of the specialization of some of these resources and the cost of the associated capital equipment, many large electronic equipment manufacturers themselves outsource those tasks to companies that have the needed specialized skills as their core competency.

Let's take the example of failure analysis. How does a failure analysis get performed and who does the failure analysis? The majority of companies have minimal product troubleshooting and questionable diagnostics capabilities. So they may remove several components to correct a PWA anomaly. The removed components are typically sent to the component supplier for verification and failure analysis. If the problem component is critical to releasing a product to production or in understanding risk in already fielded products in a timely manner, many OEMs (both large and small) rely on independent failure analysis laboratories.

Specifically, what are smaller and midsized companies to do? Small companies can use the topics presented in this book as a menu. They should pick and choose only those items that are required to meet the reliability requirements for their chosen marketplace. Small companies typically have low budgets for reliability tools, analysis, and testing, so they have to do a limited number of smart things cost effectively, such as FMEA and HALT during product development. For those reliability tasks that require specialized skill sets and expensive equipment, they need to consider the very viable option of using an outsource service provider. The obvious benefits are lower cost (than the costs of manpower and purchasing infrequently used equipment); whereas the outsource provider can spread the costs across a wide customer base and staff content experts. This allows the small to midsized OEM to focus on core competencies while simultaneously allowing the service provider to focus on its core competencies, creating a true win–win solution.

The following are some things that the small to midsized companies can do to mitigate the fact that they don't have unlimited resources.

1. Network with local peer companies to share ideas, methods, and resources.
2. Join industry associations where peers can discuss common issues and exchange ideas.
3. Use some of the readily available tools such as automated MTBF prediction algorithms, warranty cost analyses, FMEA, acceleration factor nomographs, and the like.

Epilogue

4. Partner with suppliers to obtain appropriate design, usage, and reliability test data details.
5. Investigate available outsource providers for services required such as independent test laboratories (regulatory, environmental, mechanical, etc.) and specialized engineering consultants.

What challenges do OEMs face in a faster, better, cheaper business environment to field reliable products? In a fast time-to-market environment where design cycles are compressed to 6 months or less the adoption of a faster, better, cheaper (FBC) mindset requires a paradigm change to the traditional product analysis, design, procurement, and production processes. In an FBC environment there is not enough time for lengthy and detailed analyses such as FMEA, signal integrity simulations, and the like. One must decide how much one is willing to accept and then perform only those tasks that match the risk, relying on testing rather then extensive simulations to prove the product and catch problems.

With this in mind, the following is a list of guidelines that have been successfully used in a faster, better, cheaper product design environment and are recommended to the reader.

> Collocate the multifunctional design team together in one location.
> Test is a vital factor in the momentum of a project. Do you have the right speed and are you going in the right direction?
>> Test early and often.
>> Use simulation supported by test. Don't put your full trust in simulation alone.
>> Spend one half of the allotted project time on development and the second half on testing.
>> A test plan needs to assume the worst and test for worst case conditions.
>> It is not possible to test software completely.
> Understand the risks early in the project and then develop plans to mitigate them. Retire each one by design, by test, by whatever means, until each has been reduced to the point where it is no longer a risk item, or the probability is so low that you don't need to worry about it any longer.
> Rely on past experience (lessons learned) and trust your judgment.
> Use a shootout to select competing design solutions.
> Make decisions promptly even if based on imperfect, limited, or even nonexistent information as well as on short notice.
> Project reviews are essential.
> If your chosen approach isn't working, change your mind.
> Improvising is an inseparable part of planning.
> Keep your successes in front of you. Celebrate success often.
> In the race for quality there is never a finish line.

- Every person on the design team needs to include cost, schedule, quality, and risk in his/her thinking.
- It is important that employees multitask between the technology side and the business side of a project.
- Don't change anything if you don't need to.
- Don't keep refining/improving the design. The output of even the most creative team must be a product that makes it out the door in time to serve its intended purpose.
- Don't gold plate everything in the product design, losing sight of the original goal. The less risk people want to take, the more they put into their designs to ensure reliability. The more things (safety nets) that are put in, the more expensive the project becomes and the longer the time to market.
- Stretch your design solution by using new components rather than already proven components and concepts.
- Tough constraints tend to be powerful motivators for putting aside familiar tools, techniques, approaches, and methods and aggressively search for new solutions using out-of-box thinking.
- Limitations (such as personnel or financial resources) induce innovation; and innovations lead to breakthroughs.

Index

Accelerated stress testing (AST), 15, 72, 157, 158–161, 204, 205, 336, 361–363
Acceleration factor (ICs), 210–215
Acceptable quality level (AQL), 5, 199, 200
Acoustic measurements, 162
Activation energy, 210, 211
Active heat sinks, 271, 273
Adversarial transactions, 181
Aerospace industry, 8, 11, 42, 83, 84, 168, 202, 204, 225, 315
Aircraft, 5
Analog ICs, 70, 71, 75, 76, 81, 124
Analog test, PWA, 331
Apple Computer, 175, 337
Application-based qualification testing, 224, 225, 229–235
Approved materials list (AML), 144
Application-specific ICs (ASICs), 62–63, 70, 198, 200, 203, 204, 304, 344, 346

Application tests, 155
Approved supplier/vendor list (ASL/AVL), 144, 201, 314
Arrhenius equation, 210–212
Arrhenius, Svante, 210
AT&T, 66
Automated optical inspection (AOI), 324
Automatic test equipment (ATE), 123, 136, 325, 354, 358, 371
Automatic test pattern generation (ATPG), 124, 125, 130, 133, 134
Automotive industry, 168, 204, 224, 225, 255, 294, 312, 319, 374

Back-driving ICs, 332, 333
Ball grid array package (BGA), 71, 96, 115, 156, 269–271, 325, 326
Bathtub failure rate curve, 6–12, 21, 22, 43, 44, 161, 208, 209, 230, 340, 350–352
Bed-of-nails fixtures, 327, 329–331
Beer, Michael, 171

435

Bill of materials (BOM), 15, 17–21, 70, 74, 80, 138, 139, 144, 145,154, 202, 321
Bill of materials review, 138, 139, 154, 202
Binomial distribution, 37, 44
Boundary scan, 124, 157, 130–137, 327, 329, 330
Build to order (BTO), 171
Built-in self-test (BIST), 124, 125, 134, 135, 137, 327, 328, 334
Burn-in, 23, 24, 116, 204, 337, 361

Cables, EMC control, 302, 305, 306
Categories of suppliers, 189–192
Cell phones, 312, 319, 336
Central processing unit (CPU, computer), 52–54
Chassis (or product enclosure), 250–252, 273, 274, 299–301, 306
Chip scale package (CSP), 71, 96, 115, 326
Chrysler, 182
Circuit design, 74–80
Cluster testing, PWA, 329, 334
Collaborative supplier relationships, 186, 321
Co-location, 175, 176
Coffin–Manson model, 213
Commercial off-the-shelf (COTS) components, 219
Commodity teams (*see also* Supply base teams), 172, 183, 184, 194, 196, 197
Common mode noise, EMI, 294, 295, 298, 306
Communication, 318, 323
Compaq Computer Corporation, 175, 363
Component defects, 198, 199, 206, 208–217, 324, 342, 346, 360, 362
Component degradation, 25, 26
Component engineering, 64, 68, 69, 73, 74, 84, 197–202, 313, 318
Component life cycles, 227, 228
Component placement, for EMC, 304, 305

Component qualification, 196, 198, 200, 201, 217–237, 242–245
Component qualification strategy, 235–237
Component screening, 204–207
Component selection, 84, 202–204
Computational fluid dynamics (CFD), 157, 158, 247, 252–254
Computer-aided design (CAD), 76–78, 101, 102, 202, 226, 248, 249
Computer-aided manufacturing (CAM), 102–108, 202
Computer-integrated manufacturing (CIM), 101
Computers, 2, 3, 10, 12–14, 30, 33, 34, 52, 54, 62, 63, 69, 82, 83, 95, 116, 168, 170, 172, 196, 202, 204, 224, 225, 227, 249–255, 267, 273, 275, 278, 279, 293, 310, 312, 317, 319, 337, 342–354, 356–357, 360, 374
Concurrent design, 63–65
Conducted EMI, 292, 293
Conductive gaskets, EMI, 295
Confidence limits and intervals, 45, 46, 48–52
Consumer products, 2, 3, 168, 196, 225, 349
Context tasks, 175
Continuous statistical distributions, 35–44
Contract manufacturers (CM) (*see also* Electronic manufacturing service providers), 143, 144, 178, 191, 310–323, 341, 370
Cooling solutions, thermal, 266–280
Core competencies, 171, 175–176, 314
Corrective maintenance (CM), 12–14
Cost-based pricing, 184
Cost of failure, 205, 206
Crossfunctional design teams, 101
Crossfunctional teams, 101, 184
Cumulative distribution function (CDF), 31, 39, 40, 41, 42, 47
Customer expectations, 1–4, 23, 171, 229, 400
Cycle time, 171

Index

DC/DC converters, 80, 81, 115, 116, 301
Defect characterization, 371–373
Defect (fault) localization, 371–373
Deming, W. Edwards, 206
Deming cycle (or Shewhart cycle), 15
Deprocessing, 371, 372
Derating, component, 26, 87–94, 411–424
Design cycle, 225, 226
Design errors, 165, 350
Design for environment (DFE), 151–154
Design for manufacturing (DFM), 24, 97–108, 312
Design for test (DFT), 24, 121–138, 155, 361
Design optimization, 142, 143
Design practices, 61–165
Design process (phases), 65–69
Design review, 72, 139–141, 154, 165, 382
Design tools, 76–78
Design verification testing (DVT), 154–156
Development cost, 171
Development testing, 154–161
Diagnostic tests, 124, 136, 359–361, 371, 372
Differential mode noise (EMI), 294, 295
Digital ICs, 69–71, 75, 76, 79, 80, 90–94, 121, 124, 126–129, 192
Digital signal processing (DSP), 192
Discrete statistical distributions, 44, 45, 53
Disk drive, 40, 41, 52, 293, 362, 364, 365, 368, 369
Drucker, Peter, 176
Dynamic random access memories (DRAM), 69, 70, 192, 196, 204, 206, 207

Electric field acceleration, 212, 214–215
Electromagnetic compatibility (EMC), 116, 119–121, 164, 280–282, 292–306

Electromagnetic interference (EMI), 81, 116, 259, 292–294
Electronic design automation (EDA), 76–78, 143–145, 249
Electronic distributors, 139, 143, 313, 314
Electronic manufacturing service (EMS) providers (*see also* Contract manufacturers), 143, 144, 178, 191, 310–323, 341, 370
Electrostatic discharge (ESD), 207, 209, 223, 283–291, 293, 299, 300, 302, 304, 332, 342, 355, 362, 370
EMC bonding, 295–297, 301, 307
EMC design guidelines, 298–306
EMC suppression techniques, 116, 120, 121, 295–298
EMC testing, 164
EMI caused problems, 293, 294
End of life (obsolescence) components, 86, 138, 144, 145, 202, 203, 228
Environmental analysis, 154, 336–354
Environmental stress screening (ESS), 15, 155, 158–161, 199, 204, 205, 220, 225, 336–354, 361–363, 369
ESD circuit design techniques, 285, 286
ESD discharge points, 290, 291
ESD handling techniques, 287, 288
ESD materials, 287
ESD PCB design techniques, 286, 287
Exponential distribution, 36, 38–42, 43

Fabless IC suppliers, 191
Failure analysis, 313, 318, 340, 341, 355, 357, 367–374
Failure cause, 370
Failure-free operation, 51, 52
Failure mode, 370
Failures in time (FITs), 5, 16, 17, 41, 215–217
Failure mechanism–driven qualification, IC, 220, 221
Failure mechanisms, IC, 24, 25, 208–211, 220, 221, 261, 263, 370
Failure modes and effects analysis (FMEA), 145–152, 425–431

Fans, cooling, enclosure, 114, 116, 250–252, 267, 273, 274, 365
Faraday cage/shield, EMI, 116, 302
Fault localization, 371, 372
FCC regulations, 293
Feedback of failure analyses, 370
Ferrites, EMI, 295, 300, 301, 306
Field data collection and analysis, 361–374
Field programmable gate arrays (FPGAs), 70, 77, 107, 108, 115, 125, 198, 203, 204
Field replaceable unit (FRU), 363
Fishbone (or Ishikawa and cause–effect) diagram, 382
Flying probe tester, 324, 330, 331
Focused ion beam (FIB), 373
Ford Motor Company, 179
Functional silos, organizational, 174, 182
Functional test, 227, 320, 324, 326–336, 359, 361

Gamma distribution, 36, 43
Gaskets, EMI, 300
Gaussian distribution, 35, 43
Goodwill, customer, 352
Grounding, 291, 296–304, 306

Hardware–software interaction, 344, 352–354, 356, 359, 361, 362
Hazard (failure) rate, 23, 24, 31, 36, 39, 40, 42
Heat pipes, 116, 259, 274–276, 279
Heat sink, 114, 116, 247, 255, 269–272, 279
Heat spreader, 115, 249, 257, 269, 270, 272
Heat transfer, 247, 248
Hewlett Packard (HP), 64, 158, 160, 206
Highly accelerated life test (HALT), 15, 72, 155, 158, 160, 161, 199, 225, 227
High-speed design, 228

Honda, 175
Humidity acceleration, 213

IBIS model, 120
IBM, 64
IC package material, 257, 258, 263, 275, 280
IC package qualification, 222, 229, 231–232
IC packages, 71, 96, 115, 232, 233, 257, 258, 280, 324–326
IC reliability, 204–217
IC test, 121–136
In-circuit test (ICT), 227, 319, 320, 324, 326–336, 361
Industry consortia, collaboration, 167, 171, 172
Infant mortality (or early failure rate), 7, 8, 9, 11, 43, 340, 350
Information technology, 163, 164, 170
Information technology equipment (ITE), 163, 164
Institute of Environmental Science and Technology (IEST), 339
Intellectual property (IP), 191
Integrated circuit (IC), 8, 23–26, 38, 41, 42, 69–81, 90–94, 96, 107, 114–136, 167, 189–192, 196–217, 220–222, 229, 231–232, 250–259, 261, 263, 267, 271, 272 275, 280, 283–291, 325, 330–336, 344, 356–358, 361, 362, 369–374
IPC, 5, 107, 110–113
Integrated device manufacturer (IDM), 190, 191
Internet, 108, 167, 170, 199, 200
Inventory, 171, 188
ISO14000, 152

Junction temperature, IC, 80, 91–93, 114, 256, 257
Juran, Joseph, 206

Latent defects, 348
Lead-free solder, 266–268

Index

Lessons learned, manufacturing, 352, 354–360
Life cycle cost model, 85–87
Linear transformation, linearization, 45–47
Line termination, 119–121, 305
Logistics, 169, 170, 195
Lognormal distribution, 35, 36, 38, 43
Lot tolerance percent defective (LTPD), 5

Manufacturing costs, 310–312, 318, 319
Manufacturing defect analyzer (test), 324, 326, 327, 329, 330
Manufacturing defects, 323–325, 327, 329, 333, 337, 340, 341, 350
Manufacturing engineering, 64, 197, 318
Manufacturing issues, 199
Manufacturing support, 201, 202
Market-focused companies, 180
Marketing requirements list (MRL), 380
Market/system requirements, product, 168
Material considerations, thermal, 259–266, 278
Material costs, 172, 179, 180
Mean time between failure (MTBF), 5, 32, 40, 47, 50–58, 257, 344, 363, 366, 380
Mean time to failure (MTTF), 5, 23, 24, 31–33, 41, 46, 48, 49
Mean time to repair (MTTR), 32, 33, 380
Mergers and acquisitions, 191, 192
Metal backplanes, for EMC, 278, 279
Microprocessor, 72, 116–118, 196, 250–257, 267, 271, 272, 334–336, 356, 361
Military/aerospace industry, 168, 202, 204, 374
Mixed signal ICs, 122
Moats, EMC, 302, 304
Moore, Goeffrey, 62

Moore's law, 114, 170
Motherboard (computer), 250–252, 254, 317

Networking, 170
New product introduction (NPI), 141, 142, 144, 315, 321
No defects found (NDF) (*see* No trouble found)
Nonrecurring engineering (NRE) costs, 179, 203
Normal (Gaussian) distribution, 35, 36, 43
Normative decision analysis, 352–354
Nortel Networks Inc., 231, 242–245
No trouble found (NTF) (or NDF, NPF), 12, 13, 354, 356–359
No problems found (NPF) (*see* No trouble found)

Obsolete components [*see also* End of life (obsolescence) components], 86, 138, 144, 145, 202, 203, 228
Optical microscopy, 375
Outsourcing, 171, 176–179, 180, 188, 190, 191, 310–323
Outsource support required, OEM, 311, 318, 319

Pacemaker, heart, 319
Part replacement rate (PRR), 12–14, 30, 363–367
Parts per million (ppm, measure of quality), 5, 198, 199, 206
Passive heat sinks, 269–271
PCB assembly technology, 71, 96, 315
PCB design and layout, 94–114, 224, 259, 284, 286, 287, 290, 291, 301, 303, 304, 364
PCB materials, 96, 197, 264–265
Peters, Tom, 176
Phase change materials, 259, 261
Poisson distribution, 37, 45
Porter, Michael, 171
Power control IC, 81, 116

Index

Power dissipation, 79, 80, 89, 91–94, 114–116, 254–257, 272
Power supply, 51–53, 80, 81, 86, 87, 250–253, 293, 301, 337, 344, 356, 362, 365–368
Printed circuit board (PCB), 74, 75, 81, 94–114, 118, 119, 156, 258, 264, 265, 286, 287, 290, 291, 301–303
Printed wiring assembly (PWA), 11, 16–21, 24, 76, 80, 92, 94–116, 122, 123, 125, 126, 157, 158, 160, 199, 224, 227, 232, 233, 248, 249, 265, 266, 289–291, 304, 305, 309–374
Probability density function (PDF), 31, 36, 37, 39–42
Probability distribution, 34–45, 48, 344, 345, 352, 353
Product change notice (PCN), 144, 145, 229
Product cost, 143, 157, 172, 311, 312, 321, 352
Product codevelopment, 235
Product development, 26, 188, 195, 225, 226, 235
Product development time, 26, 188, 225, 226
Product improvement, 337
Product life cycle, 26, 62, 171, 225, 227, 349
Product safety testing, 162, 163
Programmable logic devices (PLDs), 70, 77, 107, 192, 198, 200, 203
Purchased components, 172, 179, 180, 186
Purchasing (or procurement), 174, 186–188, 197, 318
PWA manufacturing, 309–323
PWA test, 123, 125, 126, 136, 137, 227, 323–360
PWA test guidelines, 137

Quality, 5, 123, 124, 136, 171, 179, 183, 191, 193, 195, 197–199, 204, 206, 207, 229, 230, 235, 311, 312, 317, 318, 352, 377–408

Qualification (*see also* Component qualification), 196, 198, 200, 201, 217–237, 242–245
Qualification costs, 218, 219
Qualification tests, 218–220, 222, 223
Qualified parts list (QPL), 196

Radiated electric field interference, EMI, 292, 293, 298, 301, 305
Radiated magnetic field interference, EMI, 293, 298, 301, 305
Reduced supply base, 172, 173, 183, 188, 201
Redundancy, 82–84
Regulatory testing, 162–164
Relationship (customer–supplier), 170, 172, 173, 178, 180, 184, 186–189, 193–195
Reliability, definition of, 5, 6
Reliability degradation, 24–26
Reliability engineering, 64, 197
Reliability (survivor) function, 31, 36, 37, 39, 40, 42, 45
Reliability goals, metrics, 5, 12–14, 16, 17, 23, 24, 31–33, 40, 41, 46–49, 52–58, 215–217, 257
Reliability growth, 23, 24, 158, 159
Reliability mathematics, 29–61
Reliability modeling, 52–54
Reliability prediction, 12, 14–21, 84–87, 362
Reliability risk, 17, 19–23
Reliability trends, 26, 27
Restructuring IC suppliers, 190–192
Risk, 17, 19, 21–23, 138, 312, 317, 354, 355, 369
Root cause analysis/corrective action, 161, 341, 342, 350, 359, 362–365, 367, 369, 370, 382
RosettaNet, 167

Scanning electron microscope (SEM), 373
Scanning probe microscopy (SPM), 373
Scan test, 124–127, 130–137

Index

SEI capability maturity model (software), 389–408
Self-certification, 195
Self-qualification, 231, 242–245
Shielding, EMC, 116, 295–297, 305
SIA technology roadmap, 254, 255
Signal integrity, 116–121, 201, 305
Single/sole source components, 84, 203
Skin effect 297, 299
Sneak circuit analysis, 138
Soft error rate, 230
Software availability, 379, 380
Software defect prevention, 385–388
Software defects, 381–385
Software design error, 11, 165
Software quality, 377–408
Software reliability, 9, 11, 377–408
Software test, 155, 384
Soldering issues, 108, 109, 325, 326, 362
Solder materials, 265–267, 268, 271
Sony, 175
Source control drawings (SCDs), 200
Sources of ESD, 289, 290
Southwest Airlines, 175
Spacecraft/satellites, 8, 11, 42, 83, 84, 299
SPICE model, 71, 77, 78, 120, 201, 226
Splits/slots, EMC, 302, 304
Spray cooling, 275–277
Stacked (3D) packages, 232, 233
Stakeholders, mutual, collaboration, 169, 180, 181, 186, 193
Statistical distributions, 34–52, 363, 365, 367, 368
Statistical process control (SPC), 381–384
Statistics terms (C_p, C_{pk}), 198
Strategic sourcing (or supply base management), 179–189
Stress analysis, part derating, 26 87–94
Stress test qualification, 198, 218, 219, 221, 228, 229
STRIFE, 155, 158, 199, 225
Sun Microsystems, 175
Supplier management, 168–197, 316

Suppliers, 84, 85, 87, 168–197
Supplier scorecards, 187, 196, 238–241
Supplier selection, 84, 86, 87, 173, 201
Supply base teams (*see also* Commodity teams), 172, 183, 184, 194, 196, 197
Supply chain, 141–145, 167–170, 313
Support costs, 85–87
Sustaining engineering, 354, 356, 358, 360
System/box build, manufacturing, 315, 316, 320
System defects, 356, 361, 367
System engineering, 115, 197
System (product) failures, 207, 208
System on a chip (SOC), 191, 203
System test, 137, 138, 348, 360, 361

TAAF, 155
Technology assessment, 69–74
Technology driver, 170
Technology roadmaps, 73, 74, 184, 199
Temperature acceleration, 210–212, 214–217
Temperature cycling acceleration, 213, 214
Testability, 121–123, 322, 361
Test, analyze, and fix (*see* TAAF)
Test/fault coverage, 329, 331, 336, 356, 360
Test development, 171
Test development costs, 171, 336
Testing microprocessor based PWAs, 334–336
Test structure, 221
Thermal airflow patterns, 249–252
Thermal analysis models and tools, 248–254
Thermal characteristics/management, 79, 80, 81, 114–116, 247–282
Thermal failures, 261–263, 332, 333
Thermal grease, 272
Thermal interfaces, 279, 280
Thermal resistance, 115, 257, 258, 261, 275
Thermoelectric coolers, 277–278
Thermography, 157, 158

Through-hole technology, 324, 325
Time to market, 171, 196, 225, 229, 316, 321, 349
Time to volume, 171, 196, 321
Transmission lines/terminations, 117–121
Transportation and shipping test, 162
Troubleshooting, 354–360

Useful life, constant failure rate, 7, 11, 43, 44, 86
Unspecified parameters (or special studies), 79, 207, 226, 231

Variability, 30, 356
Vectorless opens testing, 327, 328, 332
Venting (thermal), 250–252, 269

Virtual companies, 171, 195, 314, 315
Visual inspection, 373
Voids, 302–304
Voltage acceleration, 212, 214–217

Wafer-level packaging (WLP), 232
Wal-Mart, 175
Warranty cost, 54–56, 85–87
Wearout, 7, 8, 11, 43, 230, 352
Weibull distribution, 36, 42–44, 47, 363, 365, 367, 368
Willoughby, Willis, 206
Wireless, 170, 310

Xerox, 182, 183
X-ray inspection, 319, 320, 324–326, 336